I will not travel beyond Glasgow's city
vehicles except my bike, for a whole ca
– Ellie Harrison, January 2016

This simple proposition – to attempt to live a 'low-carbon lifestyle of
the future' – put forward by an English artist living in post-industrial
Glasgow cut to the heart of the unequal world we have created. A
world in which some live transient and disconnected existences within
a global 'knowledge economy' racking up huge carbon footprints
as they chase work around the world, whilst others, trapped in a
cycle of poverty caused by deindustrialisation and the lack of local
opportunities, cannot even afford the bus fare into town. We're all
equally miserable. Isn't it time we rethought the way we live our
lives?

In this, her first book, Ellie Harrison traces her own life's trajectory
to examine the relationship between literal and social mobility;
between class and carbon footprint. From the personal to the
political, she uses experiences and knowledge gained in Glasgow in
2016 and beyond, together with the ideas of Patrick Geddes – who
coined the phrase 'Think Global, Act Local' in 1915, economist
EF Schumacher who made the case for localism in *Small is Beautiful*
in 1973, and the Fearless Cities movement of today, to put forward
her own vision for 'the sustainable city of the future', in which we
can all live happy, healthy and creative lives.

ELLIE HARRISON was born in the London borough of Ealing in 1979.
She moved north to study Fine Art at Nottingham Trent University in
1998. In 2008 she continued northwards to do a Masters at Glasgow
School of Art and has been living in Glasgow ever since. She has
previously described herself as an artist and activist, and as 'a political
refugee escaped from the Tory strongholds of Southern England'.
In 2009 she founded Bring Back British Rail, the national campaign
for the public ownership of our railways. As a result of thinking
globally and acting locally during *The Glasgow Effect* in 2016,
she is now involved in several local projects and campaigns aimed
at making Glasgow a more equal, sustainable and connected city.

The Glasgow Effect

A tale of class, capitalism and carbon footprint

ELLIE HARRISON

Luath Press Limited

EDINBURGH

www.luath.co.uk

First published 2019

ISBN: 978-1-912147-96-0

The paper used in this book is recyclable. It is made
from low chlorine pulps produced in a low energy,
low emission manner from renewable forests.

Printed and bound by Martins the Printers Ltd, Berwick-upon-Tweed

Typeset in Sabon by Main Point Books, Edinburgh

Front cover illustration by Neil Scott, Glasgow

Photograph on back cover taken by Ellie Harrison on 28 October 2015,
featuring chips bought and eaten by the artist (at her own personal
expense) from the Philadelphia chippy, Great Western Road.

Cover design by Neil Scott, Ellie Harrison and Maia Gentle

Discussion notes and further information about the artworks referenced
in this book can be found at www.ellieharrison.com/book

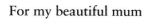
For my beautiful mum

Extract from 'Neoliberalism' by Loki

Since the youngest age I was always threatening to run away,
but social mobility isnae what they say.
Stuck at a red light on my tricycle thinking 'fuck them!'
While a bunch of Westenders whizz past me in the bus lane.
How come it's bright in this posh part, but in Pollok it's dark?
Probably cos the sun isnae shining out of everyone's arse.
Vegan liberals lecture me to 'buy local',
while they're sneering at the Glesga dialect and my vocals.
…

Apart fae that, nothing's going on here locally,
Except an off sales masquerading as a grocery.
The local shops are shutting since the new Tesco's getting built.
Thought I'd check it out. Went in sceptical,
had to walk 20 miles just to get some bread and milk.
Then got diverted at the checkout by a clothing section.
Ended up flipping out, bought some pens, Pepsi, Lilt,
a Craftmatic adjustable bed, a lamp, a feather quilt,
a pair of stilettos, salt and pepper, a kilt, a pair of stilts.
First person to mention my free will is getting killed.
Luckily there's a pretty woman on the end of the till,
offering me instant loans to help me with my credit bills.
…

While I look across the river at you sipping your Pimms,
wishing I could get hooked on yoga, swimming and gyms,
as I can to chicken dippers, Lidl's crisps and 70 million minute SIMS.[1]

Reproduced with permission of Darren McGarvey (aka Loki the Scottish Rapper)

Contents

Preface

I STARTED TO write this book – for therapy, for clarity, for closure – in summer 2018, having been thinking about it for nearly a year. There seemed a beautiful symmetry in it being exactly 30 years since I first visited Glasgow, exactly 20 since I left my family home in Ealing and exactly ten since I moved to Glasgow to live; a good point to stop and reflect. My aim was to have it finished before I turned 40 in March 2019, a task in which I almost succeeded (it went to print on 29 July 2019).

The one thing nobody seemed to consider at the time *The Glasgow Effect* 'chips hit the fan' in January 2016, was that this was a project that could only have been dreamt up by a single woman in her late 30s, a 'mid-life crisis' you could say. I had chosen a title addressing mortality because it was, and still is, a major preoccupation. I loved my family and wanted to be nearer to them and I was angry at an economic system which had torn us apart; arbitrarily scattering us in different parts of the British Isles.

When nearly all my heterosexual friends and family were 'settling down' with a house, a car, a big telly, the latest smartphone, frequent holidays and some kids, I had a niggling feeling these lifestyles were neither ethical nor sustainable. I found myself resisting conforming to this social norm. Instead, I decided to devote my life to work, to undertake the most epic and the most public artwork of my career; one that did not only last a year, but which has shaped my thinking, action and life course ever since.

Reading Darren McGarvey's *Poverty Safari* in March and April 2018 inspired me to start writing. His book, in turn, had been partly inspired by *The Glasgow Effect* project. As I sat alone, late at night, turning page after page, I came to realise the intimate connection a book enables between author and reader, away from the inherent agitation of the computer screen. A book demands patience to allow a person's stories and ideas to unfold slowly over time, offering a more complete picture of how their personal history and lived experience have shaped their politics; their thinking and action in the world. From very different backgrounds, Darren McGarvey and I have arrived at similar critiques of the city, the society and the economic system within which we live.

I can't say that this book has been an easy undertaking, but it was a necessary one. I have emerged stronger and with more conviction than ever that in order to address the 'climate emergency' we must urgently reduce the amount we travel and the amount of energy we consume. Not least because a happy, healthy and sustainable life can and should result from committing and contributing to the community where you live.

Ellie Harrison
July 2019

Ellie Harrison (right, aged nine), with her sister (left, aged ten), at Glasgow Garden Festival in 1988.
(Family photograph)

Introduction

I FIRST CAME to Glasgow 30 years ago, in summer 1988 when I was nine years old. Margaret Thatcher had had a plan. Her government thought that by imposing 'Garden Festivals' and the associated twee middle-class values onto five of Britain's most deprived post-industrial cities, she could miraculously raise the standard of living for everyone. It was symbolic of how her vision of 'trickle-down economics' was meant to work. You only need look at vast swathes of derelict and vacant land still left in Glasgow (9 per cent of the city),[2] including much still around the 'Festival Park', and at Glasgow's persistent and worsening inequalities of wealth and health, to judge this policy's effectiveness. What it did succeed in, however, was luring more middle-class cultural tourists on 'city breaks' to see the sights and spend their money. And this was indeed why my family came.

My memories of the trip are hazy. I remember having a 'Coke float' in Deep Pan Pizza Co. somewhere in the city centre (which I now assume was Sauchiehall Street, though Deep Pan Pizza Co. has long-since gone bust).[3] I remember this weird reconstructed house thing (which must have been the Mackintosh House at the Hunterian, completed in 1981). I remember the thrill of riding to and from the West End on the newly refurbished Glasgow Subway (public transport got me excited back then too). And, I remember I wasn't allowed on the Coca-Cola branded rollercoaster because I was too short. That was it. The holiday was over. Armed with our Garden Festival merchandise (my sister had the t-shirt and I had the keyring, which we both still have – how's that for 'legacy'?), we returned to suburban London to tell teacher about our trip. I had no cause to return to Glasgow for another 20 years. And once again, it was 'culture' that led me back.

Glasgow's Garden Festival could be seen as the beginning of the city's post-industrial renaissance as a global capital of culture. It was promptly followed by Glasgow being awarded the mantle of European City of Culture in 1990, and a slew of investment in the arts sector: Tramway opened in 1990, the Gallery of Modern Art in

1996, The Lighthouse in 1999 when Glasgow was named UK City of Architecture, and the newly refurbished CCA: Centre for Contemporary Arts in 2001.

The city had many high-profile artists nominated for Britain's Turner Prize, which led the international star curator Hans-Ulrich Obrist to 'parachute in' one day in 1996 and proclaim all this as 'the Glasgow miracle'.[4] What I only came to realise after five years of living in the city myself, in 2013, was that this 'miracle' was much more about how the city appeared to people living elsewhere – to potential tourists, students or other transient residents – than it was about the quality of life for the majority of Glasgow's citizens.

'Glasgow is kind. Glasgow is cruel', wrote Scottish novelist William McIlvanney in 1987.[5] And so this book is about my love-hate relationship with this city, where I have now lived for well over a decade. It is my story. It is about how I ended up here, by following the absurd and lonely career trajectory of the 'conceptual artist'. It aims to do what much of my art work has done before – that is to use my own personal history and lived experience to illustrate the impact that social, economic and political systems at local, national and global scales have on our individual day-to-day lives. Specifically it's about the forces of globalisation, unleashed by Thatcher and her fellow free-market ideologues in the '80s, which decimated industrial cities across Britain and sent millions of 'economic migrants' like me on the move. It could be about any post-industrial city attempting to fill the vast voids and regenerate its economy with 'culture', but it's not. It's about Glasgow and Glasgow, as we know, is special.

The phrase 'the Glasgow effect' began to emerge in the field of public health around the time I moved to the city in 2008. It was used to describe a then unsolved mystery. Why did people die younger in Glasgow than in similar post-industrial cities in England, such as Liverpool and Manchester? Why did Glasgow and West Central Scotland have the lowest life expectancy in Western Europe? In 2011, Glasgow Centre for Population Health (GCPH) published a report exploring 17 different hypotheses for why this could be, including: diet, other 'health behaviours' (such as alcohol and smoking) and 'individual values', 'boundlessness and alienation', lower 'social capital', a 'culture of limited social mobility', inequalities, deindustrialisation, 'political attack' and, of course, our terrible weather.[6]

GCPH concluded their 2011 report by saying that 'further research is required to fully understand why mortality is higher' in Glasgow.[7] It wasn't until May 2016, when they published their epic follow-up *History, Politics & Vulnerability: Explaining Excess Mortality in Scotland & Glasgow*,[8] which synthesises research into 40 potential causes, that they finally claimed to have solved the mystery of 'the Glasgow effect' (their findings are discussed in detail in Chapter 7). By then, I was five months into my year-long 'durational performance' named after this phenomenon, which had become one of the most controversial publicly-funded artworks that Scotland had ever seen.

Part psychological experiment, part protest, part strike, for the whole of that year, I had vowed not to travel beyond Glasgow's city limits, or use any vehicles except my bike. As stated in the application I submitted to Creative Scotland in summer 2015 to fund the project, my aim was 'to actively address the contradictions and compromises' which had been building in my own lifestyle 'as my career as an artist/academic has progressed' and to slash my own carbon footprint for transport to zero (the results can be seen in my Carbon Graph on pages 138–9).[9]

I had chosen the title *The Glasgow Effect* to dismantle the myth of 'the Glasgow miracle',[10] and to throw the spotlight back on the real story of this city – glossed over in recent years with a pervasive 'People Make Glasgow' PR exercise – where a third of children live in relative poverty,[11] and 20 per cent of our households use food banks.[12] Sacrificing myself to the social media trolls as a symbol of the detached 'liberal elite' I too had come to despise, I aimed to encourage people to make the connections between the social, environmental and economic injustices in our world, and to realise the need to fight for holistic solutions – at an individual and policy level – which address all three at once.

As Darren McGarvey acknowledges at the end of *Poverty Safari*, during *The Glasgow Effect* I was not just researching and criticising, I was actually doing; through my action that year and the way I had chosen to live, I was articulating

> what might come next... beginning to reimagine the
> society that had left so many in [his] community feeling
> excluded, apathetic and chronically ill.[13]

By undertaking *The Glasgow Effect*, I aimed to find out what happens if you make a stand against the forces of globalisation by localising your existence; what happens if you attempt to live a 'low-carbon lifestyle of the future' in a city, in a society and in an economic system with infrastructure and values still stuck in a carbon-intensive past (this is described in Chapter 7). Two years on from that intense and stressful experience and 30 since my first visit to Glasgow in 1988, this book aims to bring together everything I have learnt in order to provide the complete context for my thinking and action, which was lost in the whirlwind of the social media storm (a whirlwind described in Chapter 5).

The book is structured in three parts, each part containing four chapters. Part 1 – *A Brief History of Neoliberalism* – provides the backstory: a personal and political history examining the way our lives are shaped by wider social and economic forces often beyond our knowledge or control, and by the privileges or disadvantages that result from a person's accident of birth.[14] Part 2 offers a behind-the-scenes insight into the making of *The Glasgow Effect* – exploring the many issues the project raised, including the relationship between art, activism and well-being. Part 3 sketches out a manifesto for *The Sustainable City of the Future*, articulating the action necessary to transform Glasgow and our other cities into places where we can all live happy, healthy and creative lives.

Part 1

A Brief History of Neoliberalism

CHAPTER 1

Thatcher's Children

Straight outta Compton

NONE OF US choose where or when we are born. It just so happens that my life began in the now bulldozed Perivale Maternity Hospital in the London borough of Ealing in March 1979.[1] It was the spring following the so-called 'Winter of Discontent', which saw the coldest weather for nearly two decades coupled with the largest strike action in Britain since the General Strike of 1926.[2] In the absence of refuse collections, London's Leicester Square was used as a makeshift landfill site. It was just a few days after the first contested referendum on the devolution of power from Westminster to Scotland and Wales (although it wasn't until I moved to Scotland myself that I discovered that history). But, perhaps most symbolically for this story, it was less than two months before Margaret Thatcher came to power on 3 May 1979.

My mum and dad had been married for just two years, and I was brought home along with my big sister to a semi-detached house in West Ealing with its 'own back and front door',[3] which they had bought for £18,750 in 1977. My parents had met at work in the early '70s, where they both taught new-fangled subjects to teenagers – my mum taught 'Communication Studies' (as well as English and French) and my dad taught 'Business' at Uxbridge Technical College, opened in 1965. It was quite a handful having two children just 18 months apart, so for the first few years of my life, my mum stayed at home to look after us and get more involved in our local community.

I have happy memories of those very early days; sitting on the living room floor eating tasty snacks and watching telly. In between *Chock-A-Block*, *Pigeon Street* and *Button Moon*, I remember seeing glimpses of this powerful looking woman on the news. She had quaffed blond hair and was wearing a bright blue suit. I didn't really know what she was up to at that time, but what I did know was the more I appeared to be in awe of her, the angrier my mum would get.

My mum was born in Bickley in south-east London in March 1944, where she had to be sheltered from the air raids during the last year of the Second World War. Her parents were both Welsh. She had one Scottish grandmother (who was actually half-Irish) and despite having a booming posh southerner's accent, it remained her pet hate to be mistaken as 'English' (could there be a worse insult?). She had always voted Labour. She was active in the teachers' union,[4] and when we were young she got involved in the Campaign for Nuclear Disarmament (CND) and the Ealing Peace Register. I have one very early memory of being taken on a coach trip. Together with a big group of families and children wearing a hotchpotch of '70s styles, we were driven out of the city to a very crowded and very muddy field. While we all scampered about amongst the trees eating homemade sandwiches from foil wrappers, the adults just stood around. It was as though they were in a very long queue which didn't seem to be going anywhere. It turns out that was Greenham Common, the peace camp set up and run by women in protest against the Cruise missiles being held at the Royal Air Force base in Berkshire during the Cold War.

In the early '80s, my mum also joined the local branch of the Campaign for the Advancement of State Education (CASE), which sought to abolish selective schools and to ensure the same well-resourced comprehensive education was available to all children no matter where in the country they grew up. The campaign is still going – battling against the 'academisation' process in England, which under the mantra of 'choice' is ensuring the exact opposite. CASE also 'rejects the hierarchical division of skills into academic and vocational subjects and affirms that well-educated children need both mental and practical skills'.[5] Needless to say, my mum was always very supportive of all our skills and abilities – arts, sports and/or academic – and we were bundled off to the local state schools to make friends with local kids. It would not be a lie to say I was 'straight outta Compton' – attending Compton First School (1983–7) and Gurnell Middle

School (1987–91), which both shut down in 1993.

It's impossible to understand the nuances of the British class system when you're not yet into double digits but I was aware there were kids at school from very different backgrounds. Ealing was one of the most ethnically diverse of all London's boroughs and these schools served the big council estate, Copley Close, which was right next door – literally the other side of the railway tracks at Castle Bar Park. I remember being taken to birthday parties in my friend's flats on the estate. Holding my mum's hand, we'd navigate the dark stairwells and balconies. I was surprised at how cramped their living rooms were. That's weird, I thought, how come the kids from Copley Close always have the most expensive trainers? There was racism and homophobia in the playground and I'll never forget the bollocking I got when I repeated one of the words I'd picked up to my mum when I got home.

In 1987, the AIDS advert featuring that monolithic grey tombstone hit the telly. It said 'there is now a danger that has become a threat to us all: it is a deadly disease and there is no known cure'. We were all terrified of catching it. Kids calmed their nerves by chasing others around playing 'it', as though they were passing this mysterious disease onto one another. Another popular insult in those days was 'NHS'. If you had 'NHS specs', then it meant that they were free or cheap, whatever: they were the worst. So much of this must have been filtering down via parents from a right-wing media hell-bent on discrediting our public services and propagating hate to enable our Conservative leaders to 'divide and rule'.

Compton and Gurnell were both secular schools. In fact, we were barred from singing anything religious. Our teachers had to be imaginative in assembly. Instead we ended up reciting pop hits from the '60s: 'Downtown' and 'With A Little Help From My Friends' (the drug references were lost on us of course), as well as 'Guantanamera' based on the poem by Cuban revolutionary philosopher, José Martí. Most of it went over our heads. We sang songs reflecting our diversity like 'Linstead Market', which we learnt in Jamaican Patois. I still know nearly all the words. We had a brilliant teacher called Mr Strong, who wrote our school anthem 'Gurnell Middle's Happy Song', and helped us put on plays which he wrote and directed. I was cast in the lead role as Wally Bottle, a small boy on a mission to clean up the planet. He lived in a landfill site and his best friend was a crisp packet. It was quite ahead of its time. The following year, I was cast as Scrooge! I don't remember being any good at acting. All I remember is the feeling of being on a stage with everyone looking at me expectantly and not being able to perform quite in the way I wanted.

One day Gurnell had a visitor. It was the Conservative MP for Ealing North, Harry Greenway, who was also one of our school governors. He was later caught up in allegations of Tory 'sleaze' in the early '90s.[6] I had a feeling he was a nasty piece of work. He stood at the front of assembly and lectured us, then the head teacher asked: 'Right, does anyone have any questions?' Sitting cross-legged on the floor in the front row, I felt my pulse start to quicken. I had to say something, so I tentatively put my hand up. Harry Greenway spotted me: 'Yes, that little boy at the front'. I was so humiliated. I forgot my question, but I did answer back: 'I'm not a boy, I'm a girl!' I was traumatised by that and many similar incidents around the time. In the year before I went to high school, I started to grow my hair into a drab ponytail which allowed me to just blend in. Little did I know that Harry Greenway's Tory government were effectively eradicating difference and enforcing conformity anyway – their Local Government Act 1988 included the infamous 'Section 28' (known as 'Section 2A' in Scotland) banning discussion of homosexuality in any state school. The Act remained in force until 2000 in Scotland and 2003 in England and Wales. By then I was 24 and had left education altogether. I'd finally cut my hair short again.

My favourite book in the '80s was *Gertie & Gus*.[7] My mum used to read it to me in bed, and in later years we entertained ourselves by subjecting it to a Marxist analysis together. It tells the story of two happily married bears named Gertie and Gus. They live in a very modest little hut by the sea. Gus is a fisherman. Every day he heads out on his little boat bobbing around with a single rod to catch a fish for them both to have for tea. One day

Gertie has an idea: why doesn't Gus catch a few more fish and then we could sell them and make a bit of money? So Gus works harder. He brings home bucket-loads of fish and they begin accumulating funds. Gertie buys him a bigger boat and hires some other bears to help him. They catch so much fish that they move out of the hut and into a posh villa and start buying fancier and fancier clothes – Gus wears a Captain's uniform with gold epaulettes. But something isn't right. It's just not the same as 'the good old days' – when they caught and cooked their own dinner and actually had time to spend with each other and enjoy the beautiful scenery where they lived. They were so much happier then. So they gave it all up and went home.

What the fuck is neoliberalism?[8]

It took me nearly 30 years to work out what that woman on the telly was really up to and then the next ten to start fighting back. The aim of this book is to share all the things I've learnt during that time in the hope of inspiring young people to join the fight against an unjust system which has already robbed them of so much. I chose to include Darren McGarvey's (aka Loki the Scottish Rapper) piece 'Neoliberalism' at the start as it touches on many of the book's themes. The word 'neoliberalism' is often bandied around to explain why we live in such a precarious, exploitative and unequal world, yet it is rarely defined. I'm sure that's what Loki is hinting at by using it as a title; another little jab at the pretension of what he calls the 'progressive left'.

To understand the significance of the neoliberal policy decisions of the '80s and '90s to the world we live in today, it feels important to start with a clear definition. In simplest terms the word 'neoliberalism' breaks down into 'neo' (from the Greek for 'new') and 'liberalism' referring back to the 'liberal' economists in the early part of the Industrial Revolution who first theorised the so-called 'free-market'. For example, the University of Glasgow's Adam Smith (1723–90) believed that this market functioned like an 'invisible hand' ensuring that the world's resources would be fairly distributed to those who needed them. The 'neoliberalism' of the latter part of 20th century was like liberalism on steroids or liberalism without any moral reflection or concern.

In her book, *This is Not Art: Activism & Other 'Not-Art'*, artist and writer Alana Jelinek also complains about the overuse of the term neoliberalism in the artworld without proper explanation and then, thankfully, offers a useful definition. She breaks the amorphous concept of neoliberalism down into three key economic principles or policies – privatisation, deregulation and trade liberalisation – which began to be implemented across the world at the end of the '70s and in the '80s when Margaret Thatcher's government came to power in the UK and Ronald Reagan's in America.[9] These three principles or policies came to be accepted as the norm for the next 30 years.

Privatisation

Shortly after the end of the Second World War, Clement Attlee's Labour government won a landslide victory with a mandate to implement socialist policy – to ensure 'the common ownership and control of those things we all need to live happily and well'.[10] They decided to do this by nationalising the 'means of production' – all our key services and infrastructure which they felt should be owned and run for the common good of the British people and not for private profit. The National Health Service was founded in 1948, the same year our railways were nationalised (then named 'British Railways'). Our energy sector – gas, electricity and coal – were all taken into public ownership. There was a massive programme of social house building. From the '50s into the '70s major industries were also made public assets: telecoms, iron, steel, even cars (British Leyland Motors was part nationalised) and aeroplanes (British Airways).

When privatisation was first proposed as a way of reinvigorating these industries in the late '70s by increasing 'competition', it appeared to many as a new and exciting idea. Thatcher's government marketed it as 'popular capitalism'. The plan was to sell shares in all these companies and that everyday folk on the street would buy them, people like Sid who featured in the British Gas 'share offer' advert I remember seeing on the telly in 1986. In the '80s and '90s nearly everything was flogged: telecoms in

A Brief History of Privatisation by Ellie Harrison, installed at Watermans Arts Centre, Brentford in March 2011. This interactive installation uses a circle of six electric massage chairs to re-enact the history of UK public service policy since 1900. Each chair represents a key service or industry and switches on when the date display reaches the year it was taken into public ownership and switches off again at the year it was privatised. (Ben Wickerson)

Health:
On 1948
Off...

Railways:
On 1948
Off 1993

Post:
On 1516
Off 2013

Gas:
On 1949
Off 1986

Telecoms:
On 1912
Off 1984

Entrance

Electricity:
On 1948
Off 1990

1984, gas in 1986, electricity in 1990, buses were deregulated in 1986, railways in 1993, water in England and Wales in 1989,[11] and much of our social housing stock was sold to individuals through the new 'right to buy' scheme.

It was a story I attempted to visualise in my 2011 exhibition *A Brief History of Privatisation* (pictured opposite). When in 2013, David Cameron's coalition government – picking up where John Major had left off – privatised the Royal Mail (which had been in public ownership of sorts since it was set up by the Crown in 1516), it was only the National Health Service still hanging on by a thread. The problem with 'popular capitalism', indeed with capitalism in general, is that it creates and intensifies inequalities[12] – the more money you have, the more shares you can buy, the more power you acquire and, therefore, the more votes you get at the table. If people like Sid did buy shares, then they probably only had a tiny number to start with and more than likely sold them on to make a quick buck. The majority of all these companies – providing the essential 'things we all need to live happily and well'– are now owned by foreign investors.[13] Of course they don't care about the quality or cost of services in a country they don't live in, for these absentee landlords it's always only about maximising their financial return. Once manufacturing industries are no longer guided by the social good of a nation, operations are quickly moved overseas to exploit cheaper labour elsewhere. It's not as if we don't use steel anymore, we just import it rather than make it in Ravenscraig (the steelworks in North Lanarkshire closed in 1992, four years after the privatisation of British Steel in 1988). This is the process known as 'deindustrialisation' – outsourcing all the carbon-intensive work (and therefore carbon emissions and pollution) to other parts of the world.

Deregulation

Then there's the principle of deregulation. While Thatcher's government were busy overregulating some aspects of society, introducing that infamous 'Section 28' – 'the first new homophobic law in a century',[14] they were deregulating in terms of the 'flow of capital around the world'.[15] In 1986, Margaret Thatcher and her Chancellor Nigel Lawson created a 'big bang' on London's Stock Exchange, with sweeping changes to the way the financial sector was regulated, or not. They moved trading in stocks and shares from a face-to-face endeavour to something done from behind the computer screen. They abolished the fixed commission on trading, which meant that people could now make more and more money by sitting on their arses pressing buttons. They ended the separation between traders and financial advisors which led to less impartiality and more dodgy deals. A flurry of company mergers and takeovers helped pave the way to the many 'too big to fail' bailouts and financial crises we've had since. And finally, they opened the doors to international investors.[16]

These first two neoliberal policies alone – privatisation and deregulation – help explain why most of Britain's utilities are now owned by foreign companies. For example, 'Scottish Power' is actually owned by the Spanish multinational Iberdrola and 'ScotRail' is run by Abellio, the commercial arm of Nederlandse Spoorwegen – the railway company owned by the Dutch state. In England, 'Thames Water' is now owned by a consortium of international investors from Canada, Abu Dhabi, Australia, China and more.[17] All of them exploit and profit from our basic needs to heat our homes, travel to work and drink water. One of the most insidious aspects of privatisation and deregulation has been the continual marketisation of every aspect of our lives and the increasing amounts of money we all now need just to survive.

Trade liberalisation

The third key neoliberal policy is trade liberalisation. Following the fall of the Berlin Wall in 1989, and what right-wing commentators claimed as the triumph of liberal democracy across the world, global trade negotiations quickened pace. The World Trade Organisation was founded in 1995 (as a successor to the General Agreement on Tariffs & Trade) to help facilitate and encourage international trade, thereby 'supporting economic development and promoting peaceful relations among nations'.[18] In her book, *This Changes Everything: Capitalism vs the Climate*, Naomi Klein shows how these trade negotiations unfolded in flagrant denial of the impact of man-made carbon emissions on global warming, which had also emerged into mainstream public discourse in the late '80s.[19] Even Margaret Thatcher herself devoted her entire address to the United Nations General Assembly in 1989 to 'the threat to our global environment' which, she said, was 'the challenge faced by the world community' that had 'grown clearer than any other in both urgency and importance'.[20]

The United Nations Intergovernmental Panel on Climate Change (IPCC) had been founded in 1988 with the aim of reaching an international agreement on emissions reductions to limit the onset of global warming and preserve a climate on earth which could continue to support human life. It held its first annual international Conference of Parties in 1995 in Berlin (COP1), with the first legally binding treaty – the Kyoto Protocol – agreed at the summit in Kyoto, Japan two years later (COP3). Meanwhile, the World Trade Organisation (WTO) continued in its completely contrary mission of helping 'trade flow as freely as possible',[21] by getting more and more countries around the world to join its club. Its ambition of signing up China was accomplished in 2001. The WTO now has 164 members and 23 observers representing a vast majority of countries in the world. But, as Naomi Klein writes, nowhere in any of these trade negotiations was the fundamental consideration:

> How would the vastly increased distances that basic goods would now travel – by carbon-spewing container ships and jumbo jets, as well as diesel trucks – impact the carbon emissions that the climate negotiations were aiming to reduce? How would the aggressive protections for technology patents enshrined under the WTO impact the demands being made by developing nations in the climate negotiations for free transfers of green technologies to help them develop on a low-carbon path? And perhaps most critically, how would provisions that allowed private companies to sue

national governments over laws that impinged on
their profits dissuade governments from adopting
tough antipollution regulations, for fear of getting
sued?[22]

And so carbon emissions have been increasing unabated ever
since – more than half of all emissions since the start of the
Industrial Revolution in 1750 have been released in the 30 years
since 1988 – the three decades in which, with the help of the IPCC,
we were meant to be urgently reducing them.[23] So, put simply,
these three neoliberal policies – privatisation, deregulation and
trade liberalisation – have facilitated globalisation: the rapid
transitioning to a world economy, and the hugely increased
movement of goods and people and the explosion in carbon
emissions that has inevitably resulted.

The New Economics Foundation (NEF) is a think tank founded
in 1986 to promote a different way of structuring our economy,
which takes into account the hidden costs of globalisation by
putting 'people and the planet first'. They highlight the three key
consequences of neoliberal policies as being:

- hugely increased social inequalities,
- catastrophic environmental destruction, and
- frequent financial instability and uncertainty

It is neoliberal policies which have created the triple social, environmental and economic crises that we now face. But the worst
outcome of all has been the falling levels of human well-being.
NEF show that 'people's well-being is largely based on how we
interact with other people',[24] and that in our neoliberal era, it's
the quality of our relationships and the strength of our real social networks (not our online ones) that has suffered the most.
NEF writes:

Social networks make change possible. Social networks
are the very immune system of society. Yet for the past
30 years they have been unravelling, leaving atomised,
alienated neighbourhoods where ordinary people

feel that they are powerless to cope with childbirth, education or parenting without professional help.[25]

In his book *A Brief History of Neoliberalism*, David Harvey explains how we sleepwalked up until this crisis point: 'it has been part of the genius of neoliberal theory to provide a benevolent mask full of wonderful sounding words like freedom, liberty, choice and rights, to hide the grim realities of the restoration or reconstitution of naked class power, locally as well as transnationally' to a global economic elite.[26] But neoliberalism is more than just class war. Just as Margaret Thatcher's chilling pronouncement in 1981 suggests: 'Economics are the method; the object is to change the heart and soul'. As much as it may have been against our own interests and our well-being in the long-term, all of us, 'Thatcher's Children', raised in the '80s and '90s, came to embody the survival of the fittest mentality on which the free-market thrives. The greatest irony of all is that it's the same neoliberal ideologues, whose mantra is 'choice', who tell us 'there is no alternative' (TINA) – that we cannot fight the logic of their system or address the hidden costs of globalisation. That's what I seek to challenge.[27] As Alana Jelinek writes:

> Economically, neoliberalism rests on privatisation, trade liberalisation and deregulation. Ideologically, the concept equates to distrust of the state as the provider of social goods such as health, education and the arts. Neoliberal ideology also places an emphasis on individuals rather than systems, imagining individuals as rational beings within rational systems where choices can be made. There tends to be a generalised distrust of authorities: authorities are equated to the state and regulation.[28]

By the early '90s the ethos of the self-interested entrepreneur had begun to permeate most areas of society, including public services once considered as 'communal labour' for social good, such as health and education. My mum and dad felt this on the frontline at Uxbridge College. In June 1991, in my last summer before high school, I remember being roped in to make placards for a bicycle blockade at the college gates. Teaching staff were protesting against the new management's plans to buy themselves company cars with the college budget, which they claimed would 'boost prestige'.[29] My mum's placard read 'cash for courses, not cars!' and mine, the product of a 12-year-old mind, said 'fairer chances, not Sierra chances!' I thought that model of Ford made a nice rhyme – they actually had their eyes on the 'executive' Ford Scorpios and some BMWs.

These self-appointed execs wanted the old guard out – those like my mum and dad who had devoted their lives to the college and who actually believed in the power of education to transform young people's lives. My dad, aged 57, was offered early retirement that same summer and so he took it. Instead he spent the '90s as a househusband – volunteering with the Parent Teacher Association at our high school, Drayton Manor (making school magazines and running car boot sales to raise funds), cooking us dinners, preparing our packed lunches and working on the allotment in Pitshanger Park, which our family had taken on in the mid-'80s.

Meanwhile, my mum carried on at the college, battling against the new managers' misuse of funding. In 1997, when they decided to 'outsource' their lecturing staff to the now defunct private agency Education Lecturing Services,[30] my mum and two other passionate troublemakers were placed on a 'blacklist'. They were not allowed to reapply for the posts they had held for more than two decades. She had no choice but to resign and to look for work elsewhere where her skills and experience might actually be appreciated. She ended up teaching English at Southall College and then in Richmond, which had not yet been so badly affected by the neoliberal bug. I was well looked after and well loved, but teenagers don't appreciate these things do they? It's only with hindsight that the huge amount of support I got became clear.

Social mobility isnae what they say

In October 2017, I was invited by an arts organisation called Create to read and respond to their *Panic!* research into inequality in the arts. It was the first time I'd properly considered how terms such as 'class' and 'social mobility' are defined and measured by sociologists. They always tend to do this in reference to your 'employment status' or occupation. 'Class' is often measured by the National Statistics Socio-economic Classification (NS-SEC), which is similar to the table Darren McGarvey reproduces in *Poverty Safari*.[31]

> The NS-SEC clusters occupations together into eight
> groups, from 1 (higher managerial and professional,
> which includes doctors, CEOs and lawyers) to VII (routine
> occupations such as bar staff, care workers and cleaners),
> while VIII is those who have never worked or who
> are long-term unemployed. The NS-SEC... gives a clear
> definition of class, which is based on employment.[32]

Because the NS-SEC classifies all jobs in the creative and cultural industries in classes I and II – due to their high levels of creativity and autonomy – it means that, by their reading, there is no such thing as a working class person in the arts. You can only ever be 'working class origin'. Indeed, when I heard Darren McGarvey being broadcast on BBC Radio 3 as part of their Free Thinking Festival or reviewing the newspapers one Sunday morning in summer 2018 on the bastion of bourgeois values that is BBC Radio 4's Broadcasting House, I had a feeling he might have crossed the line. But it's clear class divides run deeper than the sociologists can quantify. In their 2019 book, *The Class Ceiling: Why it Pays to be Privileged*, authors Sam Friedman and Daniel Laurison illustrate the extent to which 'fundamental inequalities in resources (economic, cultural and social)' originate 'from a person's family background'.[33] Inequalities are regional too, and over the last 40 years we have seen the increasing 'spatial polarisation of the UK',[34] as wealth has been centralised in certain locations – an issue I'll come back to shortly. But first, one final definition.

Darren McGarvey defines 'social mobility' as the 'gentrification of class consciousness'.[35] It is certainly a concept which has only come to the fore as inequalities have intensified. The sociologists working on the *Panic!* research define and measure 'social mobility' quite precisely as the difference between what you do for a living now (your 'destination'), and what your parents did when you were 14 years old (your 'origin').[36] In 2016, David Cameron's Conservative government set up the Social Mobility Commission, as part of the controversial Welfare Reform & Work Act 2016, which also imposed widespread changes to welfare including introducing Universal Credit and the 'benefits cap' – policies which have been responsible for pushing many more people across the UK into poverty.[37] It's hardly as if we needed a whole new 'commission' to tell us how cutting social protection for the poorest in our society might affect their ability to get a 'good job'.

The one thing the Social Mobility Commission has done is publish an index of all the local authorities in England showing the 'differences between where children grow up' and their chances of 'doing well in adult life'. In 2017, Ealing was ranked number ten out of 324 council areas, and of the top 20 all are in London or the south-east.[38] According to their Index, even if you're born to working class parents in Ealing, then you're in the tenth best position in England to move into a 'middle class' profession yourself (as measured by the NS-SEC). Perhaps this explains why the local comprehensive which I attended in the '90s, Drayton Manor High School, has many ethnically diverse alumni who have been most 'mobile' – BBC Economics Editor Kamal Ahmed (born 1967), Turner Prize- and Oscar-winning artist Steve McQueen (born 1969) and Iranian-born comedian Shappi Khorsandi (born 1973). That's the effect of having access to the metropolis, where the global 'economic elite' live and where so much power, wealth, skills, investment and opportunity has been centralised. For example, public spending on public transport in London is 20 per cent more per capita than in Glasgow and bus fares are only £1.50 (compared to £2.50 on First Glasgow). This means young people are actually able to travel to take advantage of opportunities all over the city. So it seems that, no matter what your background, being a Londoner does, quite simply, make you more privileged. Does that mean that anyone with southern accent should be treated with resentment and contempt? Something to ponder later perhaps...

Waste not, want not

When I was 14 years old, my mum was a 'further education teaching professional' (Class II) and my dad was retired and therefore 'not classifiable' in the NS-SEC.[39] He was born in October 1933 and was 45 by the time I arrived. He grew up in Nuneaton, a town in Warwickshire now ranked 296th on that same Social Mobility Index.[40] He was one of the 'lucky ones' who passed the eleven-plus and made it into the Grammar School, from where he was able to get a free university place after doing his compulsory National Service in the Royal Air Force. He was the first person in his family ever to go to university. After graduating, he started working as an accountant in London, but didn't like the ethos of the corporate world and so decided to go into teaching instead, which is where he eventually met my mum. He made it to the giddy heights of Head of Department of Business & Professional Studies at Uxbridge College (Class II or perhaps even Class I), before escaping when the new neoliberal management team swept in with their plans to 'restructure'.

I was always mortified having a dad who was a pensioner and who seemed so much older than everyone else's. I remember the trauma of being taunted at school: 'Is that your granddad?' the kids would say. He'd had two hip replacement operations (due to arthritis) before I'd even left school. From an early age, this installed a heightened anxiety about mortality, which has haunted me ever since. But having an 'old parent' also opens your eyes to different experiences and ideas. At the start of the Second World War, when my dad was six, he was evacuated to a small village called Rochford, having to leave his parents behind to face the bombs being aimed at nearby Coventry. Both my parents grew up under rationing, with a 'waste not, want not', 'make do and mend' mentality, which I absorbed as a child. As I grew up and began to fuse this ethos – based on the careful conservation of resources rather than mindless consumption – with environmentalism, I adopted it wholeheartedly as a guiding principle for every decision I made.

The irony was that the more I began to incorporate the 'waste not, want not' ethos into all of my thinking and action, the more I began to notice the glaring contradictions in the way my parents behaved. As much as my dad professed to be a 'socialist', proudly recalling that he'd voted Labour in every single election (bar one) since 1955, he didn't seem to think twice about buying a holiday home, investing in stocks and shares and indulging his shop-a-holism by supporting many exploitative multinationals (such as Primark, Matalan, Wilko, Tesco and more recently online at Amazon) filling both houses with unnecessary crap. Perhaps this behaviour is just the sort of backlash to expect from any prolonged period of state-imposed restriction. But it was the 'cognitive dissonance' that always wound me up: saying one thing and doing the other. Talking the talk, but not walking the walk. But I also wonder whether it's the difference between an 'upwardly mobile' person, like my dad, and a 'downwardly mobile' person, like my mum. It's easier to reject something that you have tasted, thought about and decided you don't like (as Gertie and Gus the bears did) than to reject something which our entire culture says you should aspire to because that's what proves you have been a 'success'.

Major setback

In 1992, there was a General Election. Most people (especially the man himself) thought Neil Kinnock was going to win it for Labour and put an end to 13 years of Tory rule (that had been almost my entire life). We stayed up watching *Spitting Image* before the results came in. I remember guffaws of laughter ringing around the living room. By the morning my mum was not so amused; struck with depression and disbelief that somehow John Major had managed to win. She cut a cartoon by Steve Bell out of *The Guardian* and put it in a clip frame on her bedroom wall. It depicted a precarious looking 'Tower of Babel' crushing the mass of people at the bottom. There were a few elite men in suits grasping huge bags of money perched at the top. Around the sides were wrapped the words:

We rule you,
We fool you,
We thicken you,

We sicken you,
We tax you,
We sacks you,
We are outrageous,
But you vote for us.[41]

By the mid-'90s, I'd got in with a 'gang' at my high school, Drayton Manor. It was mainly kids from middle class backgrounds who'd gravitated together, but the naughty ones into drinking, smoking and generally disobeying our parents. I remember the first time we were allowed 'into London' alone, when we were 13 or 14. We bought child Travelcards, caught the Central Line and hung about in Covent Garden, making a nuisance of ourselves by attempting to busk. That was an exciting place. On one of these trips 'into town', we heard about a big demonstration which was about to happen. 'Kill the Bill: The Tories are the Real Criminals' it said on the flyer. I didn't really know what a 'Bill' was at that time, but it seemed important and looked like it was going to be fun. It was 9 October 1994, the third in a series of demonstrations that year against the Tories' Criminal Justice & Public Order Act (which had been the Criminal Justice Bill in its provisional parliamentary form, hence the 'Bill'). This legislation threatened to make 'non-violent protest a criminal offence',[42] and to clamp down on rave culture by banning, specifically, the use of music in public places that had 'a succession of repetitive beats'.[43]

The demo itself was like a party, loads of 'repetitive beats' and dancing. There were hundreds of police around, wearing hardcore riot gear. As we made our way towards Hyde Park, things started to get rowdier and there were a few scuffles. I remember a sharp intake of breath and this burning on the back of my throat which made us all gasp for air. Someone shouted 'gas!' and we covered our mouths with our sleeves and fled towards the park gates until we could breathe safely again. Inside, it felt like a battle scene. A huge line of police horses facing a crowd of protesters, including me and my three school friends. The protesters advanced, so did the horses. Several times they charged. I remember things being thrown and the terrifying thud of horses' hooves galloping towards us as we ran for our lives.[44] We returned home safely about 9pm. My mum was there, as she always was, waiting anxiously for our arrival: 'Where have you been pet?' 'I've been in a riot!'

I gave my mum and dad far too many sleepless nights in my teenage years, which I now feel awful for. Drinking was what we did. Every weekend without fail, we would hit the vodka, the cider, the Tennent's Super, down in the local park before we got our fake IDs sorted and could get into some of Ealing's many pubs. We gradually made our way up from one, to two, to four cans of Strongbow Super, at which point I would generally start vomiting. Nearly all our friendships were based on how we related to each other in a drunken state – that was certainly the case with my first boyfriend who I barely ever saw when I was sober. We were always too drunk to bother with 'safe sex', no matter how many AIDS adverts we may have sat through on the telly.

When I passed my driving test at 17, I just carried on. Sometimes I 'borrowed' my mum's car when she wasn't there, sometimes I asked permission. One night in November 1996, I took it to drive my friends home after a few too many cans of cider at the local fireworks display. Three of us set off towards Hanwell. 'You're Gorgeous' by Baby Bird was playing on the radio and we were singing along in unison. I was going too fast, I couldn't judge the bend in the road properly in my inebriated state. 'Look out Ellie!' my friend Caro said. It was too late. The little Ford Fiesta wrapped itself around a lamppost and we all hurtled forward. Two of us had our seatbelts on, but Caro hit the windscreen. Then the noise started. A hideous roar coming from the engine. It was loud and unrelenting. Caro had blood on her face but she was still conscious. We all jumped out thinking the car was going to blow up like it always does in the films. It didn't. But 'Shit! What do we do?' After some panicked deliberations, we went to the nearest payphone and I sheepishly called my mum. By the time my parents arrived, the police were there and so was the ambulance. Caro was taken to Ealing Hospital to be treated for concussion, and my parents had to watch while I was bundled in the back of a 'meat wagon' and driven to Ealing Police Station.

That was the first and only time I was grounded, which I was almost actually grateful for. I certainly knew I deserved it. My behaviour was totally out of control and it was a miracle no one had been seriously hurt. Shortly afterwards I caught glandular fever, so I had to stay at home for two weeks anyway. Spice Girls' 'Say You'll Be There' was playing on loop on *The Box* cable channel on the telly and I had lots of time to reflect. The next summer I was meant to be taking my A-Level exams: Art & Design, Mathematics and Physics. I had dreams of being an astrophysicist, perhaps even going into space. It was my mum that suggested I go to art college instead. Apparently, there was this thing called a 'Foundation Course', which I could do at a local college in Hounslow and stay at home for another year. Maybe that would give me a chance to channel my energy in more positive directions and to finally grow up.

CHAPTER 2

Creative Decade

Things can only get better

I WAS 18 in 1997, just in time to vote in my first General Election. Everyone was excited about Tony Blair. His campaign's theme tune, 'Things Can Only Get Better' by D:Ream, captured a sense of optimism and hope for a brighter future. He was elected Prime Minister on 1 May 1997 with a landslide victory, finally putting an end to 18 years of Tory rule. Shortly afterwards, the dream started to go sour. He introduced tuition fees in 1998, the year I was planning to go to university. My mum was angry at what she saw as a betrayal. It was an inconvenience for her as she was still slaving away to support us in a variety of teaching jobs, but there was still never any question that I wouldn't be able to go.[1] My mum was the second generation in her family to go to university, my dad was the first in his. It would have been strange for me not to go – it would have been seen as 'downward mobility'.

By then I wanted to be a 'famous artist' anyway. I had been along to see the exhibition *Sensation* at the Royal Academy of Arts during the first few months of my Foundation Course in autumn 1997. It surveyed the work of the new generation of 'Young British Artists' (YBAs) like Damien Hirst, Tracey Emin, Michael Landy, Sarah Lucas, Sam Taylor-Wood, Gillian Wearing, Rachel Whiteread and others. With the entrepreneurial attitude encouraged by the neoliberal policy of the '80s, they had attracted the attention of the art collector Charles Saatchi. The *Sensation* exhibition ultimately just became a big advert for this Tory millionaire's own private art collection – hugely inflating the 'exchange value' of all the artworks he owned. But I didn't understand the intricacies of the art market at that time. I just thought it was fucking cool.

I knew I wanted to study Fine Art, so my parents took me on a road trip. We went to Hull, then to Nottingham, Loughborough and Leicester, and also to Kingston just down the road, to look round different art colleges. When we got up north, I started to feel a little uneasy. Was I really going to move this far away from home? But then, when we reached Nottingham, things changed. A final year student took us on a tour of their studios. There was a huge homemade banner in the main space saying 'Fine Art Party this Saturday!' Then she showed us her latest creation: a pair of pants that pant! I thought that was hilarious. This was definitely the place for me. What I liked about Nottingham Trent University the most was that you didn't have to 'specialise' into any particular medium, you could follow your own interests and ideas and essentially make and do whatever you wanted. I applied, went for an interview but didn't get in. I felt robbed. I wouldn't take no for an answer and so all summer I pestered Claire Simpson, the course administrator, sending her letters and photos of my new work. Then, a week after the official start date, I got a message on my pager from my mum to say they had found me a place. I was just unpacking my belongings in the

Ellie Harrison in December 1998 with her *Take Aways* installation, one of her early experiments as a first year Fine Art student at Nottingham Trent University, inspired by 'student life'. (Ellie Harrison)

breezeblock clad halls of residence at De Montfort University in Leicester. My mum and dad drove up, helped me pack up everything again and move the 30 miles north to Nottingham where I had found a room in Sandby Hall, right behind the college.

I do believe in love at first sight. The first year studios at Nottingham Trent University were in a big vaulted room with a mezzanine. I had been allocated a small space on the ground floor – my name pinned to the wall on arrival. On my first day settling in, I remember looking up and seeing this boy on the balcony. He had dark black curly hair and the sweetest face I'd ever seen. He looked lost and afraid, a bit like how I felt. I remember thinking I wanted us to get married – that's what you're meant to do, right? I crept upstairs and looked at the name labels pinned up in the balcony spaces. That must be Jon Burgerman. I would have to track him down. We became best friends and eventually got together after an excruciating year of not really knowing what was going on between us. He knew I was 'bisexual'. It didn't seem to matter; we stayed together for nine and a half years. It was a loving partnership based on building our own individual careers.

I learnt a lot from Jon. He was one of the first people I knew to have a computer in his room at halls – a huge grey beast of a thing, bought for him by his mum and dad from Time Computers in Birmingham. At the weekends, while I was out with the rest of the students drinking and dancing to drum 'n' bass at Dubble Bubble, The Bomb or The Social, he was working away at home. He loved what he was doing so much – doodles, paintings and making lots of little digital collages and animations – that he didn't see the point in wasting time getting drunk. By my final year, it had started to rub off on me. I was totally obsessed with my work and had started to look after myself a bit better by going to the gym. I'd begun using my art practice as a way to answer questions I had about the world. Bringing in what I'd learnt in Physics at school, I began exploring the theme of 'energy'. What was it? How could it be converted from one form – say, food – into another – say, motion?

In 2000, I created an installation called *Superfluous Consumption*, which visualised the energy content of the different salty snacks we would shovel into our mouths whilst watching telly, and another piece in which a carrot and an éclair raced around two adjacent toy train tracks at speeds proportional to their calorific value – the éclair going three times faster than the carrot. In March 2001, just a few months before I graduated, I decided to set myself a challenge to see if I could document what one person consumes over the course of an entire year. Wouldn't it be amazing to see what that amounted to? Digital cameras were just appearing at that time. So I got one for my 22nd birthday and for a project named *Eat 22* (pictured overleaf), I recorded the details of all the 1,640 meals and snacks I ate over the next year.[2]

We take the facts of our lives for granted and it's only in reflection that it's possible to see how much they've helped or hindered you. When I graduated from university in 2001, I moved back home. I didn't have to pay rent; my mum was just pleased to have me back. When 9/11 happened, I was sitting in the same living room where I'd been 20 years before, this time watching the World Trade Centre crumble to the ground. My mum and dad were both there, and I felt so much relief that we were all together and safe as it appeared the world beyond was falling apart.

Being in London certainly helped my art career in those early days. I had the time to continue with my *Eat 22* project and benefited from being able to take up low-paid jobs in the arts. I worked as an invigilator at Royal Festival Hall and on the family activity programme 'Start' at the newly opened Tate Modern. It was fun. I met nice people, started to build a network of contacts and a few small opportunities began to arise. But I missed the support structure of the art college – the facilities for making stuff (particularly the band saw) and the intellectual stimulation.

So I applied to study a Postgraduate Diploma in Fine Art at Goldsmiths College, where many of my YBA idols had been before me. I remember having a heated debate with Gerald Helmsworth, the programme leader, in the interview, but remarkably they still offered me a place. I was awarded a grant from the AHRB (now the Arts & Humanities Research Council) which covered my tuition fees (£2,870) and a maintenance

0033 (16 March 2001, 7.46pm)

0106 (28 March 2001, 7.32pm)

0288 (27 April 2001, 7.36pm)

0455 (1 June 2001, 7.58pm)

0604 (30 June 2001, 2.49pm)

0889 (7 September 2001, 11.41pm)

1001 (5 October 2001, 12.57pm)

1342 (31 December 2001, 2.42pm)

1355 (1 January 2002, 11.02pm)

Ellie Harrison eating chips aged 22. Nine of the 1,640 photos taken between
11 March 2001 and 11 March 2002, as part of *Eat 22*, for which she documented
everything she ate for a year. (Ellie Harrison, Jon Burgerman)

grant for expenses across the year (£5,230). I used some of it to buy an annual Travelcard, known as a 'Gold Card' (for £912), which meant I could take any bus, train or tube within the capital. I was commuting from home to New Cross – three hours each day, back and forth across that colossal city. I had lots of reading time. And I became aware of the huge distance I was clocking up, so I decided to document it for a project called *Gold Card Adventures*. It worked out to be 9,210km in total over the course of the year, the same distance as travelling from Ealing Broadway to Shanghai.

Those were anxious times in London in the wake of 9/11 and in the run-up to the Iraq War. I remember being totally bemused at why a manhunt for Osama bin Laden who was supposedly hiding out in the mountains of Afghanistan had turned into a campaign against Saddam Hussein in Iraq. Wasn't it a bit hypocritical for the countries who already had the nuclear weapons (the UK and America) to be telling the others they weren't allowed them? Surely it would make more sense for everybody to disarm? I went on all the Stop the War Coalition demonstrations with my mum and other family and friends, including the big one on 15 February 2003 attended by more than a million people. When Tony Blair refused to back down and instead joined American President George W Bush in launching an attack on Iraq, I vowed never to vote Labour again.

I didn't get on very well at Goldsmiths College. I felt totally out of my depth. It left me with a huge inferiority complex and opened my eyes to how much I still had to learn about the world. Perhaps I'd get more out of education if I was a bit older, with more experience in the 'real world'? I also clashed with some of the tutors who were far more commercially-minded. I had just made a small online piece called *My Head's Swimming*, which captured everything I was thinking about – from the personal to the political – whilst swimming lengths at my local pool. I remember one tutor telling me not to bother making anything for the internet as nobody would ever see it, and how would I ever make any money?

The knowledge economy

Conceptual art first began to appear in the '60s as a reaction against the art market and as resistance to the commodification of art – that is art being sold, circulated and accumulated as 'designer objects' or 'investments' – a process which completely nullifies any political agency that it may once have had. Conceptual art was also a reaction to consumer culture and the detrimental impact it has on our environment. In 1969, American artist Douglas Huebler said 'the world is full of objects, more or less interesting; I do not wish to add any more'.[3] His peer, artist Lawrence Weiner, followed up with:

> Industrial and socioeconomic machinery pollutes the en-
> vironment and the day the artist feels obligated to muck it
> up further art should cease being made. If you can't make
> art without making a permanent imprint on the physical
> aspects of the world, then maybe art is not worth making.
> In this sense, any permanent damage to ecological factors
> in nature not necessary for the furtherance of human
> existence, but only necessary for the illustration of an art
> concept, is a crime against humanity.[4]

As always, it was the critical artists who were the trailblazers. Not only did they call the devastating impact humans were having on the environment in the '60s and actively try to do something about it but, in the 'dematerialisation' of their art, they prefigured the way in which the economy would be transformed in the neoliberal age: away from the manufacturing of goods, to the delivering of services or experiences. (Economists handily define 'goods' as something you can drop on your toe and 'services' as something you can't). The birth of the 'knowledge economy' in the '80s was the perfect example of capitalism catching up, as it always seems to do, and finding new ways to commodify that which had previously been thought of as uncommodifiable and to co-opt any resistance to its system. As well as the conceptual work of these once 'radical' artists now being bought and sold for vast sums by commercial art galleries like Lisson in London (who now 'represent' Lawrence Weiner), so too was the 'knowledge economy' born from the privatisation of ideas.

In 1986, the same year as the 'big bang' on London's Stock Exchange, the precursor to the World Trade Organisation (the General Agreement on Tariffs & Trade) first agreed its Trade-Related Aspects of Intellectual Property Rights (TRIPS) rules.[5] Now built into all of the WTO's international trade deals, TRIPS 'plays a critical role in facilitating trade in knowledge and creativity'.[6] It sought to fill the vast voids left in post-industrial economies by creating new service industries which could not so easily disappear to other parts of the world in search of cheaper labour. The commercial value was now inside our heads. When Tony Blair and his Chancellor Gordon Brown came to power in 1997, they saw an opportunity for economic development in the arts and coined the phrase the 'creative industries'. Over the next ten years, they pumped loads of public money into the arts sector in what is now referred to as New Labour's 'Creative Decade'. It's uncanny how this period of sustained investment mirrored almost exactly the first ten years of my art education and career.

Not only did New Labour want more people to see the arts as a career opportunity – to become 'arts professionals' – but they also thought that the arts could be a panacea for social injustice. As art critic Claire Bishop writes in her book, *Artificial Hells: Participatory Art & the Politics of Spectatorship*, the New Labour government asked 'What can the arts do for society?' answering: 'increase employability, minimise crime, foster aspiration – anything but artistic experimentation and research as values in and for themselves'.[7] Arts and culture were being used to paper over the huge cracks – increasing social inequalities, environmental destruction and financial instability and uncertainty – caused by all the neoliberal policies which New Labour had wilfully embraced. Publicly-funded artists became complicit in this prolonged PR stunt. As artist and writer Morgan Quaintance puts it,

> neoliberal capitalism has created an existential
> framework in which compromise and complicity are the
> new original sins...[8]

The 'knowledge economy' and the birth of the 'creative industries' (and now the 'gig economy' too) have also helped increase levels of self-employment – individualised workers competing against each other for scarce resources and opportunities. These new industries have been major contributing factors to the increasingly fragmented and precarious labour market we have now. Well over 15 per cent of the UK's workforce is now self-employed.[9] A largely non-unionised group for whom risk has been privatised – pensions, sick pay and the rest have become the individual's responsibility. In 2011, suffering from the negative side effects of the 'freelance' lifestyle myself, I staged the *Work-a-thon for the Self-employed* event in London and in Newcastle, inviting a group of people to join me in setting a world record for 'the most self-employed people working together (on their own individual projects) in the same place at the same time, over the course of a normal 9-to-5 day'.

The call-out for participants read:

Lone Workers Unite!
• Are you one of the growing number of self-employed people working in the UK?
• Do you often feel lonely or isolated whilst working?
• Do you find yourself working long and unregulated hours?
• Are you often 'too busy' with work to socialise?
Well, why not take part in the Work-a-thon for the Self-Employed?[10]

One of the things I want to examine in this book is the consequences of dematerialisation and digitisation, and the trading in ideas. Humans are material things. We cannot be transmitted in an instant to all corners of the world as our ideas now can. While advances in communication technology potentially reduce the need for travel, the reality has been the opposite. Producers in the 'knowledge economy', spend their lives playing catch-up, desperately attempting to chase this immaterial knowledge around the world. You only need to look at the number of international academic conferences and art biennales, to see there is an elite operating within this 'knowledge economy' who travel far too much. It's a vicious circle. They find themselves trapped in a desperate and unending search for meaning in their lives, precisely because they spend so much time on the move. Chasing opportunities from one city and one country to another, they feel little or no connection to the real material places where they happen to be 'based' at any time. Meanwhile the majority of the population are trapped at home, excluded access from this world at all.

A golden age

When I left Goldsmiths in summer 2003, I decided to move back to Nottingham to live with Jon. It had been a struggle to keep our relationship going for the two years we had been apart. I had access to the 'bank of mum and dad', which had been topped up with inheritance from my mum's parents and we were able to buy a flat in a former lace mill turned 'gated community' near Sneinton Market. I remember the big day when Jon and I sat at the desk filling in the forms, with my parents on one side and his on the other. It was the closest experience to a wedding I think I'll ever have. Quite a few of our friends from university were still living in Nottingham and had set up artists' studios above a junk shop in an old building close to the station (a building now demolished to make way for Nottingham's expanding tram network). I rented a small room for £40 per month and began working with their art collective Reactor on several performances and events.

In those days – in the middle of the 'Creative Decade' – Arts Council England had offices in each of its nine regions and the one for the East Midlands was in Nottingham city centre. The East Midlands was seen as the region with the least amount of cultural activity and the smallest number of working artists, so it was a top priority for 'investment'. We all got professionalised. I remember going to a workshop about 'Becoming self-employed' at the Inland Revenue offices (now HMRC), even one about 'Pensions for Artists'. I started 'signing on' for Jobseeker's Allowance at the start of 2004, but stopped a few months later when I registered as self-employed myself that April. I was no longer an unemployed statistic, making the government look bad. I was a 'sole trader'! I started to keep meticulous records of my incomings and outgoings then. We had other help from 'small business advisors'. Reactor was getting loads of money out of Arts Council England, so Jon and I decided to give it a go.

I remember the day I wrote my first funding application on Thursday 5 February 2004. Jon and I had egged each other on. We had a two bedroom flat which was basically just two offices. I was in my office and he was in his. For the whole day we sat bashing out words into the Arts Council's proforma, justifying our ideas and dreaming up 'budgets'. We didn't emerge until around 7pm when we started to get hungry. By then, we'd both nearly cracked it. A little bit of proofreading and neatening up and that was it. We trotted along to the Arts Council offices to hand them in the next day. We were both successful. I was awarded £4,965. I can't remember what Jon's project was, but mine was to start researching and developing a big group exhibition. I wanted to bring together a group of artists who 'collect, list and database the data of everyday life'.[11] The money was to invite ten artists to attend an 'exhibition development workshop' at Angel Row Gallery, which was then the main publicly-funded contemporary art gallery in the city (before Nottingham Contemporary opened in 2009). The 'exhibition development workshop' was a clever way of doing it as, not only did I manage to convince the gallery to help produce and launch the exhibition, but the initial pot of money was essentially there to help create a proposal for a much bigger grant. At the end of 2004, I found myself writing that funding application with help from gallery staff. We were awarded close to £34,750 towards a £56,850 total budget and the exhibition *Day-to-Day Data* launched in summer 2005 and toured to Portsmouth and then London in 2006. I then wrote another small funding application for £3,345 to stage a symposium at the ICA in London in March 2006.

This had all happened so quickly and involved such a huge amount of work that I barely had time to think critically about what I was doing. It was then that I got into the habit of working seven days a week. I had taken on so much that I had to cut back my alcohol consumption so that hangovers would no longer affect my productivity. As well as all this 'freelance' stuff, I had two other jobs. I had managed to pick up some teaching at Nottingham Trent University, who were impressed by the fact that I now had Goldsmiths College on my CV, and I was also an usher at Broadway Cinema. I began to notice the way people treated me differently depending on which 'hat' I was wearing and how much power I had at the time. It was particularly fascinating when my students came into the cinema and saw me taking tickets and picking up litter. Sometimes they didn't even notice me.

By summer 2006, I'd begun to reflect more on the systems I was operating within, and was concerned about the divergence of my art and my politics. Why was I not using my art to express how I really felt about the world? This was around the time I met the American artist Joanna Spitzner, when we were commissioned to work together on a project called *Part-time*. Her work was centred on making visible the hidden economies of the arts, especially in America where they have less public funding and rely much more on 'philanthropy' from wealthy individuals or corporations. She aimed to make people think about where this money actually comes from and who has the power to control where it goes, so she set up the *Joanna Spitzner Foundation, Inc.* She got a job at Walmart where she worked evenings and weekends – every spare moment around her other life/work commitments. All the money from the Walmart job went into the *Joanna Spitzner Foundation, Inc.* account, where it began to accumulate. After a year or so, she had acquired the power to begin awarding a series of small grants to other radical artists she admired.

Joanna and I, along with Liz Kearney, another artist from Liverpool, founded the *Union of Undercover Artists* in 2006 in order to make visible hidden labour in the arts. We went on a trip to Blackpool together where Joanna had been working in a B&B over the summer. We bought red cowboy hats, went to Funny Girls cabaret and pretended to be on a 'hen do'. It gave me an idea. Wouldn't this be a great format for a networking event for female artists, writers and curators – having a relaxed weekend away in a seaside town? I started work on another funding application and was awarded £12,508 from Arts Council England to put on a pilot in Bexhill-on-Sea. The first *Hen Weekend* event took place in March 2007 at the De La Warr Pavilion. As we sat eating delicious meals and quaffing the odd glass of Prosecco, all funded by the taxpayer, we were oblivious that this marked the end of an era. That same month Tony Blair made a speech in Tate Modern where he said 'the past ten years had been a golden age for the arts. Imagine... Britain's cultural scene without the Labour government's doubling of cultural funding since 1997'.[12] It was a warning that the life we had all become accustomed to, the only working life I'd ever known, was about to come crashing down. The glory days were over.[13]

Technologies of the self[14]

My next three funding applications were all unsuccessful. Opportunities in Nottingham were drying up. I was forced to rethink. I took the failure personally, putting it down to my own stupidity, so I began an intense period of self-education. When I was studying at Goldsmiths so many weird-sounding names were bandied around: Hegel, Marx, Nietzsche, Freud, Bataille, Foucault to name but a few – who the hell are all these dead blokes and what relevance do they have to me? I couldn't quite grasp it at the time, but now I wanted to try to work it out. I started to collect them, to write the names down in the back of my notebook and then to look them up. It turned out they were philosophers and theorists who, over the last few centuries, had offered some profound reflections on the way humanity operates. I found it fascinating and addictive. I wanted to try to capture and visualise this knowledge so, over the course of a year, I made an enormous five-piece colour-coded wallchart, with all the names arranged in chronological order from the thinkers of ancient Greece right up to the present day.

There was Socrates (470–399 BC) who declared 'the unexamined life is not worth living', that only in striving to come to know and understand ourselves and the world around us do our lives have any meaning.[15] Then there was Diogenes the Cynic

(412–322 BC). He thought that people had been 'corrupted by the false beliefs of civilisation, such as the belief that the most important thing in life is to win success and status'. He set about undermining these beliefs or 'debasing the currency' as he called it, by living in a barrel and masturbating and defecating in public without a care in the world.[16] And Thomas Hobbes (1588–1679) in England who argued that life in a 'state of nature' without a civilised society would be 'solitary, poor, nasty, brutish and short'.[17] As part of my research, I bought a tiny pocket-sized book called *The Rebel's Guide to Marx*.[18] It was the first time I'd really thought about how the economy shapes our society, and how we might be able to fight back if we come together. 'Workers of the world, unite!' proclaimed Karl Marx and his pal Friedrich Engels in *The Communist Manifesto* of 1848.[19] Marx was also critical of the thinkers who had come before him: 'The philosophers have only interpreted the world in various ways; the point, however, is to change it'.[20] It was a rousing call to not just sit on our arses drawing attention to the world's problems, but to get out there to try to solve them. The more I learned, the more I realised there still was to learn. It was an overwhelming and never-ending task – a life-long project.

I began to realise the interconnectedness of mental and physical well-being. The more I looked after my body, the better my mind would function, the better I felt and, therefore, the more likely I was to continue making time for exercise. It was swimming I got 'hooked' on. My mum helped get me into it. She'd taken me and my sister for lessons at Gurnell Leisure Centre in the '80s – the brand spanking new 50m pool which opened in Ealing in 1981 (and which is now scheduled for demolition). I stopped going when I was a teenager and found other less wholesome ways to spend my time. But when I was living at home again those two years in my early 20s, she inspired me to take it up again. A couple of mornings a week she would get up early and head off to Gurnell for a swim. 'Why don't you come along too, El?' she'd ask and so I did. It became addictive. I started to notice how my thoughts were clearer and I was happier and so much more productive on the days I'd been to the pool. My body started to change shape too; broader shoulders and more muscular arms.

As a motivational structure to keep me going with this healthy habit, I started to record all the lengths I did adding them together to see if I could swim the 5,400km distance across the Atlantic over the course of my life. In 2004, I made a little online programme called *Trans-Atlantic Challenge*, which constantly recalculates my estimated finishing age based on my current rate of progress (about 72 at present). The thought of that age continually creeping upwards, so it becomes less achievable, still makes me get up and go. When I moved back to Nottingham, I knew that I wanted to live close to a swimming pool; by then I saw it as essential for my mental and physical health. That pool was Victoria Leisure Centre. I swam there a couple of times a week for five years. One of the regulars was a woman called Margaret. She had been swimming in that same pool all her life. She must have been in her 70s by then, but was still ploughing up and down doing front crawl. Margaret was an inspiration. I knew I wanted to be like her when I got old.

Then on 8 February 2008, Nottingham City Council made an announcement that it was proposing to bulldoze our 150-year-old swimming pool as part of its 'Leisure Transformation' programme. No! What about Margaret? What about me!? Local swimmers quickly mobilised and my neighbour Mat Anderson and I founded the Save Victoria Baths campaign within a few days. I was able to use the web design skills I'd acquired at Nottingham Trent University to put together a logo and get our website up and running. It was an intense month with three public demonstrations on 19 February, 1 March and 10 March 2008. We met regularly to plan actions and used the handy new platform of Facebook to mobilise thousands of people across the city. It was the first time I'd visited a councillor's surgery. It just so happened that the councillor for St Ann's ward where I lived was the head of Nottingham City Council, Jon Collins. It was a productive meeting. He said that the final decision would be made at the council committee meeting on 18 March 2008 and if campaigners came along we would be able to attend.

On the day, a huge crowd amassed at the Leisure Centre ready to march to the council building in Old Market Square. We got to the doors and they wouldn't let us in. 'This is an outrage,

Jon Collins said we could!' I demanded to see him, and finally got past security and ran quickly up the stairs. I challenged him about what he'd said at the surgery and he backed down. A small group of us were then allowed up the stairs to make our views known. The Council did a complete U-turn and instead decided to invest £7 million in rebuilding a new Leisure Centre on the same site. It was an empowering experience. I made lots of new friends in the city; people who had lived there all their lives. It inspired me about the possibilities of a participatory democracy and helped me to identify some of the levers of social change.

Community vs career

There was no doubt the Save Victoria Baths campaign had taken over my life. It didn't end on the 18 March 2008, but continued for another four years to hold the Council to their promises until the new facility was finally open in March 2012 (I continued to update the website for Mat after I'd left Nottingham). But activism wasn't something you could do full-time – surely you need to make some money somehow? And what about my ambitions of becoming a 'famous artist'? Would I just put those on hold? Against this victory and the new friends and community spirit that had been fostered, I had another quite contradictory impulse – to get the hell out. I had a poster on my office wall by the artist Mark Titchner; a sort of motivational slogan, which read: 'If you don't like your life you can change it!' I stared up at it every day. I had a massive 'seven-year itch': not only in my heterosexual relationship, but also with the city where I lived. The altruistic activist was clashing with the self-interested artist. It was community vs career. Human instinct vs the demands of the economy.

I'd begun applying for Masters courses in 2007 – to the Royal Academy of Arts and the Royal College of Art in London. I didn't even get an interview that year. My escape plans had been thwarted. I thought I'd better take it a bit more seriously next year. So in 2008 I applied again to the Royal Academy of Arts and the Royal College of Art and also to The Slade and Chelsea in London, as well as to Piet Zwarte Institute and the

Rijksakademie in the Netherlands and to Glasgow School of Art. I spent months just working on all the applications. This time, I got three interviews: one at The Slade, one at the Royal College of Art and one at Glasgow School of Art, which took place on 13 March 2008. I was offered two places. It now looked certain I was going to leave – but where to: back to London, or to Glasgow?! I didn't know what to do. I applied for scholarships at both places, but I wouldn't find out until later in the summer. I didn't want to be in a worse financial position after two years' studying, so I decided to hang on and see. It wasn't until 29 August 2008 that I finally got a letter from Glasgow School of Art to say I'd been awarded the Leverhulme Scholarship – an £18,000 bursary over two years. It would cover the annual fees of £3,315 and leave £5,685 for modest living expenses each year, which I could top-up by doing a bit of 'freelance' work over the summer.

It was an offer too good to refuse. I had just over two weeks to 'get on my bike' and leave. It was the attitude Margaret Thatcher's Employment Secretary Norman Tebbit had espoused: 'I grew up in the '30s with an unemployed father. He didn't riot. He got on his bike and looked for work, and he kept looking till he found it' he said in 1981.[21] I had been programmed through my education and the competitive environment into which I'd graduated to think that once I'd exhausted all the local opportunities a city had to offer, when it felt as though my career was standing still, I had no choice but to move on. What I didn't stop to question was the increased carbon emissions and the social isolation that comes with moving yet further away from home (shown in my Carbon Graph on pages 138–9). Nor the fact that my ability to live this 'mobile' lifestyle, that allowed me to chase opportunities around the world, was because I had a 'bike' to get on in the first place. My parents bought it for me and taught me how to ride it, literally and metaphorically. At least I could leave knowing that plans for the new Leisure Centre were now well underway and that I'd played a small part in helping make Nottingham a better place than when I'd arrived.

CHAPTER 3

Welcome to Scotland

Dark clouds

I 'PARACHUTED IN' to Glasgow on Monday 15 September 2008.
I remember the day clearly. I had hired a Transit van. The driver
arrived early and we loaded up all my worldly possessions, which
had been carefully boxed up over the previous few days. It was a
sad day, leaving my old life in Nottingham behind. But much worse
for Jon, who was left alone in the domestic silence of an empty
flat. Our nine year relationship was now entering a new phase.
I was off to seek my fortune, in search of 'the Glasgow miracle'.
I remember driving up the motorway, particularly the bit when
you pass the sign that says 'Welcome to Scotland' and then
seem to be going through huge swathes of wilderness before
you reach city life again. Was I really moving this far away from
home? I was heading for a ground floor flat on Dudley Drive
in Hyndland, sharing with a man from Costa Rica called Juan,
who was also about to enrol on the Master of Fine Art (MFA)
course. I knew nothing about Hyndland. I knew nothing about
Juan. This really was an adventure. From 15 September 2008,
life would never be the same again.

Early that same morning, news started to break that the
fourth biggest investment bank in America – Lehman Brothers
– had gone bankrupt. Its $600 billion holdings disappearing in a
puff of smoke. This event came to be seen as the dramatic start
of the global economic crisis – the most severe since the Wall
Street Crash of 1929 and the Great Depression of the '30s that
followed – a crisis we're still experiencing the repercussions of
now. Deregulation in the '80s and the 'loads of money' financial
risk-taking that it encouraged had finally come back to bite
us, and bite hard it did. Suddenly other global banks started
to teeter, it looked like the whole financial system could come
crashing down, with all our bank balances wiped to zero. So

American President George W Bush stepped forward with a $700 billion bailout plan to settle all these banks' 'toxic debts',[1] largely caused by flogging 'subprime' mortgages which enabled people on low incomes to get on the 'housing ladder'. Our Prime Minister Gordon Brown (who had taken over the reins from Tony Blair on 27 June 2007) came forward with what amounted to a £850 billion gift[2] from the public purse to UK based banks: Royal Bank of Scotland and Lloyds Banking Group (which had swallowed up HBOS, the bank formed in 2001 through the merger of Halifax and Bank of Scotland). Suddenly the modest £1.6 billion allocated by the UK Government that year to fund its entire department of 'Culture, Media & Sport',[3] which includes over 50 public bodies including Arts Council England,[4] was put into perspective (it is 0.18 per cent).

Within my first few months in Scotland, the dark clouds of recession had begun to settle over the UK. They weren't the only dark clouds in Glasgow though, that's for sure. I had arrived totally unprepared. Nottingham, I realised in hindsight, is one of the driest places in Britain. Slap-bang in the centre of England, it is most protected from coastal storms. I'd come with my trusty bike, which had been my main mode of transport since I was a child.[5] Not only was it a shock how far apart everything was compared to life in Nottingham – where I was a 15 minute stroll from the city centre, my studio, the station and most other places I needed to go – the weather was beyond belief.

Within my first few days I had to get a complete waterproof suit – jacket and trousers – which I never went anywhere without. I quickly realised that even if it looked sunny when I left in the morning, it was very unlikely to still be when I returned. I experimented with the public transport – perhaps that would be a more civilised way of getting around the city? I remember an early trip into town to buy a kettle from Watt Brothers and to get a few bits from Woolworths on Argyle Street (in its last few months in operation before the whole chain collapsed in autumn 2008). I stood in the rain at the bus stop in Hope Street waiting for a First Bus to come along. Overcrowded, windows all steamed up, I squeezed on clutching my purchases. Why was it so bloody expensive? And how come you couldn't use the same simple ticket on Glasgow's trains, Subway and buses – First Bus or the many other private operators? That's something which has been straightforward in London since 2003 when Transport for London launched the Oyster card, and has encouraged bus patronage to grow (whereas in most other places in the UK it has declined).[6]

Glasgow's public transport was so complicated, frustrating and expensive that I decided to persevere with my bike. I found it hard work. Potholes everywhere and drivers who seemed to have little or no empathy for other more vulnerable road users. I remember one night on Byres Road trying to get home in the pissing rain when a van pulled up alongside me. The passenger took the trouble to wind down the window just to hurl out some abuse as the rain, and the tears, rolled down my face. I eventually got used to what life is like for a cyclist in Glasgow, where you're in a tiny minority and therefore viewed as a nutter. I now realise this is because Glasgow has significantly fewer people cycling on a daily basis than many cities in England, just 1.6 per cent of the population[7] (though this is now fortunately on the increase). In London it's 3.7 per cent and in Amsterdam it's 60 per cent![8]

The flat which Juan had found before I arrived was a ground floor tenement. My bedroom was at the back with a window overshadowed by the railway embankment. There was no mobile phone reception and no internet to start with either. I felt totally disconnected from everything I had known before. Sleeping alone in a bed for the first time in over five years was profoundly isolating. It was damp too. It used to take a full seven days for my washing to dry on the clothes rack in my bedroom. I just thought that's what Glasgow life was like. I guess you could call it 'culture shock'. At least I was only from England, not Costa Rica! Juan couldn't hack it and so decided to leave after the first year of the course to head back to a life in Canada instead.

Creative education

But none of that really mattered, I was here to learn. I had been given this amazing opportunity: two years' study paid for with a bursary. I would not have to worry about money. I could just focus on my work. I was desperate to research and find out what was really going on in the world and to try to find a way of fusing together my art, my politics and the way I lived my life. Within the first week, I discovered that as a Glasgow School of Art student, I could also access the facilities at the University of Glasgow. It was just £40 for a year's swimming pass, so I started going four times a week at 8am before heading to the studio. I was also allowed to sit in on any of the university lectures. I got in touch with the philosophy department and, as well as the taught MFA programme, I put together my own multi-disciplinary timetable attending several university courses. First I did 'Introduction to Marx' and 'Environmental Ethics'. Then in the second semester I studied 'Hegel's Philosophy of Right', alongside two MFA electives: 'Psychoanalysis in Art & Culture' led by Laura González and 'Contemporary Art: 1960 onwards' led by the late great John Calcutt (1951–2018). In my second year, I studied 'Political Philosophy' and 'Meta-Ethics'. I was attempting to work out how all these things fitted together.

Again my art practice became a way to answer the questions I had about the world. My first works focused on the economic system: why do we keep getting these periods of crisis and recession? Inspired by Spanish philosopher George Santayana (1863–1952) who said 'Those who do not remember the past are condemned to repeat it', I looked to history as a way of understanding the present. I researched the number and frequency of financial crises over the last century and thought about fun and engaging ways to communicate this history to an audience. The popcorn making machine I'd won at a quiz in Nottingham the previous year came to mind. I remember the manic experience of overloading it and turning it on in our kitchen – the loud whirring noise, erratic popping, corn quickly filling the bowl and watching helplessly as it flew out all over the floor. That would be a nice metaphor for a financial crisis I thought. So I made the installation *The History of Financial Crises* (pictured opposite), using 11 popcorn making machines (each bought second-hand on eBay) to retell this history from the Wall Street Crash of 1929 up to the Global Credit Crisis of 2008, over the course of a day. When I could see that the majority of the machines went off in

The History of Financial Crises by Ellie Harrison, installed at Mejan Labs, Stockholm in April 2009. This installation uses a row of 11 popcorn making machines to re-enact the history of financial crises over the last century. Over the course of a day, the machines are programmed to go off at the time corresponding to the year in which the crisis occurred. (Ellie Harrison)

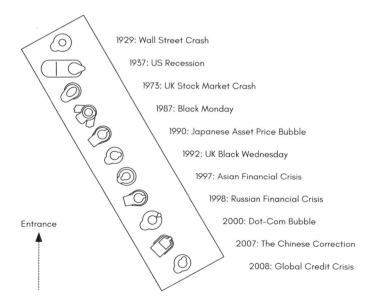

1929: Wall Street Crash

1937: US Recession

1973: UK Stock Market Crash

1987: Black Monday

1990: Japanese Asset Price Bubble

1992: UK Black Wednesday

1997: Asian Financial Crisis

1998: Russian Financial Crisis

Entrance

2000: Dot-Com Bubble

2007: The Chinese Correction

2008: Global Credit Crisis

the last hour, it suddenly became clear what was happening. The frequency of these crises was increasing – our economy was becoming more and more unstable.

It's all very well saying that capitalism – and the accelerated version that neoliberal policies have created – is 'bad', but clearly it wasn't that simple. I wanted to acknowledge and take account of our own complicity in feeding the machine. So I made a little piece called *Transactions* to be shown alongside *The History of Financial Crises*. This time, I used the 'Dancing Coke Can' that my sister and I had fought over in the '80s as our favourite toy. Designed to be placed next to your boombox, the Coke can has a small audio sensor inside so it can dance along to the tunes. I placed it on a small plinth next to an old Nokia mobile phone. Then for the duration of the exhibition, I sent a text message to the phone, each and every time I bought something out in the 'real world', causing the little Coke can to jiggle with joy. I was interested in the word 'transaction' and how in a society dominated by money where nearly everything we need to survive has to be paid for in cash, 'transactions' are increasingly replacing 'conversations'. The 'I give you this and you give me that' attitude overrides real human connection.

That same year I made my *Vending Machine*, which was re-programmed to vend out crisps only when news relating to the recession made the headlines on the BBC News feed. I always tried to use humour as a levelling force – to draw people from all sorts of backgrounds in and to make them think about the world around them in a different way; questioning the things they may have previously taken for granted. The question of whether or not it was 'art' they were looking at became irrelevant. I also found that humour in the form of 'satire' allowed us to laugh at the absurdities of the world we humans have created and our (self-)destructive urges.

As well as the start of global financial crisis, the political backdrop to my first year in Glasgow was the run-up to the United Nations Climate Change Conference in Copenhagen in December 2009 (COP15). I remember the moment I first realised fighting climate change was not a question of 'saving the planet', it was a question of saving ourselves. The penny dropped a couple of years before I left Nottingham, when Jon and I were watching telly. Louis Theroux was visiting Chernobyl (in Ukraine) for one of his documentaries. Chernobyl had been the site of a huge nuclear power disaster in 1986. The whole area around the town had been evacuated and left deserted because of dangerous radiation. The cameras toured around all the old Soviet housing blocks, which had by then been standing empty for 20 years. In that relatively short space of time, they had been reclaimed by nature. Huge trees had taken root in what had previously been bedrooms, greenery cascading down the walls. It was clear to see: the planet is going to be fine if and when humans are gone.

Many scientists were warning that the Conference in Copenhagen could be our 'last chance' to reach a global agreement on emission reductions which would prevent us from reaching the dreaded 2°C increase on average global temperatures from 1990 levels. It was at 2°C when 'runaway climate change' would be triggered, causing the rapid acceleration towards a planet largely uninhabitable to humans by 2100: rising sea levels flooding many of our big cities (including Glasgow and London), widespread desertification of land, fresh water shortages, crop failures and the displacement of millions of climate refugees. It was a future which was unlike anything we had known or become accustomed to in the western world. It was a future very different from the one 'promised' when I set out to follow the career trajectory of my YBA idols at the start of the new millennium.

Studying Environmental Ethics made me start to seriously question the ethical implications of continuing to make art – pursuing a career in the production of pointless objects in a world where the production and circulation of pointless objects was one of the major causes of our carbon crisis. We looked at the famous essay 'Killing & Starving to Death' by philosopher James Rachels (1941–2003) – about the moral differences between causing harm through your own conscious action, compared to failing to act on the knowledge that you have acquired. He writes that:

> There is a curious sense, then, in which moral reflection
> can transform decent people into indecent ones: for
> if a person thinks things through and realises that he

is, morally speaking [in the wrong]... his continued
indifference is more blameworthy than before.[9]

The more we learn, the more necessity there is to act on our
knowledge. My first essay on the MFA was called 'How Can We
Continue Making Art?'.[10] It questions whether it would not be
more ethical just to give up art altogether and devote all our time,
energy and ideas to fighting climate change instead. Art emerges
battered and bruised. I conclude that as much as we need to
fight climate change we also need art as the antidote to consumer
culture – as a forum for free ideas and discussion away from the
marketplace. Without any art, our free museums, public galleries
and the other places where we make or encounter it, would
quickly be subsumed by capitalism and turned into shopping
malls instead. There would be no space for resistance at all. But,
in the age of accelerating social, environmental and economic
crises, we need art that draws attention to the injustices of our
world and inspires people to fight back. We need art that is 'free'
in both senses of the word: inclusive and accessible to everyone
and intellectually autonomous and uncorrupted by commercial
(or other) interests.

We also need art which minimises its own material impact on
the world. So, in February 2010, with a cheeky reappropriation
of corporate jargon, I launched my own 'Environmental Policy'
on my website. Under the headings: Diet, Energy, Transportation,
'Reduce, Reuse, Recycle', Banking and Continuous Improvement
– it outlined all the day-to-day actions which I vowed to take
responsibility for in order to limit my own carbon footprint.
Under Transportation, the stated policy was for Ellie Harrison to:

make all her local journeys by bicycle, on foot or on
public transport; to only use taxis on occasions when
large amounts of luggage or equipment need to be

The global warming projection graph (opposite) was used in the introduction to
How to Reconcile the Careerist Mentality with our Impending Doom, Ellie Harrison's
MFA dissertation, in May 2010. It is extracted from the UK Government's official
response to the United Nations Climate Change Conference in Copenhagen in
December 2009.

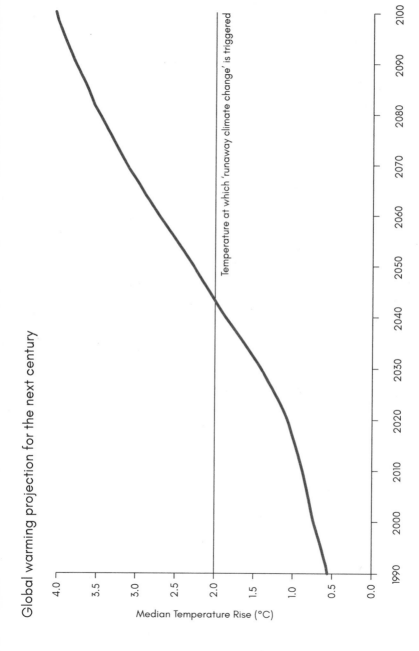

Global warming projection for the next century

Median Temperature Rise (°C)

Temperature at which 'runaway climate change' is triggered

transported or when it may be unsafe to travel by other
means; to make all long distance journeys within the UK
by train, coach and occasionally by car (but never by
plane); to attempt to limit international flights to one trip
per year, and to make all shorter journeys to northern
Europe by train or coach.

This Environmental Policy became the foundation on which all
my other activity could be built. And in my MFA dissertation
completed in May 2010, *How to Reconcile the Careerist
Mentality with our Impending Doom*, I explored how the skills
and characteristics acquired through an art school education –
what artist and economist Hans Abbing describes as our 'flair,
self-assurance and… sense of audacity',[11] and our 24/7 work ethic
could be 'reconfigured' and put to more altruistic ends, rather
than the bolstering of our own egos or the furthering of our own
careers.[12] I concluded by advocating 'an approach to practice in
which the artist becomes adept at switching between different
"hats" – moving between direct political action and more frivolous
artworld spectacle in order to raise questions about and challenge
different aspects of our current political and economic order'.[13]
The art school education and the freedom it allowed to pursue my
own research had given me an eclectic mix of important life skills:
critical thinking, practical skills, confidence and self-motivation.
Now it was time to put them to good use.

Bring Back British Rail

In my first year in Glasgow, I found myself spending a lot of time
on trains. I had been naïve in thinking that I could just move –
cut myself off from my old life – and that would be that. I would
always have commitments back where I had come from: to visit
Jon and my family and for work. The summer before I moved,
I was offered an opportunity as 'artist in residence' at Plymouth
College of Art of all places, a mere 830km away from my new
flat in Hyndland.[14] I had to travel down there once a month. I
just got used to these longer journeys and normalised the seven
or eight hours I would now travel for work. It was a real insight

into just how bad the railway network had become since it was privatised by John Major in 1993. To get to London, I now used Virgin Trains – why on earth was billionaire Richard Branson involved at all? To get to Nottingham I had to take my chances with two, three or sometimes four of the 17 different private Train Operating Companies, depending on my route. Why was it more than six and a half hours to get to Nottingham, when Glasgow to London was only four and a half despite being 203km further? Where was the 'guiding mind' ensuring that investment was fairly distributed around the network to overcome these inequalities? Did it really make sense to have so many different 'competing' entities running different parts of an essential public service that is totally dependent on cooperation?

At Christmas 2008 I went back to stay with Jon, bringing with me many of the exciting things I'd been learning about on the MFA, including an old 2002 documentary by Adam Curtis called *The Century of the Self*. It tells an alternative history of the 20th century, showing the impact that Sigmund Freud (1856–1939) and his theory of psychoanalysis had on the rise of consumer culture. PR specialists had used Freud's theories to work out how to tap into our most primitive urges – 'unconscious sexual and aggressive drives' – to sell us stuff we didn't really need. In the process they created a culture obsessed with self-gratification. In the final part of four, he looks at the rise of the 'focus group' in politics, honing in on the way that Tony Blair changed the Labour Party in the '90s. It was an eye-opener for me. I'd never heard of 'Clause IV' before and what it stood for in the labour movement. Nor the fact that Tony Blair had taken it out of the Labour Party's constitution in 1995 in order to appeal to the whims of the 'people sipping wine in Kettering', rather than to fight for the interests of the oppressed working class.[15] Clause IV was first defined in the 1918 Labour Party constitution, and was seen as the party's commitment to socialism. Without democratic public ownership and control over our key services and infrastructure, we were destined to be exploited. It read:

To secure for the workers by hand or by brain the
full fruits of their industry and the most equitable

distribution thereof that may be possible upon the basis
of the common ownership of the means of production,
distribution and exchange, and the best obtainable
system of popular administration and control of each
industry or service.

As I sat with Jon on the sofa sipping tea in Nottingham, suddenly
the penny dropped. New Labour was not actually representing
'labour' at all. No wonder there was no public debate about the
renationalisation of our railways, because it was no longer the
policy of any of our country's elected politicians. Who, beyond
the inward-looking trade unions, was representing passengers'
interests and forcing the common-sense idea of reunifying our
railways under public ownership back into mainstream political
debate? I found myself doing a little doodle on a scrap of paper. It
was a sketch of a t-shirt with the words 'Bring Back British Rail'
emblazoned across the front. Two days later, I set off to catch
the train from Nottingham station. It was a tiny little service
run by East Midlands Trains (a brand name of Stagecoach, the
company owned by Scottish billionaire Brian Souter who funded
a massive homophobic campaign to keep 'Section 28' in Scotland
in 2000 and has also donated millions to the SNP).[16] The little
train left late and continued at a very leisurely pace as though
it didn't have a care in the world. I did! I had a connection with
another Train Operating Company, which I inevitably missed.
My letter of complaint to National Express in January 2009 read:

> I am writing to claim a full refund for my disrupted journey
> between Nottingham and Glasgow… The journey got off
> to a very bad start when the train leaving Nottingham at
> 13:34 was 20 minutes delayed causing me to miss my first
> connection in Grantham. Following this, ALL three of the
> subsequent trains I had to catch were ALL subject to delays.
> My final journey on the 19:15 from Edinburgh did not
> arrive in Glasgow until 20:35 – over an hour and a half after
> I was meant to! Bring back British Rail I say!

I finally got back to the flat just after 9pm. Juan was still away in Canada for the Christmas break and the heating had packed up. I called the landlord and he didn't seem that fussed, even though it was freezing. When I threatened to withhold the rent, he eventually brought round a little oil-filled radiator, which I huddled around for the rest of the week until he bothered to get the boiler fixed.

The 'Bring Back British Rail' doodle loomed over me like a

Bring Back British Rail screenprint by Ellie Harrison, enlarged from her sketch of an idea for the campaign, made in Nottingham on 2 January 2009. (Ellie Harrison)

spectre. I knew it was big, but I also knew I had to do it. So when the first year of the course was complete, our Interim Show had taken place and our essays were handed in, I began setting up the campaign. I worked with Fraser Muggeridge – a designer with an interest in typography who had worked on the *Day-to-Day Data* publication in 2005 and on my *Confessions of a Recovering Data Collector* book that April.[17] Fraser designed Bring Back British Rail's logo, which is like the old British Rail logo except that the famous double arrow icon is flipped, so that it is now pointing 'backwards'. I set up a Facebook page and a petition on the UK government's No.10 website.[18]

Bring Back British Rail struck a chord with people. Within a few months the Facebook page had six or seven thousand supporters. In November that year, the government announced that National Express, the company running the East Coast mainline from London to Edinburgh and onwards to Aberdeen and Inverness, was not making enough profit and so had decided to walk away.[19] There was no option for Gordon Brown's government but to renationalise the route to keep services running. On 26 November 2009, sooner than I'd expected, I found myself on national radio as the campaign's spokesperson doing my first live debate on BBC Radio 5 Live.[20] At that time, there was near unanimous support from all the pundits for wholesale rail renationalisation, but yet none of the main political parties were advocating the policy. The new publicly-owned East Coast ran the route successfully, until it was reprivatised on 1 April 2015 (despite our extensive campaign against reprivatisation in autumn 2013, pictured opposite).

In autumn 2009, from an empty office space in the Barnes Building at Glasgow School of Art, I set about developing Bring Back British Rail merchandise and a membership scheme and reaching out to politicians – the late Tony Benn (1925–2014) and the then leader of the Green Party of England & Wales, Caroline Lucas, both agreed to become our 'patrons'. The Greens then became the first major political party to put rail renationalisation in their election manifesto in 2010, promising to: 'Return the railways, tube system and other light railway systems, including both track and operations, to public ownership'.[21] By the time that Labour finally announced it was reinstating rail renationalisation as party policy 20 September 2015,[22] Bring Back British Rail had morethan 150,000 supporters across the country.[23]

I came to public transport campaigning from an environmental perspective. With a growing understanding of the implications of climate change and the absolute urgency of reducing carbon emissions by whatever means possible, I began to see the privatisations that took place in the '80s and '90s as huge barriers to creating and delivering a strategic long-term plan for decarbonisation. As the old adage goes 'you can't plan what you don't control, and you can't control what you don't own'.[24] The most carbon-intensive sectors of our economy – transport, energy and to a large extent housing – were no longer under public control. Because of the neoliberal imposition of a so-called free-market onto sectors which are 'natural monopolies',[25] such as the railway network and the national electricity grid, there were now huge barriers preventing individuals living more sustainable lives. In public transport, rail tickets have risen well above inflation

Keep East Coast Public petition hand-in at the Houses of Parliament, London on 18 October 2013. Left to right: Cat Hobbs (We Own It), Lilian Greenwood MP (then Shadow Rail Minister), Manuel Cortes (General Secretary TSSA), Bob Crow (1961-2014, then General Secretary RMT), Caroline Lucas MP and Ellie Harrison (Bring Back British Rail). Ellie Harrison's dad is also pictured far right. (Action for Rail)

since privatisation forcing many people to buy cars as a more affordable option, despite the consequences for carbon emissions and air pollution.[26] The reason the UK has the highest number of domestic flights in the whole of Europe is because our railways are so messed up. And in the energy sector, there was no long-term strategic plan for switching to renewable sources of energy on the scale that was necessary, something that was simple in the post-war period under public ownership with massive investment building hydroelectric power plants in Scotland.

Bring Back British Rail and its sister campaign Power For The People (demanding the public ownership of our energy production, distribution and supply), launched in 2013, were all about climate change.[27] But I saw little point even mentioning it. Climate change was such an enormous and intangible concept that it made people feel totally disempowered to try to do anything to stop it. The last thing they wanted was a load of privileged people ranting on about it. I realised that the key was to connect decarbonisation with the struggles people do feel in their everyday lives, where these broken systems hit them personally in the pocket or make their lives miserable in many other ways – the frequent train delays or the frustration caused by trying to buy a simple train ticket on our ridiculously overcomplicated fare system (the only 'innovation' privatisation has delivered), or the extortionate energy bills which result when companies switch you to their most expensive tariff without your knowledge. To use a quote from the activists that ran Climate Camp from 2006–11,[28] we need 'system change, not climate change'. And so Bring Back British Rail became part of a movement of anti-capitalist campaigns originating from climate campaigners – campaigns such as UK Uncut (founded in October 2010) to fight the tax-dodging multinationals who continue embezzling money out of our economy, money which should be invested in public services and sustainable infrastructure instead. Over the last ten years, together we've helped to shift the mainstream discourse back towards socialism – awakening people from the spell of neoliberalism.

Hedonism vs asceticism

I didn't know much about Glasgow before I 'parachuted in' in 2008. The myth of 'the Glasgow miracle' had permeated down to Nottingham in news and reviews in the artworld press of supercool places like the CCA and Transmission gallery – by then immortalised in the Franz Ferdinand song, which refers to its 'arty' parties. But I'd also heard rumours about alcohol and violence.

In 'The Demon Drink', a chapter in her book, *The Tears that Made the Clyde: Well-being in Glasgow*, Carol Craig uses the theory of 'dislocation' drawn from the book, *The Globalisation of Addiction: A Study in Poverty of the Spirit*, to attempt to explain the historical context for Glasgow's problem with alcohol. The theory holds that all humans have two somewhat contradictory 'psychosocial' needs: the first is the need to express our 'individuality' and to distinguish ourselves from the crowd. The second is the need to feel as though we are part of something; part of a larger group which we can identify with: a community, a class, a nation – a need for 'belonging' (or affiliation and community). If we fail to fulfil both needs, then 'dislocation' occurs and with it a deep yearning for something that is not there. We attempt to quell this yearning with various addictions. In a globalised world, it has become increasingly difficult to fulfil both needs. Either we're part of the 'knowledge economy' or the 'creative industries' which are all about us expressing and commodifying our individual 'brilliance' or we're locked out of that world altogether as a member of what Darren McGarvey would call the 'underclass'.

Glasgow as a city grew incredibly quickly in the 19th century with its population peaking at well over one million people in 1939.[29] Living conditions were among the most squalid and overcrowded in the western world. As a result people, whilst having a strong sense of class identity, were robbed of their need to be seen as individuals – to stand out from the crowd. Many quelled their feelings of 'dislocation' by turning to alcohol. In a city built by immigrants from the Highlands, Ireland and elsewhere, everyone suffered from 'dislocation' and so was born 'a culture where the middle and working classes alike used every occasion as an opportunity for drinking'.[30]

By the age of 29, I'd finally begun to moderate my drinking. When I started to get busy with work, I noticed the effect that hangovers had on my productivity and on my mood. I couldn't think straight, recall information and synthesise ideas – and that, after all, is what being a 'conceptual artist' is all about. I was concerned that moving to Glasgow might set me back to the old ways of my teenage years, especially after my brief visit to look round the art school on 17 January 2008. I came up on the train with my friend Niki from Nottingham who was also planning to apply for the MFA. We'd booked beds in the Youth Hostel near Kelvingrove Park. Two students, Emmett and Conor, took us on a tour of the campus. We had brief introductions to the teaching staff and then we headed to The State, just off Sauchiehall Street for a well-earned rest. In the pub were many more MFA students who answered our questions about the course over several pints. By the end of the evening, not only were most of the course's students there, but also the entire teaching team. We then proceeded up the steep hill to The Vic, continued drinking and dancing before rolling back down again in the early hours of the morning and straight into the Blue Lagoon. That's when I first got to sample the Glaswegian chips and noticed the thick congealed saturated animal fat left on the paper. That was something completely alien to me, as all the chips I'd ever eaten in England had been cooked in unsaturated vegetable oil. As we stumbled back to the Youth Hostel, I remember thinking how insane it was that you had to cross a six-lane motorway just to get home.

When I arrived back in Glasgow to study in September 2008, I just picked up where we'd left off. Drinking was a good way to get to know people and to feel part of a community. There was also so much free booze on offer at artworld openings, it was almost impossible to say no. But I really found it conflicting with my desire to work hard and learn more about the world. I'd relapsed back into old binge-drinking habits and was embarrassing myself on a regular basis. After one terrible incident where I declared my (unrequited) love for one of my course mates, got thrown out of the SWG3 gallery for accidentally breaking an artwork and then ended up in a gutter where a kind passer-by picked me up, put a ten pound note in my pocket and got me in a taxi, I decided enough was enough. From 1 January 2010, I quit drinking altogether. I had just six months of my course left and a 9,000 word dissertation to write. That had to become my priority.

As I was approaching the end of the MFA, in March 2010, I did a talk at the CCA called *Hedonism vs Asceticism: A Control Freak's Guide to the MFA*.[31] It aimed to explore the conflicting motivations and desires I had experienced whilst studying and my tendency for 'all or nothing' behaviour; to swing between extremes. I'd become interested in the concept of asceticism – severe self-discipline and the deliberate abstinence from all forms of indulgence. Had I known more about Glasgow's history at the time, I would have seen that I was actually following the city's long tradition of temperance. *Hedonism vs Asceticism* was highlighting what Sarah Lowndes describes in her book, *Social Sculpture: The Rise of the Glasgow Art Scene*, as one of the fundamental 'oppositions which… govern the unstable character of Glasgow', which she identifies as:

temperance/dissolution
wealth/poverty
Protestant/Catholic.[32]

Going it alone against what felt like an art school culture hell-bent on boozing, little did I know I was actually following in the footsteps of women like Mary Carmichael of Cowcaddens,[33] highlighted in Elspeth King's book *The Hidden History of Glasgow's Women*, who were key players in the temperance movement in the early 20th century which was rooted in socialism and women's suffrage as well as the church. It was an austere new life I had made for myself – alcoholism had become work-a-holism as a way of dealing with my 'dislocation'. But not drinking was also very cheap. In 2010, austerity was in the air.

Austerity politics

Even before the General Election in May 2010, David Cameron was promising a 'new age of austerity'. He called it 'a whole new, never-been-done-before approach to the way this country is run. Because the world has completely changed'.[34] The Tories blamed Labour for the entire global financial crisis. It was a bit of an over exaggeration, though as New Labour, they had wilfully embraced neoliberalism and done very little to reverse the damaging policies implemented under Thatcher and Major, which had been the root cause of the crisis. David Cameron saw the crisis as an opportunity to cut back state spending by an enormous amount: 'the age of irresponsibility is giving way to the age of austerity', he said.[35] Writer John Lanchester points out that the genius of the Conservatives was to simplify the national economy by equating it to how an individual or a household runs its affairs. Their favourite slogan was to say Labour 'maxed out the nation's credit card'.[36] As John Lanchester writes, austerity itself 'is a heavily loaded term, taking a personal virtue and casting it as an abstract principle used to direct state spending'.[37] It was what philosophers call a 'fallacy of composition' – assuming that the whole (in this case the national economy) has the same properties as its constituent parts (individuals or households).

Whilst my own ethos was becoming more ascetic, for the sake of my own well-being (mental, physical and financial health) and for the sake of the environment, there could not have been a time when we needed state spending more. We should have been spending on low-carbon infrastructure – world-class public transport, renewable energy, insulated homes and decarbonised heating – which would create meaningful jobs allowing people to contribute to the transformation of our society and so enabling more sustainable lifestyles for us all. This is what China did – in the ten years after the financial crash they built tens of thousands of kilometres of high-speed rail, with the aim of connecting up the entire country with low-carbon transport links for centuries to come. In some Chinese cities like Shenzhen, in order to address their air pollution crisis, 16,000 buses were replaced by electric ones almost overnight, now saving 440,000 tonnes of carbon every year.[38] In Britain we took a very different path driven by neoliberal ideology and a desire to further shrink the state. It was clear the burden of Tory austerity would fall 'much more on the poor than on the better-off'.[39] It was wealthy and privileged people imposing austere lives on the poorest; the ones with most right to aspire to something better. But, it was shocking the extent to which this fallacy was bought by the electorate.

The General Election on 6 May 2010 was Gordon Brown's first opportunity, since he had been handed power by Tony Blair behind closed doors three years before, to see whether anyone actually wanted him to be Prime Minister. It was a test to see what people thought about his handling of the financial crisis and the massive bailouts he'd given to all those corrupt banks. There was a lot of speculation in the run-up to the election that it would result in the first 'hung parliament' since 1974. There was a fear that, following the vote, the country could be left in a total mess. I wanted to use my work to draw attention to this key turning point for Britain and to question the effectiveness of our representative parliamentary system in its present form with its outdated 'first past the post' voting system. So I came up with a crazy idea, especially crazy for someone who had just quit drinking.

General Election Drinking Game would be an 'endurance performance' taking place live on election night as an 'alternative commentary on the results'. It was inspired by a game I played as a teenager (around the time I lost my driving licence) called One Hundred Minutes. Each player has in front of them a plastic 35mm film container – the sort which were ubiquitous before digital cameras. It didn't look like it would contain a lot of liquid. That was the point. The rules were: every minute, on the minute, for one hundred minutes, the players would drink one film container full of beer. Sounds easy, right? But it's the relentlessness of the shots that gets you, which turn out to be far more frequent than anybody would normally sip a beer. Very few players make it to the end. On the occasion I played with all my school friends in 1996, only two of us were left standing at the end and one of them was me. Firstly because of my dogged determination to complete any challenge that I set myself, and

General Election Drinking Game by Ellie Harrison, performed live at the Star
& Shadow Cinema in Newcastle on 6 May 2010. Left to right: Oliver Braid
(Conservative), Ellie Harrison (Labour), Harriet Plewis (Liberal Democrat) and Paul
Knight (Other/Independent). (Ilana Mitchell)

secondly, because I puked in a vase halfway through. It was
strange the way this all came back to me during my first few
months of being sober. I guess I was interested in exploring the
potential backlash after any (self-)enforced period of prohibition.

The plan for *General Election Drinking Game* was simple. On
a spot-lit stage four 'players' would draw straws to see who they
were going to represent: Labour, Conservative, Liberal Democrat
and 'Other/Independent' (which at that time only made up 4 per
cent of the seats including the SNP).[40] Then, as the votes came in
live throughout the night, each player would drink one shot of beer
for each seat in Parliament their party won. Drunkenness would
be used as a measure of 'success'. I wanted the event to take place
in Glasgow, but none of the venues I approached were interested
(Transmission gallery or The Vic), so we had to travel down to
Newcastle. My friend Ilana Mitchell offered to produce the project
at their community-run Star & Shadow Cinema. Despite my
teetotal vows there was never any question that I wouldn't take
part – how could I ask other people to humiliate themselves for the
sake of my art, if I wasn't prepared to do it myself?

I drew the 'short straw' of Labour. Oliver Braid (my friend from the MFA) picked Conservative, our friend Paul Knight from Australia (who had been the year above us on the MFA) picked the 'Others' and Harriet Plewis, an artist based in Newcastle who just happened, quite serendipitously, to be wearing a yellow t-shirt, picked the Liberal Democrats. It's crazy to think how popular the Liberal Democrats were then – I'd even been duped into voting for them myself. The event lasted five hours and 35 minutes. By the end, Oliver had disappeared to have a joint and had been replaced by some actual Geordie Conservatives from the audience. After a few heated arguments culminating in a tirade delivered from atop a chair, I found myself vomiting in a big green bucket on the floor. Paul and Harriet looked on in relatively sober horror. Artist Tony Kemplen, watching our live webcast of the event, said: 'The drunken shambles seemed a perfect comment on the outcome!' The actual election had resulted in a 'hung parliament' and back room discussions which gave rise to the Conservative–Liberal Democrat coalition. When Nick Clegg signed the deal with David Cameron, I vowed never to vote for the Liberal Democrats again either.

It was a shock living under a Conservative government again for the first time in my adult life. We were graduating from the art school and going back out into the 'real world' at the very same time. When I looked at the map of the UK election results and saw a sea of blue south of the border and just one blue seat in the whole of Scotland, it felt like there was no way I could go back. Who were all those nasty selfish people down there? I liked being in Glasgow, amongst people with similar politics. You could say I became a 'proud Scot'. For the first few years after I graduated from the Masters I described myself in my artist's biography as 'a political refugee escaped from the Tory strongholds of Southern England'. Maybe I'd be holed up here for quite a while, until the map down there started to go red again (or better still, green). Then it would be safe for me to return.

Plus Glasgow was a great city to be an artist. Living costs were considerably cheaper than London (if a bit pricier than Nottingham). In summer 2009, when Jon and I finally went our separate ways (he ended up moving to New York), I was able to take advantage of the 'comparative cheapness of Scottish property',[41] and the fact Jon's dad was a wheeler-dealer mortgage broker in Birmingham, to buy myself a flat. This time it was in the attic of a converted Georgian townhouse near Kelvinbridge. It wasn't damp at all and my clothes now dried in 24 hours. I spent two months over the summer doing it up on my own – using all those practical skills acquired through my art school education and the drill and jigsaw I bought to install my first Degree Show in 2001. With the help of fellow MFA graduate Amelia Bywater, we installed my new kitchen. Once the flat was fixed up, Oliver Braid – my friend from the MFA who was five years my junior – moved in as my lodger.

With all that investment in the 'creative industries', there were loads of studio blocks in Glasgow and I got myself a space in one in Dennistoun run by Wasps (Workshop & Artists Studio Provision Scotland). If anywhere sums up the city's abrupt transformation to a 'knowledge economy' it's their building on Hanson Street – once Wills cigarette factory with its entrance on Alexandra Parade, it now accommodates 80-odd individual artists, makers and creative businesses. In many ways it's also symbolic of the growing divide, or the 'gulf of experience' as Darren McGarvey calls it, with one class of people being moved out of the space and another class being moved in. I loved having a quiet space where I could think. It was a 20-minute cycle from my flat and once used to the weather, I came to appreciate the gentle exercise I got every day whilst travelling. I always arrived at my desk fresh faced. I signed up for Working Tax Credits – the benefit introduced in 2003 when Gordon Brown was Chancellor to subsidise people on low incomes (most self-employed artists did), and I got a discount on my studio rent for doing jobs for Wasps as a Studio Rep. Life was good.

Once I was out of the art school bubble and in the 'real world', I began to try to get my head around Scottish politics and what it meant for me being here as an English-sounding émigré. During those first two years in Scotland, my perspective on the UK had shifted dramatically. You could say it was turned upside down, with a growing understanding of the resentment people in the north felt for the distant and aloof Westminster. In December 2010, I was invited to do a Skype broadcast to the Whitechapel gallery in London as part of an evening event featuring 'transmissions' from different parts of the world. In *Transmission: Glasgow to London*, I bigged up all the things I loved about my adopted home: from its free dental care to its radical past. I told the tale of when the workers of Red Clydeside descended on George Square in 1919 and raised the red flag, a whisker away from a revolution.

The following May, I got to vote in my first Scottish Parliamentary election, the one where the SNP gained the

Stills from *Personal Political Broadcast* by Ellie Harrison, webcast live for Bloc digital arts on 5 May 2011. Duration: 14:38. (Ellie Harrison)

majority that was thought to be impossible. I think I even voted for them myself – that's what 'proud Scots' do, isn't it? I certainly remember being excited at the results: that'll show them down there! That was also the same day as the UK-wide referendum on voting reform for the UK Parliament and I did another live webcast on election day commissioned by the Welsh digital arts organisation Bloc. In *Personal Political Broadcast* (pictured on page 81), I used a variety of different regional accents (one bad generic Scottish, one slightly better South Wales, which I had perfected as a child amongst all my mum's relatives in Carmarthenshire, and one over-the-top Cockney). The broadcast highlights the massive discrepancies in how voting works in the Scottish Parliament, Welsh Assembly and the UK Parliament (the former both have 'proportional representation' and the latter still has 'first past the post'), as well as the fact that England has no parliament at all.[42]

My *Personal Political Broadcast* also looks at one of the 'unintended consequences' of the pro-union Conservatives' policies of privatisation. That is in breaking up and selling off all the national infrastructure, industries and services which were nearly all 'British' or nationwide: British Rail, British Gas, British Telecoms and British Steel, the Central Electricity Generating Board and the National Coal Board, they were effectively erasing any coherent sense of national identity and paving the way for devolution and also, potentially, Scottish independence. To see yourself as 'British' was now totally out of date. I was trying to work out what other choice I had as an English-sounding, half-Welsh person living in Glasgow. I was trying to work out how and where I fitted in to this place, but something always seemed missing.

Long-distance love

During my first year in Glasgow, I made friends with an Israeli woman called Shelly Nadashi who was the year above us on the MFA. She said something which has haunted me ever since: 'You will never find love in this city... there is no love in this city!' Shelly was a romantic. That was her analysis of Glasgow, having lived here for just over a year. It reflects the findings of the chapter 'Love Actually' in Carol Craig's book, *The Tears that Made the Clyde* – that the principal causes of the poor mental health which drives premature mortality in Glasgow 'is not brain disease or disorder but poverty, exploitation and lack of love'.[43] Shelly left in 2010, after just three years, to move to Brussels, then Paris where she eventually did find love. But her words chimed with my experience at that time. The further away from home I moved, the less love there seemed to be. The warmth of life was missing, literally and metaphorically. I still found sleeping alone in a bed after all those years with Jon profoundly isolating. It was a powerful feeling which would wake me up at night. A sudden panic that I was so far away from anyone who really loved me or who I really loved, unconditionally: my family.

Families should always have your back, be there when you need help or fall ill, provide the support structure necessary to make us 'resilient' as individuals. Along with 'neighbourhood' and 'community', families make up what feminist economist Neva Goodwin defines as the 'core economy'. The unconditional, unpaid labour that provides the 'operating system' on which everything else in our society is built:

Family, neighbourhood, community are the Core Economy. The Core Economy produces: love and caring, coming to each other's rescue, democracy and social justice... territory lost to the commodification of life by all sectors of the monetary economy, public, private and non-profit.[44]

But families are not always easy places to be. As Carol Craig's book shows, they can also be deeply oppressive. It's impossible to understand why without knowing the historical and cultural context. One of my mum's favourite poems was the classic 'This Be The Verse' by Philip Larkin about how we inherit flaws from our parents. My mum and dad had a strange relationship. They were never affectionate with each other yet they stuck together for the sake of raising my sister and me. But they were both the products of a far more oppressive society. My dad was 34 before homosexuality was decriminalised in England in 1967 (it wasn't decriminalised in Scotland until 1981). In the '70s, my mum was desperate to get married, because her younger sister already was and she didn't want to be seen as the 'failure' of the family. Growing up with that and the 'don't ask, don't tell' atmosphere fostered in state schools by 'Section 28', it's no wonder it took me until I was 31, in summer 2010, to 'come out' to my parents.[45] It didn't go down that well. My dad never mentioned it again and my mum seemed to hope it was a phase that would go away. But my mum and I kept talking – talking was what we both loved to do. It took a couple of years, but eventually she got used to the idea and changed as a person as a result – less concerned with 'keeping up appearances' and more concerned with living her values.

But while I was coming to terms with all that, it suited me to be far away from home, far away from anyone who had known me in my previous heterosexual life. Coming to Scotland was like 'running away' from the past. But it turned out Scotland wasn't even far enough. The first woman I fell in love with was in Australia. I had been chatting to Paul Knight, star of the *General Election Drinking Game*, who had also 'come out' later in his life. He said I reminded him of his friend Kim Munro who lived in Melbourne: 'In fact, you guys might really get on'. On 14 February 2010, he introduced us on Facebook and thus began an obsessive correspondence. Within the first two weeks, we had written back and forth more than 10,000 words. It was completely absurd seeing the main reason I had quit drinking was to focus on writing my dissertation and I'd barely put pen to paper on that. We started Skyping and speaking on the phone most days. Within a month or two, we were officially 'in love'. She watched the live webcast of our drinking game from Melbourne and texted us throughout. It was totally addictive, for both of us.

The longer the digital communication went on, the clearer it became that we had to meet, somewhere, somehow in 'real life'. Kim was six years older than me: an OWL (older, wiser lesbian). She had also travelled a lot: to Europe several times and round Asia and Japan, like many of the 'dislocated' white population of Australia do. It would be 'too easy' for her to come here. On the other hand, I had never expressed any interest in going to Australia, a place where many of my school friends had gone on 'gap years' after finishing high school and where three of them now lived, with Australian husbands and kids. And yet now I found myself, the day after all my work for the MFA course was finally finished on 3 July 2010, hopping on a plane (we decided to split the cost of the ticket). It was the longest plane journey I had ever taken (to Bangkok), to change for the second longest plane journey I'd ever taken to Melbourne (you can see the environmental impact of this trip in my Carbon Graph on pages 138–9). We were 'together' for about six months, meeting on three occasions: once in Australia and twice in the UK, but it was not going to work. Both of us were too old for working visas (only available if you're under 30), and I was not up for a civil partnership just yet. I was looking for something a little more practical, and a little more sustainable: for 'local love'.

By the end of 2011, Oliver and I had been living together for two years and had created a neurotic, homonormative little home of sorts. Kim and I had parted company at Heathrow airport on 9 January that year and I had flung myself back into work. I was offered some more teaching back at Nottingham Trent University, and began making the six and a half hour commute every two weeks. It was a really full-on year, with many projects, exhibitions and events. I went for the first five months without a single day off, weekends included. It was these working conditions and my extreme self-exploitation which gave me the idea for the *Work-a-thon for the Self-Employed*, which I staged twice that year in London on 13 June and in Newcastle on 3 November 2011.

By December, I really felt like I needed a break from all the travelling. Many people I knew had left Glasgow after the MFA course – most international students had been forced out by immigration and others had just had enough and followed the trail of artworld 'opportunities' elsewhere. It was a very different city without the art school community.

My aim for 2012 was to 'have a more social year',[46] and to try to make some more friends in Glasgow who weren't just transient residents. I joined a reading group, and over the next two years we read two volumes of Karl Marx's *Capital* with the help of David Harvey's lectures on YouTube.[47] I never would have thought I'd be able to read a book like that, but it was relatively easy with the motivational structure provided by the group. I began to see the world in a whole new light, and started to realise that being a landlady – that is collecting rent from a poor tenant on account of my privilege in owning the property in the first place – was not exactly in line with my anti-capitalist beliefs. Or as Marx put it himself:

> You may be a model citizen, perhaps a member of the Society for the Prevention of Cruelty to Animals, and in the odour of sanctity to boot; but the thing you represent face-to-face with me has no heart in its breast.[48]

The landlady–tenant power dynamic became central to my relationship with Oliver. We wanted a forum to explore this further, so we set up our own radio show, the *Ellie & Oliver Show*, which we broadcast live from our living room every Friday lunchtime throughout 2012. It was the perfect structure to require me to stay put. We were both trying to make sense of our lives and work through new ideas. Our shows were always based around a theme (a plural noun), which we chose alternately. For our fourth show on 27 January 2012, I bucked the trend by picking the only singular noun we discussed that year: 'Isolation'. Introducing the theme I referred to a 2011 loneliness survey, which showed that most people in the UK 'feel lonely'. 'One in three people in the UK actually live on their own and that being lonely can be as bad for your health as smoking'. I said to Oliver: 'I can't distinguish isolation – or how I feel in my life – from being in Glasgow. But for me, this radio show is an amazing tactic for breaking through this isolation... this is our reaching out to the world'.[49]

At the start of 2012 I'd been invited by Glasgow Women's Library to be one of the 21 artists to make a new work for the *21 Revolutions* project celebrating their 21st birthday.[50] I had also decided to take up a hobby, which ended up taking over my life. I first heard about Women's Flat Track Roller Derby when I was in Newcastle for the Wunderbar Festival in 2011. The

Newcastle Rollergirls had done a 'skate skills' demonstration at the *Desk Chair Disco* which I'd staged in an empty office building. People were invited to come and whizz around to music on ordinary office chairs. The Rollergirls blew my mind. These were the coolest women I had ever seen. They told me there was a great team in Glasgow and that I should check it out. I got a cheap pair of roller skates for Christmas and in January 2012, I went along to the 'fresh meat' training session at Firhill Complex. I loved it. The skating was so much fun and the people were great. It was also very queer. I became interested in the history of Women's Flat Track Roller Derby, a grassroots sport, which had spread across the world in a very short space of time (the first team in the UK, London Rollergirls, was founded in 2006; Glasgow Rollergirls was founded one year later in 2007). Organised by women from the ground up: 'by the skaters, for the skaters' was the motto.[51] It was real collective labour, not for the money, but for love.

I was spending one day a week in Glasgow Women's Library learning about women's history – specifically the Women's Liberation movement of the '70s. I was struck by the similarities in how these women-led movements from different points in history had been organised: an older generation who had helped win the massive gains for women's rights and who had help to set up institutions like the Women's Library, and the younger generation who were so passionate about this new and empowering sport. I thought both groups would benefit from getting to know one another. So I set up the *National Museum of Roller Derby* at Glasgow Women's Library, to archive the history of the sport in the UK, and to bring a whole new, young audience into the library to encounter more women's history and politics. I did a performance talk in the library on 22 March 2012 where I explained more about the ideas and the NMRD was officially launched on 14 June 2012.

I also got a brief glimpse of another Glasgow in 2012. On Saturday 3 March, I went to an evening event at Kelvingrove Art Gallery. I was with a group of people, none of whom I knew very well, as was often the case, so social anxiety was rife. Across the crowd I spotted this woman. Her face jumped out to me in that way Jon's had done 14 years before. Wherever I went in the room I kept seeing her out of the corner of my eye. She didn't look very gay. There's no way she would be interested in me. It was too depressing. I made my excuses and skulked home alone; drowning my misery with a disgusting midnight feast. A few days later I was in my studio and an email popped up in my inbox. It said:

> This is bizarre even for my standards but I saw you in the Arches on Friday and then again at Kelvingrove on Saturday... I wondered if you might be interested in a beer some time?

What! That's so strange? I was at the Arches on Friday for a meeting. Do I have a stalker? What if, just what if, it could be that woman? Whoever it was, I admired their boldness – if only I could be more like that. I didn't want them to suffer an anxious wait, or worse still not get a reply at all, so I just dashed off a quick message. What did I have to lose? We arranged to meet that Thursday at a bus stop on Great Western Road. What better location for a blind date? I was so nervous, anyone could have turned up... But then, as if by magic, that very same woman came walking along. Wow! This is meant to be. Now I really was in love. Sarah-Jane was actually Scottish and she lived about five minutes from my flat. For one blissful month, this city suddenly did feel like home. I remember walking back and forth along Great Western Road and it was no longer cold, bleak and isolating, suddenly there was warmth, light and love. It was short lived. It turned out she already had another girlfriend and I was just some weird experiment. I was heartbroken.

On the rebound, I found myself in another long-distance relationship with Becky 'Lawless' Bengtsson who was one of the top skaters in the Stockholm Roller Derby team, who had come to play a bout against Glasgow on 14 April 2012. We were 'together' for nearly a year. But that was difficult too. I'd become so weary of 'long-distance love' after my relationship with Kim that I could hardly believe I'd fallen into it again. I found it harder, in many ways, than being alone. Skype could offer the illusion of intimacy for the time you had it running; to make it feel as though there is someone there, only to be plunged further into the harsh reality of isolation when it's time to hang up. Long-distance love can make the bleakness of your actual existence, sleeping alone in a bed night after night, not having someone there when you most need them, ever more apparent.

Reality check

I remember the day austerity politics impacted on my own lifestyle. I was in the kitchen listening to Money Box on BBC Radio 4, as all good capitalists like to do. There was a piece about George Osborne's changes to the Working Tax Credits system, which would soon affect all self-employed people. From the following year, we would only be able to claim credits for a low income for the first year of our 'business'. After that, HMRC would assume that you were making the minimum wage, regardless of what your actual income was, and so would not give you any additional credits. The rationale being that if your 'business' wasn't making any money, surely you'd cut your losses and stop? That's one of the problems with classing an art practice as a 'business' in the first place, and in 'professionalising' artists. Most of us aren't doing it for the money, but to give our lives a sense of meaning. Whatever the case, I realised my cushy lifestyle – where I'd received £6,476.70 back from HMRC over the last two and a half years – was not going to be possible for much longer.

George Osborne's policy had worked. I was cast back out onto the 'labour market', to instrumentalise my education and experience to earn a crust.[52] I had added myself to the 'job alerts' for all the art colleges in Scotland in the hope of picking up a bit of teaching a bit closer to where I lived. That would be a great job. I'd loved working with students at Nottingham Trent, being able to use all my knowledge, experience and enthusiasm to inspire young people to learn. But English universities were becoming ever more elitist establishments since the Coalition government voted to raise tuition fees to an eye-watering £9,000 per year.[53] Teaching in a university in England was now like teaching in a private school. At the heart of my 'refugee' status in Scotland was the desire to work in an education system that was still free at the point of use. It felt like less of an ethical compromise to be employed as a teacher within a system that still 'believes access to education should be based on ability to learn, not ability to pay'.[54]

In May 2012, a job popped up at Duncan of Jordanstone College of Art & Design in Dundee. Artist Gair Dunlop (who I'd met on a residency in Bristol in 2002) worked there and had invited me to do a talk to the students in November 2010. What I liked about Duncan of Jordanstone was that it was the only one of Scotland's 'big four' art colleges (alongside Gray's in Aberdeen, and Glasgow and Edinburgh) where you didn't have to 'specialise' in any particular medium at undergraduate level. Like Nottingham Trent, students could follow their own interests and ideas and develop a diverse range of skills. They were also encouraged to study philosophy (just as I had done in Glasgow) with a unique Art & Philosophy course. I went for the job. I remember being casually confident, preparing my presentation just the morning of the interview and going along wearing my *Ellie & Oliver Show* branded t-shirt as visual aid. I was offered a three-month contract to cover another member of staff's 'research leave'. I would be teaching two days per week from September to December 2012. Moira Payne was the head of the Contemporary Art Practice programme at that time. She really looked out for me. During those three months she encouraged me to apply for a permanent lecturing post.

I loved working in the art school. I was passionate that a creative education with its eclectic mix of essential life skills – critical thinking, practical skills, confidence and self-motivation – really had 'the potential to offer young people the best possible start in life'.[55] It shouldn't just be those wanting to be 'artists' that get to go to art school, everyone should! As artist Bob & Roberta Smith says, 'All schools should be art schools!' Creativity is a fundamental part of what it means to be a human being. But I remember thinking a permanent job would be a big step. Did I really want to be tied down forever? But I had to think about the security, not having to worry about money in an increasingly unstable and uncertain economy – plus Oliver had dropped the bombshell that he was moving out in February. The income from my lodger would be gone. I had to quit Roller Derby as I could no longer make all the training sessions (it also didn't look very professional presenting a lecture covered in huge black bruises). The fun was over.

CHAPTER 4

Socialist Dystopia

Turning point

IS THERE A key 'turning point' in a person's life when they just 'naturally' become more conservative and start to betray their idealistic former self? That was the central question of my *Transmission: Glasgow to London*, broadcast in December 2010 when I was approaching the age of 32. It's a phenomenon summed up in a quote, which I first heard in the German film, *The Edukators*, in 2005,[1] but which I think can be traced back to the former MP for Dundee Winston Churchill (1874–1965). He himself defected from the Liberals to the Conservatives in 1925:

> Under thirty, not liberal, no heart.
> Over thirty, still liberal, no brains.

I wondered whether this supposed psychological change could actually be attributed to gaining more financial responsibility. In getting a mortgage and helping to fulfil Margaret Thatcher's dream of creating 'a nation of home-owners', was I also subconsciously compromising my politics and becoming less likely to take risks? Polling data shows that home-owners are far more likely to vote Conservative than people who are renting.[2] It was what American philosopher RJ Wallace has described as 'the bourgeois predicament', 'in which affirmation of certain projects that we have come to build our lives around',[3] lead us to entrench social inequalities in the world, to which we're in principle opposed, despite the fact that 'we continue to view those inequalities as objectively lamentable'.[4] In other words, we become hypocrites.

If I had to pinpoint a 'turning point' in my life, it would be 2 February 2013. That was the day Oliver moved out of the flat. He was heading to France to do an artist's residency in

Marseille and was never coming back (not to my flat anyway). On 13 December 2012, I'd had an interview for the permanent lecturing post in Dundee – a gruelling day-long process for which the shortlisted candidates had to give a 20-minute presentation to the entire academic team, then have lunch together and finally be grilled by an interview panel from across the whole university which the art college is part of – mathematicians, scientists and philosophers to boot. Somehow I got through it and was offered the job. My fate was sealed. My contract was due to start on April Fools' Day 2013. For the first time in my life I would have a proper salary. I could afford to live on my own. It meant I'd no longer have to deal with that fundamental compromise of being a landlady. My living costs would be almost double, but it would be worth it, right? No longer would I have to put up with the mess, the queues for the bathroom, the disturbances late at night or the occasional whiff of marijuana from the room next door. I could do all those home improvements I'd been fantasising about: clear out the loft (crammed full of the last few occupants' junk) and have it insulated courtesy of the Scottish Government's Green Homes Cashback scheme. It was going to be great!

I'd never lived on my own before, except for my one year in the halls of residence in Nottingham in 1998–9. In hindsight, I now recall a similar feeling of bleakness during that time – as though you are that tree in the forest that nobody hears fall over. I was used to having people around. I'd grown up with my family in Ealing, where we also had various lodgers over the years including my friend from school who'd fallen out with her mum and so moved in with us for a year. Then after halls, I lived in a big shared student house at 319 Woodborough Road in Nottingham, where five art students and one fashion student cohabited for two years. Then I lived with Jon – we slept alongside each other every night for seven of the nine and a half years we were together. Then I had my flatmates Juan and then Oliver in the room next door. Then suddenly on that day in February the door shut and I was left alone. This is what 'successful' people do isn't it, 'grown-ups' with 'proper jobs'?

'Single person households' are on the rise across the globe and have more than doubled since the '60s – with the highest percentages in the most 'developed' countries in Europe and Asia, Australia, Canada and America. In 2018, it was predicted that 'more than a million Scottish households will be inhabited by just one person within the next decade'.[5] As lovely as it was to wake up in the morning, go for a swim, have my coffee, listen to the radio and plan my day with no disturbances whatsoever, there was something vital – about what it actually means to be a human being – that was no longer there. Not only is it clear that living alone in a two bedroom flat is not the most sustainable thing you can do (no matter how much insulation you have in your loft), this massive demographic shift is what's fuelling an 'epidemic of loneliness'.

They say loneliness can be as damaging to health as smoking 15 cigarettes a day.[6] This is something I've observed and now understand viscerally. I've watched my dad's brother grow old. Ever since I was born, he has always lived on his own. He had a heart attack in the '80s and has been on medication for high blood pressure ever since. He often has a 'bad turn' in the night and calls an ambulance or my dad – he's terrified of his heart giving up again. I know this feeling. I have experienced it often: a sudden waking in the night with a racing heart. It's a fight or flight response. Your body knows that you are lying alone hundreds of miles from anyone who really loves you. It's a primitive instinct to remain alert, conscious of potential threats even whilst asleep. It wears you down. The problem with social isolation and the loneliness it causes is that it becomes a vicious circle. The more isolated you become the less confidence and ability you have to change your circumstances. The harder it is to start new relationships. It becomes profoundly alienating.

In *Capital*, Karl Marx names 'alienation' as one of the key consequences of a capitalist system of production, which divides our society into 'specialisms' or 'professions' and disconnects us, as individuals, from the fruits of our own labour.[7] French Marxist philosopher Henri Lefebvre developed this theory in the book, *Critique of Everyday Life*. Writing in the '40s, on the cusp of the consumer boom which was to transform the way we live, he described the predicament of advanced capitalist societies:

> Human life has progressed: material progress, 'moral'
> progress – but that is only part of the truth. The
> deprivation, the alienation of life is its other aspect.[8]

In 1972 the Glaswegian trade unionist Jimmy Reid made a famous speech about 'alienation' at the University of Glasgow when he was made its Rector.[9] He began by defining what he meant by the term:

> It's the cry of men who feel themselves the victims of
> blind economic forces beyond their control. It's the
> frustration of the great mass of ordinary people excluded
> from the processes of decision making. It's the feeling
> of despair and hopelessness that pervades people who
> feel, with every justification, that they have no real say in
> shaping or determining their own destinies.[10]

Jimmy Reid was speaking about the swathes of people whose livelihoods were threatened by the closure of the Clyde shipyards.[11] But he is clear that the phenomenon of alienation is experienced by everyone living under capitalism and that it is the so-called forces of economic 'progress', which are actually disempowering us.

In *Critique of Everyday Life*, Lefebvre describes what he sees as the paradox at the heart of the 'bourgeois intellectual'. For 'bourgeois intellectual' you can read 'conceptual artist' or anyone else who's ended up working in the 'knowledge economy'. Alienation becomes a powerful motivating force, as work steps in to fill the void. But the paradox is that it is the activity which the 'conceptual artist' does in order to connect to and communicate with society (ie their work) which is the very same thing which isolates them from that society as they spend days locked away in flats and studios on their own, trapped in a world of ideas:[12]

> nowadays we are still struggling with this deep – in other
> words everyday – contradiction: what makes each of us a
> human being also turns that human being into something
> unhuman.[13]

I found myself working because I was lonely and that often I was lonely because I was working. So many of the people that I'd met when I first came to the city had now moved elsewhere, or were always to-and-froing, going on an artist's residency here or there. They were not around. It felt like work was all I had left. Lacking real human contact, I found myself turning to Facebook for comfort. It was then that it dawned on me that the lifestyle I'd had since moving to Glasgow more than five years before had only been made bearable by the illusion of connection that technology provides. But it actually, potentially, made things worse. Because all the time channelled into digital communication prevents us from meeting real people nearby. I was spending too much time stalking the Facebook profiles of my ex-loves, dreaming of the multiple lives that I could have had if I wasn't stuck here.

Facebook wasn't making me less lonely, it was just making me *feel* less lonely. I needed real social networks, not online ones! So in December 2013, I made the decision to 'deactivate' my Facebook profile – seven years of photos, comments, connections and memories now out of sight and out of mind. I realised that all this digital admin which was supposed to be 'social' actually just felt like work. So, from now on, I would only use the medium for work. I made a new ghost profile with no 'friends' at all, and used it simply for updating all the project and campaign pages for which I was the 'admin'.[14] No more newsfeeds of babies and weddings to sift through on a daily basis, just a new 'interest bubble' of public transport policy, renewable energy and art. I watched my increasing social isolation unfolding, but I felt powerless to stop it. But at least I had my artwork to help me fight back. And that became darker as a result.

You are what you eat

In summer 2013, John O'Shea, an artist based in Manchester (who I'd met on an artist's residency in Hull in 2010), invited me to work on a project with him. I remember how grateful I was that he'd reached out. John was an interesting artist, who used conceptual propositions (like thought experiments) to make people think critically about their rights and responsibilities as human beings. In 2012, he proposed a piece called *The Meat Licence*. It was about a parallel universe in which, just as we are now required to have driving licences to prove our competence at handling such dangerous machines, people who want to eat meat are required to undertake an examination to prove they are capable of killing an animal before being allowed to eat the meat. This new bureaucratic hurdle would enable everyone to understand the consequences of what we shovel into our mouths.[15]

John wanted to collaborate on a project at MediaCityUK in Salford Quays. His idea was inspired by the weather map used on ITV's *This Morning* in the '80s and '90s. The plan was to invite artists to use the succinct format of the weather forecast to describe other complex global systems to a lay audience. I thought it was a great idea and said I'd help produce it, coming up with the title *The Other Forecast*. We used the University of Salford's greenscreen studio and invited four other artists to join us in making live forecasts, which were then broadcast outside on the BBC's 'big screen'. I used my 'Other Forecast' (pictured overleaf) to bring together what I saw as several interconnected 'growth' trends, which were leading us towards climate catastrophe – our growing dependency on technology, our growing alienation as individuals, growing levels of obesity and the growing energy consumption and carbon emissions that result from these changes to our lifestyles. Donning a fat suit for extra impact, I hoped to create a warning of the dystopian future we're heading towards if our economic system remains unchecked – a world with a lot of 'very fat, very lonely and very sweaty people'.

I began to see over-consumption as an inevitable consequence of a capitalist system of production predicated on always making and selling us more, more, more. I started to look more critically at my own relationship with food, questioning why it had endured as a focus for my work for more than 15 years (since I was studying on my Foundation Course in 1997–8, throughout university and into my project *Eat 22*).

I was 12 when I first began to realise the consequences of the everyday act of eating and the power we have as individuals to make ethical choices about what we choose to eat. That last

Stills from *The Other Forecast* film by Ellie Harrison, shot live in front of a green-screen at MediaCityUK in Salford Quays on 25 October 2013. Duration: 05:29. (Ellie Harrison)

summer before I went to high school, we went on a nice middle class family holiday to France. While we were away we got the news that our cat, Friday, had been run over by a car and killed. That was my sister's and my first experience of loss. We cried for days, feeling powerless that we were so far away and with the guilt of never knowing whether it would have happened had we not left the poor thing home alone. That same trip I remember staring up at one of those sticky fly paper things which was hung up in the kitchen and watching the poor creatures which had come into contact with it. Their wings were stuck, but they were still very much alive, legs whirring manically in a futile attempt to escape. My 12-year-old mind just couldn't bear that. So I spent the rest of the day trying to save them, bathing their wings in water to free them from their ordeal. I decided then that I wasn't going to eat meat again and my sister followed suit. Our parents were both meat eaters, but they humoured us. We had vegetarian dinners from then on. In 2009, when I learnt more about the environmental impact of the livestock industry – that the farming of cows, sheep, pigs and fowl for meat and dairy products is the single greatest contributor to climate change, producing 18 per cent of the world's greenhouse gas emissions (compared to the 13 per cent produced by all global transport)[16] – I thought I'd better try to go vegan instead. I liked having a structure and a set of rules and principles to live by, knowing I was not causing unnecessary harm. I could live and reaffirm my values on a daily basis, every time I had a meal.

But my fixation with food wasn't necessarily all that healthy. Social isolation made it much harder to moderate or control my eating habits – there was no one around to tell me what I was doing was disgusting. To paraphrase the pathos of a well-known internet meme: 'I wasn't hungry, I was just empty inside'. I found myself in an endless cycle of bingeing and dieting and so began to reflect on the origins of this 'boom and bust' behaviour. For as long as I could remember my mum was always on a diet of some sort, whatever fad was next: The F Plan, the 'cabbage soup diet', Atkins or Weight Watchers. Often I'd join in or devise my own. When I was about 13, I was inspired by the Slimfast adverts I'd seen on the telly. They said you could replace two of your meals with 'Slimfast shakes' and just have one 'normal meal' in the evening. I didn't have enough pocket money to buy the real stuff, but I realised that if I drank one can of Diet Panda Cola (which was then only about ten pence) for lunch and one for breakfast, then I could get through the whole day on just two calories and 20p! I kept that up for a week, before nearly passing out in the school common room. My mum's body obsessions had rubbed off on me. When in 2007, my work *Eat 22* ended up in the Wellcome Collection museum in London in a section called 'obesity' – peeping out from behind a huge obese blob sculpture by John Isaacs – I found it hilarious, insulting even. But suddenly, in 2013, it all started to make perfect sense. Obesity was just one of the many increasing 'diseases of affluence' (as well as heart disease, cancer and mental health disorders)[17] in our western world which *The Other Forecast* aimed to highlight.

It is the global food industry driven by the profit motive which is responsible for the dual crises of the obesity epidemic (now affecting a quarter of UK adults) and the scandal of food waste which sees more than 15 million tonnes end up in landfill every year.[18] I didn't want to point the finger at individuals for their own weight problems, but instead, to illustrate that the very fact that it's an 'epidemic' affecting millions proves that it's the broken system that's to blame. That year, I had been working with an organisation based in London called Invisible Dust, who paired artists up with scientists to make works in response to our environmental situation. I got to meet Kevin Anderson, the renowned climate scientist who has spent his life demanding radical policy on an intergovernmental level, to move away from 'growth' driven economies towards 'planned recessions'. This was the only way to reduce carbon emissions to the extent that we need to. If you look at the graphs on page 98, it's impossible to deny that carbon emissions, energy consumption and, indeed, instances of obesity fell immediately after the financial crash of 2008.

I had an idea for a 'revolutionary new diet' programme. I wanted to use the format of the support group (a group of people coming together to help each other), which has been hijacked and monetised by multinationals like Weight Watchers, to help radicalise overweight people to fight back against the system. The *Anti-Capitalist Diet Plan* would 'riff on the aesthetic of existing diet plans' except it would advocate 'a very different medicine'. I started attending Weight Watchers at Maryhill Central Hall as 'research' in May 2013. My mum was pleased as she was a great fan of Weight Watchers and had been a volunteer at her local branch since 2011. She believed the only way to keep her weight under control was having a group of people bear witness to your progress every week – to have a 'big other' watching you get

Still from the film documenting *Anti-Capitalist Aerobics* by Ellie Harrison, performed live at the Ways of Seeing Climate Change conference in Manchester on 30 October 2013. Duration: 07:09. (Emma Crouch)

weighed. I lost a stone and was awarded my Gold Member card. The funding for the *Anti-Capitalist Diet Plan* fell through, so I scaled back my plans slightly. Instead, I channelled the same spirit into a one-off performance: *Anti-Capitalist Aerobics*, which I set to the trance track 'More' by RokCity. In the performance I took on the persona of a bossy aerobics teacher who, whilst leading the routine, delivers a polemic about capitalism:

> We're living beyond our means.
> We waste money we don't have on things we don't need.
> We're over-consuming left, right and centre.
> Fattening foods everywhere we turn,
> the sugar highs of consumerism.
> Everything always needs 'charging up'.
> That's it, get your smartphone out and keep on charging it up!
> There's an endless supply of energy out there.
> God put the coal in the ground to be burnt,
> so come on Siemens, let's do it!

The system makes you feel as though it's all your fault.
Anxieties are reproduced because anxieties help sell.
I'm fat because I'm a failure,
I am never good enough!
...
The trouble is we don't know what's good for us.
We need to start joining up the dots.
On the one side, we need to burn energy because we're
too fat.
And on the other side, we need to save energy to reduce
our carbon emissions.
We should be looking for easy ways to incorporate
exercise into lives, which enable us to reduce our energy
consumption elsewhere, like cycling or walking to work
rather than driving in our cars! Sometimes the simplest
solutions are actually the best.
...
What's the point of wasting time, money and energy
consuming more than we need and then wasting more
time, more money and more energy cleaning up the mess
it makes?![19]

System change, not climate change

Also in summer 2013, through my work with Bring Back British
Rail, I was awarded a place on the New Economics Foundation's
Campaign Lab course, run by its new organising arm, the New
Economy Organisers Network (NEON). NEF was a think tank that
had come to understand the necessity for reflection and action.
It was a nine-month programme based around one intensive
workshop a month. The first session was a residential weekend in
Devon, the rest were in London. I managed to get the university to
pay for my travel expenses and every month I got on the train to go
back 'down south' (you can see the increased 'Rail Miles' for 2014
on my Carbon Graph on pages 138–9). I was the only participant
based in Scotland and the only one who was not being paid to
campaign (an 'amateur activist' you could say).

The main aim of Campaign Lab, described as an 'economic

justice' campaigning course, was to help us understand neoliberalism – not just what it is and the consequences it has had for our society (described in Chapter 1), but also the organising strategies that had been put in place to make it into the dominant ideology of our times. Neoliberalism had begun as a fringe idea in the late '40s. A tiny group of wealthy people including economists Friedrich Hayek and Milton Friedman met in the Swiss Alps to found an organisation called the Mont Pelerin Society. Its aim was to make the idea – that the market knows best and that government intervention should be minimal – the norm. In 1947 Friedrich Hayek declared 'It is the beliefs which must be spread'.[20] For 30 years, they worked tirelessly to build a network of think tanks across the world funded by big business to lobby politicians. It took persistence and hard work and, by the late '70s, the politicians they'd won over (such as Margaret Thatcher in the UK and Ronald Reagan in America) began to seize power. Our aim was to learn from what they had achieved and to start organising in a similar way – building a movement around the world demanding a 'new economy' which puts people and the planet before profit.

It was all about systemic change. Campaign Lab helped explain the meaning of the word 'radical', which relates to the Latin word for 'root'. It means that in order to solve the recurring symptoms of neoliberalism – from rising inequality and rising carbon emissions to obesity and social isolation – we must change the economic system that is their *root cause*. It is a huge undertaking – much more difficult than 'firefighting' the symptoms with temporary measures – and it will be a long slog to achieve. That's why it was essential for us to plan a strategy and work together in a coordinated way, just as the Mont Pelerin Society had done. The course vindicated a lot of what I had been advocating and doing with Bring Back British Rail and Power For The People, but which I didn't quite have the language to express.

It was from the New Economics Foundation that I began to understand the interconnectedness of our social, environmental and economic problems and to realise the need to fight for holistic solutions – at an individual and policy level – which address all three at once. Unless we do this, it's the poorest that will continue to suffer:

> When banks collapse, when temperatures and sea levels rise, when food and energy prices soar, people who are poor and powerless bear the brunt. It happens in rich as well as in poor countries. Those in lower income groups are less cushioned against risk and hardship, less likely to have savings or insurance, and less likely to have access to credit at affordable rates. They are often the first to lose jobs, homes and livelihoods. And they are more vulnerable to shortfalls in public services.[21]

This is what is known as 'climate injustice'. It's something Glaswegian activist and anti-poverty campaigner Cathy McCormack is well aware of. In her book *The Wee Yellow Butterfly*, she tells the story of the floods in Easterhouse in July 2002, which took weeks to sort out. Nobody had insurance and tenants were left at risk of electrocution from fallen power lines (managed by the privatised Scottish Power). There were too many companies and organisations passing the buck; none wanting to take responsibility. She makes the stark contrast with the twee village of Boscastle in south-west England, which flooded in 2004 and was 'treated as a national emergency'.[22] The New Economics Foundation's 2009 manifesto *Green Well Fair: Three Economies for Social Justice*, makes clear that 'we cannot allow carbon reduction schemes to penalise the poor; instead, we want them to help narrow inequalities'.[23] Suddenly I began to realise that public transport provision and energy production and supply – the two issues that had been preoccupying me – were the clear points where social, environmental and economic justice met.

We were given a reading list to keep us focused in between the Campaign Lab sessions, which included many inspiring texts. There was American activist Saul Alinsky's *Rules for Radicals* written in 1971 (shortly before his death), which featured 13 rules for successful community organising. The Rules include: 'Power is not only what you have but what the enemy thinks you have', 'Never go outside the expertise of your people',

'Ridicule is man's most potent weapon', 'A good tactic is one your people enjoy', 'The threat is usually more terrifying than the thing itself', 'If you push a negative hard and deep enough it will break through into its counterside' and 'Pick the target, freeze it, personalise it, and polarise it'.[24]

There was a text describing the Montgomery Bus Boycott in 1955 in Alabama, which showed how oppressed peoples – in this case the black population of a city in pre-civil rights America – could bring an institutionally racist bus network to its knees by collectively organising and removing their custom en masse.[25] We also listened to a recording of civil rights leader Martin Luther King's 1968 sermon on 'The Drum Major Instinct'. I discovered that King's argument was similar to the one made in my MFA dissertation, *How to Reconcile the Careerist Mentality with our Impending Doom* – that human motivation is complex, and can be simultaneously altruistic and self-interested. We must acknowledge that we all have 'those same basic desires for recognition, for importance... a desire to lead the parade'. But that if we can learn to channel this 'Drum Major Instinct' as King calls it – this ambition, this 'will to power' (as German philosopher Friedrich Nietzsche describes it) – towards positive and altruistic goals, then it can become a powerful force for social change. We must 'seek greatness, but to do so through service and love'.[26]

Being surrounded by all those 'professional activists', I began to make some interesting observations. At one of the sessions we undertook a Power & Privilege Workshop with community organiser Isis Amlak. This required us all to line up in the centre of the room and either step forward or step backwards in response to a series of questions determining how privileged you are. The question 'If there were more than 50 books in your house when you grew up, take one step forward' really jumped out at me, as that was something I'd always taken for granted growing up with two teachers for parents.[27] When we ended the exercise, all standing predominately at one end of the room, it became clear we were a largely privileged, white, middle class bunch. We were those who weren't 'At the Sharp End of the Knife' as Cathy McCormack would call it[28] – having to 'firefight' the symptoms of neoliberalism on a daily basis in our own lives, struggling to put food on the table or heat our homes, and so we had the luxury of the time and education necessary to be able to take this 'big picture' approach.

As Loki says in his *Poverty Safari Live* show, 'for middle class people... life is just like one big TED talk'.[29] French President Emmanuel Macron came to realise after the 'gilets jaunes' (yellow vests) protests in 2018 that 'some people worry about the end of the world, whilst others worry about the end of the month'.[30] Middle class people have the luxury of the time and education to ask questions about the world we live in and to write self-indulgent books aimed at providing the solutions. It seemed most activism was fuelled by middle class guilt. Unless we are acutely aware of this privilege and take a holistic approach to solving social, environmental and economic problems (as described above), artist Laura Coleman says, we run the risk of speaking 'with a language embedded with violence'.[31] Middle class do-goody-ness – banging on about climate change and 'better futures for our children' – completely ignores the struggles, injustices and violence the most marginalised people in our society are experiencing right now. It also fails to acknowledge that it's the lifestyles of middle class people that are responsible for the majority of environmental destruction in the first place. We're the ones that need to change.

Being surrounded by 'professional activists' also made me appreciate my amateur status and the many 'hats' I had by then acquired, which enabled so many different perspectives on the world. Although I was one of the few who couldn't 'clock-off' at weekends (that was when I got most of my campaigning work done), I also began to appreciate my dual role of artist and activist. The 'art' was what offered the light release from the hard slog of trying to change the world. This dual role enabled me to come at the same problems from these very different perspectives. It was the artist's duty to 'zoom out' – to 'stand back and view the world objectively' (as I described in my MFA dissertation) and draw attention to the absurdities of the social, economic and political systems we humans have created. Whereas, it is the activist's obligation to get 'stuck in' – attempting to transform these dysfunctional systems which are perpetuating inequalities in our world. The two roles represent our 'passive' and our 'active' responses to the crises that we face.[32]

But with the 'art' came some interesting ethical problems. Does making 'socially-engaged' work about any political issue ultimately always end up serving the ego of the artist more than it empowers the participants or affects any meaningful social change? In our current artworld (described in Chapter 6) that felt like an irreconcilable contradiction. The only way I could deal with it was to separate off my activism from my art and to persist, in a sort of Jekyll and Hyde manner, as 'good cop, bad cop', where:

the activist	proposed solutions and was optimistic about the possibility of social change, and
the artist	asked questions and was pessimistic about humanity's capacity for altruism (a misanthrope).[33]

Asceticism and the spirit of capitalism[34]

My recent work had certainly become more misanthropic, fuelled by alienation, anger and frustration. The more socially isolated I became, the more intolerant I was of other human beings, the more the thought of living with one again seemed impossible, but yet, at the same time, there was nothing I craved more. I felt totally trapped. I had resolved one fundamental contradiction in the way I was living (being a landlady) and another had arisen in its place.

It's clear that in order to reduce our energy consumption and to improve the quality of our relationships, and therefore our well-being, we need to be moving to more communal ways of living. Yet I had found myself doing the exact opposite. I wanted to draw attention to this contradiction as a way of, hopefully, starting to resolve it. So for Glasgow Open House Festival in April 2014, I put on an event in my flat called the *Transition Community of One* (pictured overleaf). Inspired by the Transition movement,

Transition Community of One by Ellie Harrison, staged in the artist's flat for Glasgow Open House Art Festival on 13 April 2014. (Ellie Harrison)

founded in 2005 to encourage communities to work together to 'transition' to a low-carbon society, I invited strangers into my flat to watch a programme of socio-political films. As described in the event blurb: 'I hoped to expose the paradox at the heart of my lifestyle – to test the intolerance caused by solo living and challenge my actually existing "socialism in one person"'.[35] I was attempting to reconcile the fact that I had ended up living in a totally unsustainable, individualistic and ultimately unhealthy way.

I started to behave ever more ascetically to try to counteract the excesses of living alone. When Oliver left, I cancelled the internet subscription and the TV licence. It was Oliver's telly anyway and I had begun to see sitting around passively ogling at a screen as a massive waste of time.[36] When I got my boiler serviced I had a big shock. The engineer had taken the front cover off and was testing that it fired up okay. I got to see this massive fireball of gas igniting in the corner of my kitchen – all those fossil fuels being burnt right there in my flat! It felt so wrong that I was causing that combustion every time I absentmindedly turned on the hot tap.

So I took to turning the boiler off at the switch at the wall. 'Off' was its normal state and I only used hot water and heating when absolutely necessary. I had a new portable thermostat courtesy of the Scottish Government's Green Homes Cashback scheme so I could carefully monitor room temperature. I realised that I could survive at 14°C with aid of a hot water bottle, a blanket and some gloves to stop the chilblains.

I pared back other aspects of my life too. I bought a set of 'uniforms' in April 2014, which were plain 'smartish' clothes that I could wear for multi-purposes: teaching, ushering (at the Glasgow Film Theatre where I was a volunteer), working in the studio, meetings, whatever. It meant that I wouldn't need to buy anything else or even go to the shops for the next few years. I took to refining my diet to the simplest ingredients possible – breakfast was exactly 50g of mixed nuts weighed out precisely and washed down with black coffee. I'd given up on all those crazy milk replacement products because of the cost and unnecessary packaging involved. It felt idiotic to be carrying home heavy Tetra Packs of various liquids, which were more than 95 per cent water (water comes out of the tap!). Lunch was a pre-prepared curry, made by soaking some lentils or beans overnight. I would do big batches and measure out portions to put in the freezer. I became fascinated with honing the everyday act of eating to hit a 'win-win-win' sweetspot of being simultaneously the cheapest, healthiest and most environmentally friendly food. I calculated my lunches were just 62p each.[37] My life became a sea of tupperwares – constantly transporting, washing and refilling.

It was hardly as if my new salaried job meant that I started spending more money. If anything, it was the opposite. I still had the same accounting system I'd set up when I first became self-employed in 2004 and had attended all those workshops to help me 'professionalise'. There were three separate bank accounts: one for the 'flat', which all my living expenses would come out of by direct debit: mortgage, council tax, factor, gas and electricity (and in the old days: internet, phone and TV licence).[38] Then I had one account for my 'business', where any money earned from any source would go: my salary and other 'freelance' fees. And finally one for my 'personal' expenses: food, drink and entertainment. I created an elaborate set of standing orders which automatically transferred exact sums from 'business' to 'personal' to 'flat' on the first of each month.

Providing there was enough money going into the 'business' account, it meant that I always only had the same allowance every month for 'personal' expenses, which worked out about £80–100 a week. It was regular and reliable, and meant that I knew I could live off around £8,400 a year. I didn't experience 'boom times' when I received a big pay cheque. As far as my 'personal' account was concerned, it always only ever had the same in it. At the start of the week, I would take out my personal allowance in cash and know that that was all I had to spend until the following Monday. I never used my debit card for my 'personal' account, so I could keep tabs on my budget, plus why would I want my bank and the shops collecting all that data about my spending habits? Given that my salary was more than £18,000, I started to accumulate a lot, which I then used to start paying off my mortgage in large chunks. Asceticism clearly had material benefits for the individual, as well as for the environment. The sooner I was rid of that financial responsibility, the sooner I would regain my autonomy and be able to start taking more risks.

Compromise and complicity are the new original sins

There was one idea which particularly influenced my thinking and action during this time, which I'd discovered when I was invited to write a text for a report exploring how art and culture affect our values in summer 2012.[39] It was psychologist Tim Kasser's theory of the universal human values which recur in different cultures all across the world.[40] Kasser had arranged these values into two opposing sides of a circle (known as the 'Circumplex', opposite). On the left side are the 'extrinsic' values such as conformity, popularity, image and financial success, which are materialistic – all about valuing how you appear to others from the outside. On the opposite side are the 'intrinsic' values such as self-acceptance, affiliation and community, which are all about valuing what you have already or what you can develop within yourself or within your real social networks:

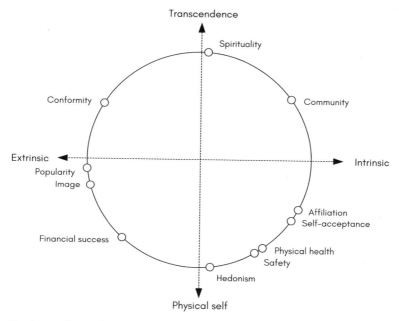

The Circumplex model of universal human values aims to show how values adjacent to each other are experienced as relatively compatible whereas values on opposite sides are experienced as conflicting. Originally published in 2005 in the *Journal of Personality & Social Psychology*. (American Psychological Association)

your 'core economy' close to home. In a healthy society, it's the 'intrinsic' values which can and should help us to fulfil our dual 'psychosocial' needs for 'individuality' (self-acceptance) and 'belonging' (affiliation and community).

Tim Kasser was able to show some 'deep connections between seemingly different issues'.[41] The more we prioritise 'extrinsic' values, the 'less happiness and life satisfaction' and 'fewer pleasant emotions (like joy and contentment)' we have, and the more likely we are to behave in 'manipulative and competitive' and 'unethical and antisocial' ways. In contrast, the more we prioritise (and successfully pursue) 'intrinsic' values, the 'happier and healthier' we are and the more likely to have 'ecologically sustainable attitudes and behaviours'.[42] This theory blew my mind. I realised that all my ascetic choices – my uniforms, my packed lunches, my tinkering with automated standing orders – had served to help create a situation where, once up and running, I wouldn't need to think about money or material goods. Instead I could prioritise 'intrinsic' things which would help to improve my 'well-being, the care with which I treat others, and the extent to which I live in an ecologically sustainable fashion'.[43] I could feel the theory held true. I was beginning to crack it in aspects of my own lifestyle, but yet I found I was being pushed to do the exact opposite at work.

I had chosen to work in the arts and in education (as opposed to being, say, a stockbroker) because I believe they are the activities which should enable us to prioritise (and successfully pursue) 'intrinsic' values. The university our art college is part of even circulated a colourful little postcard proclaiming its 'core values', such as 'valuing people', 'working together' and 'integrity' (all of which are 'intrinsic'). But there was a stark difference between what the university preached and what it practised. In reality, it clearly prioritised the exact opposite. When I began my contract on April Fools' Day 2013, I was signing up to a three and a half year 'probation' with numerous criteria attached, including fundraising. One read that I must 'write and submit a significant research grant application'.[44] These criteria were the same as those issued to scientists and mathematicians across the university. It might have made sense to apply for lots of money if I was trying to collaborate across the globe, in order to, say, cure cancer, but did it really make sense for me to do that to fund my art?

What I liked about having a part-time teaching job in the first place was that I could subsidise all my other work. I didn't have to worry about it being profitable or compromised by the agendas (hidden or otherwise) of various funding bodies. In my campaigning and in my art work I could do and say whatever I wanted. I had the autonomy necessary to 'speak truth to power' knowing that my teaching work would cover the bills. What's more, these so-called 'research projects' we were encouraged to get funding for were about as far away from what I believed the purpose of art to be: that is to critique and to challenge the status quo. They often had million pound budgets, normally funded by the European Union – the bigger they were, the more meaningless they became: totally disconnected from the human scale or from any specific locality, the outcomes lost forever within academia.[45] But worst of all, these so-called 'research projects' were actually sucking staff resources away from teaching the students, as 'successful' staff were given 'research leave' to complete them.

When studying the first module of my Postgraduate Certificate in Teaching in Higher Education in 2014 (another mandatory element of my 'probation'), I began to critique this situation, writing then that:

> I see no greater absurdity than that of a university which, in the ruthless pursuit of funding through a dysfunctional system, completely sidelines the educational experience of its students: its essential raison d'être.[46]

It was clear that the spell of neoliberalism had captured the entire higher education sector,[47] pushing universities to continue to prioritise 'extrinsic' goals: popularity, image and financial success. Our managers were obsessed with our ranking in the 'league tables', offering pep talks on how to play the system. There was a culture of dreaming up ever more complex and absurd 'research grant applications' with ever larger budgets in a futile attempt to break even. The Research & Innovation Services department was on hand to help us 'researchers' create a funding 'applications treadmill'. 'Successes' were celebrated. I remember my shock when it was announced with some fanfare that a colleague who was employed as a full-time 'researcher' (what sociologist David Graeber might describe as a 'bullshit job')[48] had been awarded £15,000 from Creative Scotland. This was someone with a decent salary and no conceivable need for that money. What benefit would that have to the students or wider public, I thought? It certainly wasn't going to make any interesting or provocative art!

A few small acts of resistance helped me cope in this oppressive environment. There were two t-shirts that I would often wear whilst teaching. The first I made myself depicting a replica of a screenprint made by students and staff of Hornsey College of Art during their occupation of the college in 1968. It said 'Bureaucracy makes parasites of us all'.[49] The second was given to me by artist Graham Ramsay (my old tutor at Glasgow School of

Art) and was a replica of a print by the conceptual art collective Art & Language from 1977, which said 'When management speaks, nobody learns'.

Jimmy Reid was made Rector of the University of Glasgow in the '70s because he understood *the true value of education*. Education was not a way of making money and joining what he called the 'rat race', moving on and leaving your community behind (what might now be described as 'social mobility'), but instead of learning to live a valuable, meaningful life and *contributing your knowledge back to your community*, especially in the absence of full-time work caused by deindustrialisation. As a call to arms against a small-minded system obsessed with 'extrinsic' goals, he said in his famous 'alienation' speech:

> To the students [of the University of Glasgow] I address this appeal. Reject these attitudes. Reject the values and false morality that underlie these attitudes. A rat race is for rats. We're not rats. We're human beings. Reject the insidious pressures in society that would blunt your critical faculties to all that is happening around you, that would caution silence in the face of injustice lest you jeopardise your chances of promotion and self-advancement. This is how it starts, and before you know where you are, you're a fully paid-up member of the rat-pack. The price is too high. It entails the loss of your dignity and human spirit. Or as Christ put it, 'What doth it profit a man if he gain the whole world and suffer the loss of his soul?'[50]

Progress trap

I started to question what 'progress' actually was. At a certain point in their lives my parents stopped riding their bikes. They got jobs that were further away from our house, they got cars so that they could drive there and they gradually got fatter and a bit wealthier as a result. They were too busy or too tired to exercise regularly, which led to the relentless cycle of dieting. Why did it always have to be 'boom and bust'? It was clear that the older and more 'successful' I got, the more my own lifestyle became

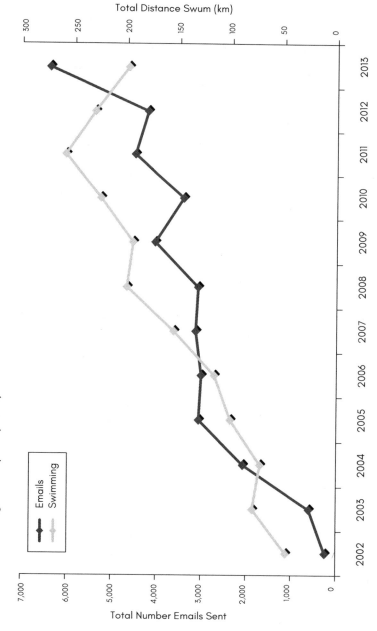

Ellie Harrison: Progress Report (2014)

riddled with contradictions and compromises. I hadn't gone as far as buying a car – that really would be the 'original sin' – but I certainly wasn't always practising what I preached. After all those years of banging on about a sustainable lifestyle, I came to realise that my work-a-holism, the fact that I was always busy, busy, busy, with no time to rest, was hardly very 'sustainable'.

So in 2014, I took the time to analyse more than a decade of personal data to make my very own 'Progress Report' (on previous page). To my horror I discovered that my own lifestyle, measured in crude metrics representing 'work' (number of emails sent) and 'leisure' (number of lengths swum) had simply been mirroring capitalism's 'growth' fetish. Continual growth could mean only one thing: crisis was imminent. Although I was also guilty of committing a 'fallacy of composition' (like our then Chancellor George Osborne had done), I found that indulging in a little comparison helped me comprehend our economic system and understand my place within it.

That year I watched *The 17 Contradictions of Capitalism* – a lecture on YouTube by David Harvey (author of the book, *A Brief History of Neoliberalism*), which highlights all the 'internal contradictions' within the system of capitalism, which inevitably lead to points of crisis. One of them is the contradiction between 'use value' and 'exchange value', which is most evident today in our private housing market. We all need somewhere warm and dry to live (that's a fundamental human need), but the contradictions of capitalism have made houses into 'commodities' which are bought and sold to enable people to make money out of nothing, artificially inflating prices to the point where people can no longer afford a home. Another contradiction is that with nature – the demand for continual and infinite 'growth' on a finite planet, which is leading us towards climate breakdown. One is inequality, which David Harvey phrases as the 'accumulation of capital by the minority' against the 'immiseration of the masses'. Another contradiction is urbanisation – more and more people moving to cities in search of a better life, which leads to more squalid conditions (something most evident in Glasgow in the 19th century). These contradictions all eventually lead to crises and David Harvey says, 'the bourgeoisie only has one solution to

Breaches to the Transportation section of my Environmental Policy (2010-5)

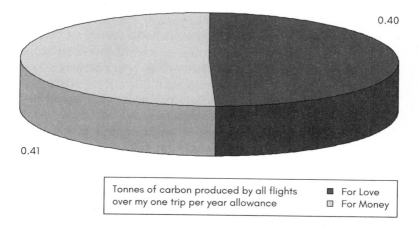

0.40

0.41

| Tonnes of carbon produced by all flights over my one trip per year allowance | ■ For Love □ For Money |

its crisis problems: it moves them around' – temporarily resolving one only to have it reappear elsewhere.[51]

It made me realise that if 'mid-life crises' do occur, it's because the 'internal contradictions' in the way we are living – between what we believe in and how we actually act – reach boiling point. I could feel the pressure starting to build.

So in 2015, I further interrogated my own behaviour and was horrified to discover that I had, in fact, breached the Transportation section of my own Environmental Policy on several occasions in the previous five years. I made a pie chart (above) to illustrate the extent of my own hypocrisy – showing that it had only ever been 'Love or Money' that pushed me into compromising my principles.

I might not have been driving, but it was clear I was still travelling far too much, particularly since I got my job in Dundee and was doing a huge commute every week – two hours door-to-door crossing the whole width of the country, which made for a 12 and a half hour day (you can see in my Carbon Graph on pages 138–9 how my 'Rail Miles' increased from 2012 onwards). The crux of the contradiction was that in order to slash our carbon emissions to the extent necessary, we must urgently reduce the amount we travel and the amount of energy we consume. Instead, we're doing the opposite. People are increasingly travelling further distances for work: longer daily commutes are becoming normalised,[52] stress and decreased well-being result.[53] It's all made possible by the larger pay packets that seem to make all this travel feel 'worthwhile'.

And it is transport infrastructure – particularly the railways – which has enabled these sorts of lifestyle to emerge. When Britain's railways were built, during the Industrial Revolution as a way of speeding up the transportation of goods and people and therefore accelerating the profit-making machine, it was seen as one of the greatest technological advancements of all times. Suddenly it was possible for the English to work in Scotland, and for the Scots to work in England and to travel between the two in a matter of hours. As Glaswegian artist and writer Alasdair Gray writes: 'After north Britain became reachable by railway and steam ship many southerners began coming here, and have been coming ever since'.[54]

In *Railways: Nation, Network & People*, Simon Bradley describes the impact this technological advancement had on people at the time. Those who were part of the generation which spanned the building of the railways, and the new lifestyles linked to a national economy (as opposed to a local one) which ensued, could never quite cope with the transformation. He says:

Many of those who had reached adulthood before the railways came, were haunted by a sense of loss, of having experienced a world that seemed secure, but which had melted away… Railways achieved a sort of revolution in the head. A sense of the forceful urgency of the present day, before which old customs and attitudes were doomed and unimportant.[55]

Not only do I feel that I am of an equivalent generation in relation to the huge technological advances of the internet and digital communications (having lived until I was 19 before even getting an email address, 21 before I got a mobile phone), but I also feel that the dramatic shifts rail travel (and now plane travel too) have encouraged, have affected more than just that Victorian generation. The essay *Reflections on Exile*, written in 1984 by the Palestinian philosopher Edward Said (who spent most of his life as a political exile in New York), describes the 'condition of terminal loss' experienced by a person forced to live away from their 'native place' (where they are born); forced to live away from what he calls their 'true home'. He questions why, therefore, this condition of the 'exile', the 'émigré' or the 'refugee' has

> been transformed so easily into a potent, even enriching, motif of modern culture? We have become accustomed to thinking of the modern period itself as spiritually orphaned and alienated, the age of anxiety and estrangement.[56]

It's the opposite of what is needed to improve human well-being and enable more sustainable lives. The 2018 book, *Situating Ourselves in Displacement* (a collection of essays published by the Journal of Aesthetics & Protest) also describes 'displacement' as the 'key paradigm of our time'. 'For who can afford not to move, to shift, to change, to develop and improve – or to be moved, shifted, displaced?' Yet they claim that it is the opposite experience of 'situatedness' – that is having real roots and connection to the actual place where you live, which is the 'key condition for solid and sustainable practices, in politics, arts, research or otherwise... Yet situatedness is not something we can take for granted today'.[57] The more transient and disconnected our lives become, the less we understand the places where we live and work, the less likely we are to fight to make them better, or even to know where to start.

The leaky bucket

It was when I finally turned my attentions to my local area – the city where I'd been 'based' for over five years – that I began to realise that Glasgow itself was also riddled with contradictions and compromises. As Carol Craig puts it in the introduction to *The Tears that Made the Clyde*, 'This is a city which likes to espouse its egalitarian values but it is deeply divided by social class'.[58] Or as Glasgow-based writer Gerry Hassan puts it 'Scotland loves talking about social justice, but it doesn't, doesn't do it!'[59]

My disenchantment with Scotland began when I got involved in a local campaign. There were plans to open a new Tesco store at the end of my street and lots of local people were up in arms. They were concerned that it would put all the local food shops in the area out of business. It would be the 'tenth Tesco in two miles' – a pincer movement of new Tesco 'Metro' and 'Express' stores (which are not subject to any planning controls due to their small size), squeezing local competitors out of the market and gradually gaining a monopoly over our food supply.

I was inspired by the New Economics Foundation's theory of the local economy, which uses the metaphor of the 'leaky bucket'. The holes in the bucket are caused by all the big multinationals like Tesco (then Britain's biggest food retailer, taking '£1 out of every £8 spent by British shoppers').[60] As soon as profit is made in Tesco, it is sucked out of the local area and redistributed to wealthy shareholders elsewhere.[61] For every new minimum wage job a new Tesco store creates, another more rewarding and empowering job is being lost elsewhere as local family-run businesses are forced to close.[62] It's a zero-sum game. To build a strong local economy we need to 'plug the holes in the bucket', and make sure wealth generated locally is retained and then re-spent in the local area.[63] With unfettered expansion, these global corporations are creating 'clone town Britain',[64] where all our local high streets end up looking the same and, therefore, all suffer the same fate when financial crises strike.[65] As NEF say:

> It represents a slow emptying of our institutions, a destruction of our way of life as tangible as any terrorist attack, a rotting away of the nation of shopkeepers – of imagination, innovation and pride at local level – to a miserable, slavish acceptance of whatever narrow aspects of life our centralised systems choose to deliver.[66]

As the 'Say No To Tesco' campaign, we launched a petition to change planning regulations so that the combined square-footage of all one company's stores in any given area would be taken into account when considering planning permission (there's a more rigorous process for large out-of-town stores). The campaign was a big learning curve. Before we went to the Scottish Parliament in January 2014, we took our petition to Glasgow City Council. That's when I had my first run-in with one of Glasgow's Labour councillors. The week before our presentation to the Petitions Committee on 3 September 2013, he invited me to the City Chambers for a 'chat'. That was a meeting I'll never forget. I was taken up into the palatial surroundings of one of the Council's wood-panelled committee rooms. He sat down opposite me and leaned back in his chair, his legs far too wide apart and his little grey suit far too tight at the crotch. When I explained what was happening: Tesco's new tactic of opening lots of small stores to get around planning regulations and how this was not going to help our local economy in the long-term, he replied: 'It's just market forces, love. Sorry'. What! That was insane. I thought governments, especially Labour ones, were there to regulate markets to ensure that they don't cause unnecessary harms. Who was this nasty little patronising man?

When I got to see the brilliant play *The Cheviot, the Stag & the Black, Black Oil* at the Citizens Theatre on 19 October 2016, I recognised him immediately. Written in the '70s (but just as relevant today), the play tells the story of centuries of exploitation of ordinary Scots: by English landowners, by American oil tycoons and also by other Scots. One of the characters is a Labour councillor turned property developer – a wheeler-dealer from a working class background who gets a taste for power and wealth and so quickly betrays the very people he'd set out to serve, in favour of enriching himself and preserving his status. He'd become that detached 'liberal elite'. I guess you could call that 'social mobility'.

These characters were rife in Glasgow's Labour establishment and the deep sense of betrayal they've caused has long-historical roots, going back nearly a century. As Carol Craig shows in *The Tears that Made the Clyde*, the nationalist movement was only founded in 1928 after the first wave of Independent Labour Party MPs elected to represent the people of Glasgow at the

UK Parliament became so enchanted by Westminster that they defected to the mainstream Labour Party and were no longer committed to home rule.[67] In the post-war period, Labour in Glasgow became more paternalistic, with an agenda lacking any imagination and ambition (other than for enriching themselves). She quotes from the book, *A Century of the Scottish People 1830–1950*, by Christopher Smout:

> Labour in Scotland became synonymous with the defence of council housing, jobs in heavy industry and sectarian schools: it had nothing whatever to do with participatory democracy, enthusiasm for socialism or hope for the future.[68]

But the Say No To Tesco campaign was not totally in vain. At the Scottish Parliament, the debate we provoked fed into the Community Empowerment (Scotland) Act 2015, which gives local people the right to buy or take back land from absentee landlords to put to good community use.[69] It also made me realise that there's no use just saying 'No' – you need to provide positive solutions and alternatives. You need to devise new systems for distributing essential resources, which do deliver the social and environment outcomes that we want to see – and that has to be done at a local level. That was when the idea of a local currency – money which can only be spent within a certain geographical location – first cropped up. It was a tool that had proved successful in other parts of the world, for counteracting the forces of globalisation, reducing carbon emissions from imports and exports and building stronger local economies. If pound sterling is too easy to be extracted as profit and whisked away to some offshore tax haven, let's design our own currency which only has value in Glasgow. At the end of the Campaign Lab course, I made a list in my notebook entitled 'Think Global, Act Local' – a phrase coined in 1915 by the great Scottish thinker Patrick Geddes (1854–1932).[70] It detailed all the local projects and campaigns I would have loved to make happen in Glasgow, if and when I ever found the time.

Pages from Ellie Harrison's Notebook 29 (April 2014) showing list of local projects and campaigns she would like to make happen in Glasgow if she had the time.

Worst inequalities in Western Europe

In terms of my disenchantment with Scotland, it was discovering the term 'the Glasgow effect' and finding out what it meant, which really tipped it for me. I was at an event called 'Cultures of Independence' at Glasgow School of Art, which was one of the many discussion events around the country in the year running up to the Referendum on Scottish Independence in September 2014. There was a presentation about 'public health' – not something I knew much about. Apparently Glasgow – the city where I'd been 'based' for over five years – had the worst inequalities in health in the whole of Western Europe. 25 per cent of the city's population are registered disabled,[71] and 32 per cent of adults have no educational qualifications at all.[72] Where are all these people, I thought? Everyone I know seems to have a PhD! Why had this story, which now appeared quite starkly as the dominant reality for so many people in this city, not been reflected in any of the art and culture I had seen until now? Perhaps we were really living in the 'two different worlds' Darren McGarvey describes.[73] 'The Glasgow miracle' and the narrative of the 'creative city' which had lured me here in the first place, had simply been used to paper over the cracks. As an artist – a 'creative industries' professional – was I partly complicit? Scotland clearly wasn't the socialist utopia I had hoped. On many counts it was actually worse than England!

What's more, it was the hypocrisy of what claimed to be a 'social democratic' Scottish Government continuing to preside over such a divided nation, which appeared to make the situation so much worse. It created a culture of complacency where people believe the Scottish Government is on their side so there's no need to fight for social change. It also created a culture of denial. The same 'it wisnae us' attitude which pervaded Scotland's relationship to the slave trade, despite the fact that Glasgow was 'the second city of the empire' in the 18th and 19th centuries with much of its wealth built on the brutal forced labour of slaves in the Caribbean.[74] Scottish politics was a blame game. Just as everyone was getting optimistic in the run-up to the Referendum, I was starting to lose faith.

I did tentatively vote 'Yes' in the end, but in summer 2014 I made an installation called *After the Revolution, Who Will Clean Up the Mess?* to explore my ambivalence to the event. It featured a line of four huge confetti canons installed inside Talbot Rice Gallery in Edinburgh, which would only be detonated in the event of a 'Yes' vote. The sense of anticipation, and then anti-climax, was captured in the stillness in the room. If the Brexit vote has shown us anything, it's that the fallout of a 'Yes' vote would have lasted for decades. To quote a man I spoke to on the train recently: 'We're now damned if we do, and we're damned if we don't'. My feeling with both referendums was always that there are much more pressing issues we need to be dealing with now – namely cutting carbon emissions and reducing inequality. We are deliberately creating massive distractions. To paraphrase an infamous troll critiquing the self-destructive introspection of identity politics: 'We're fiddling with our constitutions, while the world burns'.[75]

Of course, the poverty and deprivation which is the key driver

After the Revolution, Who Will Clean Up the Mess? by Ellie Harrison, installed at Talbot Rice Gallery in Edinburgh from 1 August to 18 October 2014. (Chris Park)

of Glasgow's poor health is partly caused by the benefits system controlled by Westminster, but there's lots that the Scottish Government could and should be doing to reduce inequality now. It has power over education, the environment, health, housing, civil and criminal justice and transport.[76] With additional powers over taxes and welfare and even the public ownership of Scotland's railways (thanks in part to my campaigning) granted in the Scotland Act 2016. Why are these powers not being put to good use?

Scotland has had power over its bus network since devolution in 1999. Yet it has done *absolutely nothing* to reregulate the buses and undo the damage caused by Thatcherism. Real socialists see the importance of providing high-quality public transport to the poorest people in society who live in inaccessible places and cannot afford cars. In fact, public transport should be an absolute priority for any government supposedly concerned with reducing inequality, addressing climate change and delivering 'inclusive growth' – the Scottish Government's favourite catchphrase.[77] But there are some powerful Scots who made it rich as a result of bus deregulation in 1986 – two of the UK's biggest private bus companies (First and Stagecoach) are based in Aberdeen and Perth – and their billionaire bosses are too in bed with our self-interested politicians. So this privatised system, which has the worst impact on our poorest and most marginalised communities, prevails.

To add insult to injury, when the Scottish Government did reform public transport in its Transport (Scotland) Act 2005, it chose to remove 'passengers' from the picture all together. Strathclyde Passenger Transport – the regional transport authority which is meant to oversee all our public transport was renamed 'Strathclyde Partnership for Transport'. 'Partnership' being the most insidious of New Labour terms that basically translates as 'pandering to private companies'. But because the SPT acronym remained the same, nobody really noticed the profound damage that had been done. Plus the Scottish Government is on our side, right? Nearly everyone I have met working in the transport sector in Scotland now (at SPT or at Transport Scotland, including two Transport Ministers) dutifully refers to the bus 'market', rather than to bus 'services' as if there really 'is no alternative' – which simply is not true! They are all still firmly trapped under the spell of neoliberalism.

It also became evident that the education system, which I'd initially been so proud to be part of, was actually, in many cases, more unequal than in England. Glasgow School of Art itself was seen as one of the most elitist educational establishments in the whole of the UK, worse than Oxford or Cambridge for the number of students from working class backgrounds being admitted.[78] No wonder the art school is the butt of many jokes for Glaswegians such as Darren McGarvey,[79] and comedian Limmy. The sketch *Glasgow School of Art Tragedy: My Thoughts*, created by Limmy following the first fire in 2014, shows a posh Scottish character bemoaning the loss of the art school and showing little concern for community centres and schools which have been deliberately bulldozed in other less glamorous parts of the city, because they haven't created artworks which have 'travelled all over the world' and 'touched a lot of people'.[80] Inequality runs throughout Scotland's education system. I remember my shock when opening a copy of *The Herald*, a supposedly left-leaning

Scottish newspaper, and a supplement on 'Independent Schools' fell out onto my lap. God, I thought, you wouldn't even get that in Ealing! Not only were kids being segregated at an early age on wealth and class grounds, many were subject to state-sponsored segregation on religious grounds. And we wonder why there's inequality?

'The Glasgow effect' was something I wanted to find out more about, to draw attention to and to start fighting to try to address. I still have the clipping I cut out of *The Guardian* on 17 April 2014, with the headline 'Glasgow counts down for [Commonwealth] Games with worst life expectancy in UK'.[81] It features an illustration showing two maps of the UK, one for men and one for women with the lowest life expectancy marked in red and the highest in dark blue. It's striking how Scotland is almost wholly reds and oranges in both maps apart from East Renfrewshire and East Dunbartonshire, two of Glasgow's neighbouring councils, which are the same pale blue as Ealing. Who lives there, I wondered?

Settlers and colonists

By the start of 2015, my dreams of the socialist utopia had been shattered; my love affair with Scotland was over. There was nothing I wanted more than to go home. My instincts now told me being closer to the people who really loved me and who I really loved, unconditionally, would be a far more natural way to live, and much more sustainable. I wondered about how to make it happen. I'd been away from London so long that returning seemed almost impossible – private housing was so expensive and I wasn't allowed on the social housing register because I was now a 'home-owner' despite the fact my flat was worth nothing in comparison to London prices. Moving south would mean a crippling mortgage and a much lower quality of life in many respects. Perhaps I could move back into my parents' house? It's only our individualistic society developed over the course of the 20th century which labels people who move back in with their parents at the age of 36 'failures'. It actually makes much more sense for younger family members to be around to care for the older ones – that's what the 'core economy' is all about. If I could have picked up my flat in its entirety and dropped it down next to my mum and dad, I would have done it in a flash. I'm sure there are millions of people all over this world who've had that same thought: a sign of how broken our housing system is (both private and publicly-owned). I had another massive 'seven year itch'. Just as had happened in Nottingham at the end of the previous decade, I wanted to 'change my life' again and get the hell out.

But is wasn't just the practicalities of the housing system that was stopping me moving home – giving up my job, my studio and the other few roots I now had in Scotland. My friend Neil, who was born in Leicester but had been living in Glasgow since 2005, had told me about the 'Settlers and Colonists' essay by Glaswegian artist and writer Alasdair Gray, which was published in the book, *Unstated: Writers on Scottish Independence* in December 2012. It was all about English immigrants who move to Scotland to take up opportunities in the 'creative industries'. They 'know or care nothing for... local achievements' beforehand except for 'rumours of gang violence and radical socialism, both of which should be forgotten'.[82] He defines two different types of immigrant: the 'settlers' who go on to make lives for themselves, embracing Scottish culture and identity and becoming honorary Scots: 'as much a part of Scotland as Asian restaurateurs and shopkeepers, or the Italians who brought us fish and chips'.[83] Then there are the 'colonists' who stay for a few years, cream off whatever rewards they can before returning to England with a little more wealth and prestige than they had before. Colonists use Scotland as a stepping stone in advancing their careers.

He was describing *so many people* I had seen come-and-go in the short time I'd lived in Glasgow, including the majority who studied on the MFA. People who'd succumb to the pressures of a globalised 'knowledge economy' to live transient, itinerant and opportunistic lifestyles, chasing work around the world.[84] He was also, potentially, describing me. As much as I wanted to leave, I did not want to be *that person* who used and abused the city where they lived without giving anything back. My thinking had moved on since my Nottingham days. I now had more of an understanding that a happy, healthy and sustainable life comes from committing and contributing to the community where you live. Instead of following my gut instincts to escape, an idea started to brew in my mind for something much more counterintuitive. Maybe I should really start to 'Think Global, Act Local'?

Perhaps I could devise a framework within which I could start to put all my learning about how to build a more equal, sustainable and connected city into practice on my doorstep. I could attempt to 'situate myself in displacement',[85] by offering positive solutions to the many social, environmental and economic problems which Glasgow clearly has. It would be the opposite of the artworld and academia's obsession with 'internationalisation', which I had come to think of as the antithesis of what we need to be doing to address climate change. They were 'thinking locally and acting globally' – a petty and small-minded attitude which sucks money and opportunities away from local communities by shipping slick artworks (and high-profile artists) to global exhibitions in glamorous and inaccessible parts of the world, like the Venice Biennale,[86] which I witnessed first-hand on a visit in May 2015. They were all obsessed with money, power and prestige.

But as I entered the final year of my 'probation' at work, I still had that dreaded 'write and submit a significant research grant application' on my list of things to do. When I returned from Venice in May 2015, it all started to come together. Perhaps I could get a £15,000 grant from Creative Scotland (as my colleague had done) to do this 'durational performance' thing where I would force myself not to leave Glasgow at all for a whole year.[87] I could easily justify that to the funding body in terms of my previous year-long performances like *Eat 22* and *Gold Card Adventures*, and with my continued concerns about, and action on, climate change. I could pitch it as a real life experiment in 'thinking globally and acting locally'. Then if the grant was awarded, it would put the university in a 'catch 22' which would highlight the absurdity of their value systems and the 'manipulative and competitive' behaviour they encouraged – either they could have the money (a £15,000 grant to list on a spreadsheet somewhere showing how 'successful' they had been), or they could have me there actually teaching the students. Then, if I continued to travel to Dundee for work, I would be breaching the terms of the grant. It was genius. I knew it was pretty hardcore to draw so much public scrutiny of my own 'lifestyle choices' and the unsustainable way I'd found myself living, but I also knew I had to do it, even if it did mean getting the sack. I had to show them the 'rat race is for rats'![88]

If I did get sacked then perhaps that would be the excuse I needed to pack up and go home. I would be following in a history of critically-engaged artists who'd also had uneasy relationships with the institutions where they worked because they felt pushed to compromise their personal principles. Artist John Latham (who founded the *Artists Placement Group* with Barbara Steveni in 1966 with the aim of placing artists into 'government, commercial and industrial organisations' to respond to and influence their policy), was dismissed from his teaching post at Central Saint Martins in London in the late '60s.[89] His crime was to check out a copy of Clement Greenberg's book *Art & Culture* from the college library – a much overused text seen to be restricting the interpretation and production of art to meaningless modernism – and to individually rip out and chew up the pages one-by-one before returning the pulp back to the library in a glass phial. Influential German artist Joseph Beuys was also sacked from the Düsseldorf Academy of Fine Arts in the late '60s when he refused to follow the official procedure of selecting ten students to study with him, and instead let all four hundred applicants join his class.[90]

I took courage from these forebears' integrity and spent the month of June carefully crafting my funding application. I spoke to my mum and dad and they really didn't like the idea. They were quite upset at the thought of not seeing me so often. But my sister helped and encouraged me. She worked in academia as well – as a 'researcher' at the university in Norwich – and saw it as a brilliant challenge to the unnecessary pressures placed on staff to fundraise; a great practical joke. When I finally submitted the application on 29 June 2015, I was so relieved. It was a massive task that had loomed over me the last two years, which I could now finally tick off my list. It was almost the end of a non-stop slog of work which had been just as relentless as my 'work-a-thon' in 2011 where I'd barely had a day-off all year. It was a slog which had caused the break-up of my last relationship. When I got dumped in New Year 2015, Isla said it was never going to work between us because she defined her identity through her relationships with friends and family, whereas I defined my identity through my work. What would it take for that to start to change?

First as tragedy, then as farce

The day after the application was submitted to Creative Scotland, I began my five weeks of 'annual leave' from the university. I could relax slightly, but it wasn't exactly a 'holiday' as I still had two more big projects to deliver in Cardiff and London as well as travelling to Basel in Switzerland on the train to do a talk about my 'love-hate relationship with data' at a conference called *Big Data in the Context of Culture & Society* on 4 July 2015. When I got to Cardiff I broke my right arm. It was all I needed – a silly accident involving a skateboard, which made me start to realise: 'I'm getting too old for these sorts of things'. It was incredibly painful and swelled up black and blue. But I'd never broken a limb before, so didn't actually realise that's what I'd done until I returned from Switzerland the following Monday and walked myself into A&E at the Glasgow Royal Infirmary.[91]

Having this temporary disability was an eye opener. I could no longer cycle around town. I went to get the First Bus again – the number 42 which ran from near the Mitchell Library to Alexandra Parade in the East End just outside my studio – only to discover that it no longer existed. What? I got in touch with First and was told bluntly: the route had been cancelled because it was not 'commercially viable'. Argh! How could this supposedly 'social democratic' Scottish Government preside over a privatised public transport system which shits on the people most in need of their support? I knew that someone had to fight back against the privatisation of our bus services, but I didn't have time right then. For now I'd just have to walk.

I continued working with my right arm out of action for the rest of the summer. I was forced to get back on my bike after three weeks, holding onto my right handlebar tentatively so as to avoid the worst of the bumps. There was just no other easy way to get around town. Fortunately I had a project called *This Is What Democracy Looks Like!* to deliver in London in July so got to stay with my parents for ten days.[92] They looked after me. My mum said I was looking 'a bit tubby' and got me going to Weight Watchers again. I discovered it's impossible to cut up vegetables with just one arm, so they made me tasty salads for my packed lunch every day. By the end of the summer, as the students were arriving back at the college for the new semester, my arm was almost back to normal. I'd lost ten pounds again and was feeling great. On 25 August 2015, I got an email from Creative Scotland to say I hadn't got the funding, but they offered some feedback on what needed further elaboration. I tweaked a few things and resubmitted the application on 1 September 2015. They had an eight-week turnaround, so I would find out by the end of October whether this insane project was going to go ahead.

On Wednesday 23 September, I left my studio about 5.20pm to cycle through town to the Glasgow Film Theatre where I had been working as a volunteer usher for the last five years. David Cameron's 'big society' (the friendly face of austerity) had made it acceptable to withdraw wages from so many jobs which had previously been paid. But I enjoyed taking tickets and picking up litter. It was worth it for all the free films. As I rounded the corner into Cambridge Street, I crossed onto this weird bit of cycle lane in the middle of the road, which must have been installed during a previous ill-thought-through attempt to encourage more cycling in the city the decade before I arrived.[93] As a cyclist you have right-of-way on this lane, but it's hard not to feel very vulnerable with traffic passing closely on both sides and only green paint for protection. I saw a white car pulling out of a side street on my left, it was coming straight for me. Surely it had seen me? Surely it was going to stop? I rang my bell: tink, tink. Whack! It was too late. I went flying, landing on the road to the right of the car. I could see the faded markings of the cycle lane below me, the majority of the green paint by then worn off. The pain was excruciating. I looked down at my left wrist. I could not believe my eyes – the bone was bulging out through the skin. This was much, much worse than the last time. I've only gone and broken my other bloody arm! I scrambled over to the pavement and a passer-by brought a chair from a nearby café. Somebody gathered my smashed up bike out the road.

Awwwwwwwwwwwwwwwwwwwwwwwwwwww! I'd never felt so much pain in all my life. What the fuck was I going to do? The car driver had got out to join us on the street corner. A woman on her way home from work at Lloyds Bank had stopped to help calm the driver down, as I continued to yelp out loud and stamp my feet up and down manically to quell the pain. A paramedic showed up, but he was in a car not an ambulance, so couldn't actually take me to the hospital. He offered me some gas and some painkillers, but I said no. I wanted to have my wits about me in case I had to make life-changing decisions about surgery. They asked me if there was someone who could come and help me. I didn't really know who to call. I had just seen my friend Neil who was helping build my new website. We had left the studio at the same time following our meeting. I felt terrible burdening him with this, but after they spoke to him on the phone he came up on his bike straight away.

Eventually, after about an hour, an ambulance showed up and took us both to A&E. They put a temporary cast over the bulging bone and said they would have to operate in the next few days. I went back to stay at Neil and Laura's that night.[94] I took all the Co-codamol I could get, but I didn't sleep a wink – writhing around in pain, plagued with a vision of this splintered bone underneath the cast popping right out and puncturing my skin. I didn't tell my parents what had happened. I didn't want to worry them, especially after they'd had to look after me over the summer. But, come the morning, I thought I'd better give my mum a call.

'Hi mumma, it's me'.

'Hello pet', she said.

I was straining my voice to sound as 'normal' as I could: 'You're not going to believe this, but...'

'What! Not again?!' said my mum.

'No, actually it's a bit worse than before... I'm going to have to have an operation', I admitted sheepishly.

There's nothing I wanted more than for my mum to be there to comfort me, but I didn't want to put any pressure on her to make the long journey up. After I put the phone down, I started to make my way back home, this time on the Subway. At 4pm I heard a text message come through. It was from my dad: 'Mum has now set off. She's looking forward to seeing you and helping. Hope the pills keep the pain at bay. xxx'. I was so relieved.

I slept so much better that evening knowing my mum was in the room next door. The next morning I was under orders not to eat anything in preparation for the operation. We got a taxi back to the Royal Infirmary. I enjoyed lying around in the hospital bed waiting. It was an enforced rest. It was a little taste of real Glasgow life. I made some interesting observations. When the nurses asked me what had happened, they acted as if I was crazy even being on a bike in the first place, maybe it served me right? When I finally got wheeled in to meet the anaesthetists who were further up the NHS pecking order, there was more interest and empathy with the injustice of the cyclist's injury. How come only the wealthier, more privileged people used this cheap and healthy form of transport? It made no sense to me... but before I knew it I was unconscious, having the most intense and vivid dreams. When I woke up, my mum was there smiling at me. She had been worried sick as the operation had taken longer than planned. She'd bought me loads of cereal bars, the only vaguely vegan snacks she could find in the hospital shop. It was the evening by then and I was starving. I shovelled them all into my mouth, one after another in a slightly deranged fashion. My sister was on her way up on the train from Norwich by then too. She came to the hospital and then the two of them went back to my flat to rest. I was kept in overnight. Maybe it was the morphine, but I just had the most brilliant time. I wasn't annoyed by the lights or noises in the ward. I liked it when the kind Glaswegian nurse came to check my blood pressure every four hours. I felt so safe there, in solidarity with all the other women in the ward with plaster-casted arms.

My mum had to leave the next day and my sister left on the Sunday. I had been given the week off work and had to try to cope in the flat on my own until Monday 5 October 2015 when I had an appointment at Stobhill Hospital to have the cast removed and replaced by a splint. Fortunately I could still type somehow, just about, with the tips of my fingers popping out of the end of the cast. I had to build a new website and devise a crowdfunding campaign for my *Radical Renewable Art & Activism Fund* project, which was due to launch at Beaconsfield gallery in London the following week. I went to try to buy a Zone Card to use on all Glasgow's public transport so I could get around without my bike. The private bus companies don't want you to know these 'multi-modal' tickets, issued by our regional transport authority, SPT, actually exist so that you are forced to buy a weekly pass you can only use on their buses instead. Because they hold all the power in the broken privatised system, they have helped to make Zone Cards incredibly expensive and complicated to get.

When I was growing up in London, bus passes and Travelcards were available from nearly every newsagent in the city so that using public transport was made as simple as possible – a normal part of everyone's lives. The only place you can buy a Zone Card is in the Buchanan Bus Station in the city centre. So either you have to pay for another bus to get into town to buy the Zone Card or you walk. I'm lucky enough to live within walking distance, so I put my coat on as best I could over one shoulder, cradled my plaster-casted arm and slowly made my way in. When I got to the Buchanan Bus Station, there was a bewildered looking woman sat at a desk behind a desperately disorientating array of timetables issued by all the different competing private bus companies. She told me that no matter what day you bought the Zone Card (that was a Tuesday) you still had to pay for a full seven days even though it would always expire on a Saturday. It was a total joke. But I'd walked all that way so I reluctantly handed over the cash. I have never bought one since. I got the Subway back home.

That was a tough week. I couldn't really cook or exercise, so started to pile on all the weight I had lost that summer. I was so depressed, that I even relapsed on my near two-year abstinence from Facebook. On 30 September 2015 I reactivated my personal profile and bleated out a little message of pity to tell people what had happened:

Dear friends, I've not been on Facebook since December 2013. Did you notice? Well, I'm back! For a bit at least... Firstly, because I need some VIRTUAL TLC having been hit by a car when cycling last week and undergoing surgery on my left arm 😔 And secondly, because in typical Ellie Harrison style, I want to shamelessly promote my new *Radical Renewable Art + Activism Fund* project which will be launching in London next week. Please give it a boost by LIKING the page and please come and see me in person at Beaconsfield on Tuesday 6th at 7pm xx

As soon as I got the cast off and got the 'all clear' – the operation had been a great success (apparently I had 'the best hand doctor in Glasgow' who was actually Polish) – I jumped straight on the train heading south again. I stayed for another ten days, working at Beaconsfield gallery and being looked after by my mum and dad. My little nephew (who was then four) had just had his tonsils out and was also staying there to convalesce. On 20 October 2015, just before I returned to work, I got the news from Creative Scotland:

> We are delighted to let you know that your Application for funding from Creative Scotland has been successful. As a result, we are making you an offer of Funding up to the maximum of £15,000 (Fifteen Thousand Pounds Sterling) (the 'Funding') towards the cost of *Think Global, Act Local!* which is a year-long 'action research' project/durational performance, for which artist Ellie Harrison will not travel outside the Strathclyde region (the 'Project').

I had mixed emotions. Yes, I'd done it! But shit, this was going to be a rollercoaster ride. On 7 November 2015, Oliver (my old flatmate who was now living in the East End) texted me to invite me out to the cinema the following weekend. I couldn't make it because I had to go back down to London for another event at Beaconsfield. At 5.50pm I replied to say that I'd have lots more time to hang out next year: 'Yeah, in 2016 I'm committing career suicide!! xx'. 'Well, if only you were doing it without making such a fuss ☺' he replied at 5.51pm.

Creative Scotland had set the deadline of 17 November 2015 for me to accept the grant. Before I could, I had to get the university to agree to let me have the time off as 'research leave'. It took nearly that whole month to negotiate an agreement. I told them that I had raised the money for them, as per the terms of my 'probation', and that I had no need or desire for it myself. Instead, I would 'donate' the whole lot to them so that they could get some high quality teaching cover for me while I was away doing my 'research'. I was walking on thin ice. I had to try to get them to agree and eventually on the afternoon of the 17 November 2015, I finally got a letter of support, which I scanned and sent to Creative Scotland just in the nick of time.[95]

I'd been working at the university for more than three years by then. I seemed to be forever looking for places in Dundee to stay overnight to reduce the amount I had to travel: colleagues' spare rooms and various B&Bs. It was a transient and disconnected existence and that November I hit rock bottom. Still nursing a broken arm, I had been up and down to London three times in one month (twice for events at Beaconsfield and once for the funeral of my school friend's dad who had died very suddenly from pancreatic cancer). I was catching up on emails in my office at the university and got booted out by security when the building shut at 9pm. I began making the lonely walk through the night streets back to the B&B. I hadn't had any dinner so stopped in at the chippy on Nethergate and bought some of those potato fritter things (basically just some chunks of potato, battered and deep-fried). It felt like a good idea at the time but, when I got back to the room, I could feel the thick grease sitting like a weight in my stomach, smears of it across my face. In desperate need of some distraction, I switched on the telly at the foot of the bed. There was a programme on about Glasgow's new Southern General Hospital (now named the Queen Elizabeth University Hospital), showing off all its state-of-the-art facilities and saying what a great place it was. I remember this feeling: I wish I was in there, around other people, being looked after.

What kind of life was this?

Tonnes of carbon produced by the personal transportation of a 'professional artist' (2019)

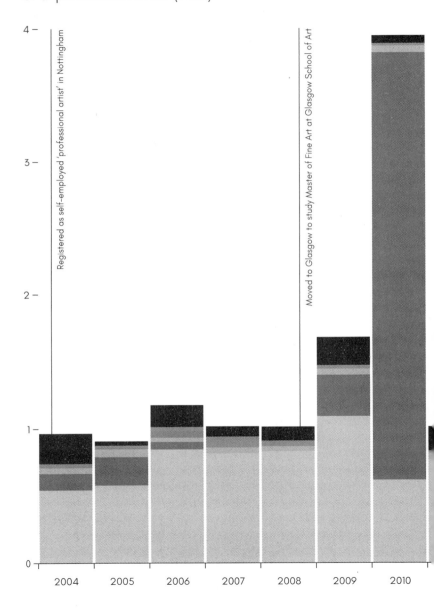

Registered as self-employed 'professional artist' in Nottingham

Moved to Glasgow to study Master of Fine Art at Glasgow School of Art

4 —

3 —

2 —

1 —

0 —

2004   2005   2006   2007   2008   2009   2010

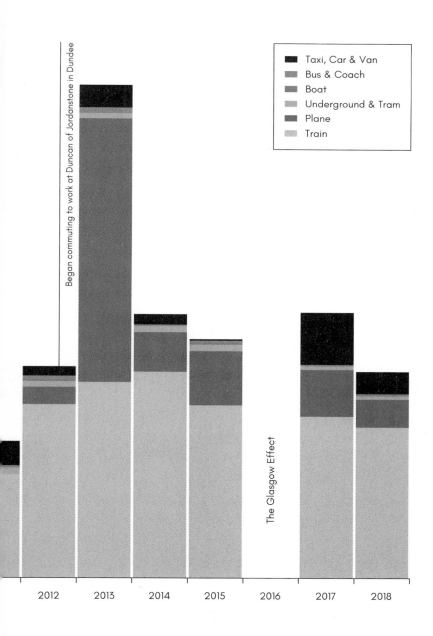

Began commuting to work at Duncan of Jordanstone in Dundee

The Glasgow Effect

Taxi, Car & Van
Bus & Coach
Boat
Underground & Tram
Plane
Train

2012   2013   2014   2015   2016   2017   2018

Table 1: Annual distances for each transportation mode (km)

	2004	2005	2006	2007	2008	2009	2010	2011
Taxi, Car & Van	1,261	174	916	420	481	1,045	313	921
Bus & Coach	743	669	668	2,811	1,099	596	602	143
Boat	0	67	446	13	0	4	8	46
Underground & Tram	1,167	1,542	835	1,142	992	1,174	1,368	1,312
Plane	1,441	2,478	664	0	0	3,290	36,877	0
Train	12,120	12,913	18,913	18,236	18,545	24,434	13,758	17,149
TOTAL	16,731	17,843	22,441	22,622	21,117	30,543	52,926	19,570

	2012	2013	2014	2015	2016	2017	2018
Taxi, Car & Van	388	940	405	50	0	2,122	924
Bus & Coach	553	936	161	201	0	258	439
Boat	107	102	0	120	0	0	0
Underground & Tram	1,124	1,102	1,175	1,272	0	684	661
Plane	1,528	22,692	3,435	4,715	0	4,090	2,453
Train	28,881	32,623	34,368	28,790	0	26,975	25,156
TOTAL	32,581	58,395	39,544	35,147	0	34,129	29,633

My Carbon Graph on pages 138–9 is the central illustration of this book. It shows my increasing amount of travel in the years running up to *The Glasgow Effect* and the dramatic impact the project had on reducing my carbon footprint for transport. The graph was compiled over the course of a month, in April and May 2019, by calculating the distance of all 3,628 journeys I made in motorised vehicles since registering as a self-employed 'professional artist' in 2004 (journeys on foot or by bike are excluded as they have zero carbon footprint). To identify the journeys, I reviewed all my records of 'travel expenses' in parallel with my old diaries and photographs, and then used the websites below to calculate the distance in kilometres of each:

- Rail Miles Engine for all UK train journeys
- Google Maps for all international train journeys, bus, car and all other land and sea transport
- MapCrow Distance Calculator between Cities for all plane journeys.

I then multiplied the total annual distance for each transportation mode (Table 1) by the most up-to-date Conversion Factors (Table 2), issued by the UK Government's Department for Business, Energy & Industrial Strategy in July 2018.[96] The kg CO_2e (kilograms of carbon dioxide equivalent) Conversion Factors take into account the impact of the seven main greenhouse gases that contribute to climate change, as defined by the Kyoto Protocol. The greenhouse gases are:

- carbon dioxide (CO_2)
- methane (CH_4)
- nitrous oxide (N_2O)
- hydrofluorocarbons (HFCS)
- perfluorocarbons (PFCS)
- sulfur hexafluoride (SF_6)
- nitrogen trifluoride (NF_3)

Table 2: Conversion Factors

Transportation mode	Conversion Factor kg CO_2e	with Radiative Forcing (RF)*
Ferry (average)	0.11287	
Coach	0.02801	
Local bus (average)	0.10097	
Domestic flight (average)	0.15780	0.29832
Short-haul flight (economy)	0.08440	0.15970
Long-haul flight (economy)	0.08610	0.16279
Black cab	0.21420	
Car (average)	0.17753	
National rail	0.04424	
London Underground	0.03760	
Light rail and tram	0.03967	

*Radiative Forcing (RF) is a measure of the additional environmental impact of aviation caused by emitting nitrous oxide and water vapour at high altitude. The Conversion Factors without RF were used in my Carbon Graph on pages 138-9, so that it is possible to see clear definition between the other modes of transport.

Part 2

The Glasgow Effect

CHAPTER 5

When the Chips Hit the Fan

Calm before the storm

ON 1 JANUARY 2016, I awoke fairly late with a hangover.[1] I had an anxious feeling in my stomach. What the hell was I doing? My one task for that day was to set up the personal GPS tracking device which I had bought to be my virtual gatekeeper all year. 'Trackimo: Caring from a distance' went its sinister sounding marketing slogan. I lazed around my flat trying to build up the motivation. There was a SIM card inside the device which would send a GPS reading of my exact coordinates to their data server every minute, once it sensed I was on the move. My plan was to set it up so it would send a text message to my Creative Scotland Officer to alert them automatically if I went beyond the predefined zone (Glasgow's city limits) or travelled at more than 20 miles per hour. That seemed an apt way of symbolising the conditions of their grant. I had already decided that it wouldn't be such a challenge to have the whole region of Strathclyde to play with (even though I have no living relatives within that zone either), and that I wanted to attempt to slash my carbon footprint for transport to zero (as well as reducing my living expenses to a bare minimum) by not using any vehicles except my bike. The backlash against the private motor car and my own personal boycott of Glasgow's privatised public transport was about to begin. The Trackimo was up and running by 4.10pm and so I was free to leave the house. I didn't go very far and was back in my bed by 9pm.

Sleepless nights followed for those first few days. Little did I know what was to come. I had these weird conflicting feelings. Firstly, of being 'cut loose' of all responsibility, no job to go to any more, less motivational structure (I would have to make my own). Then secondly of being 'trapped'. A self-imposed exile which panicked me to think how far away I was from my family,

Pages from Ellie Harrison's Notebook 35 (June 2015) showing proposed branding for *The Glasgow Effect*.

some of whom were pretty elderly by that time. There was my great aunty Rhianydd in Cardiff – the last remaining from my grandparents' generation – who was due to celebrate her 90th birthday that July. There was going to be a big family gathering in Wales, which I would have to miss. Then there was my uncle who was going to be 85 in September, my dad who was going to be 83 in October and my mum, the spring chicken of the bunch, was approaching 72.

Other than set up the tracking device, what else did I need to do? In many ways this was a 'wildcat strike', the whole point was to withdraw labour. But it was never going to be that simple. There were all the things I'd said I would do in my application to Creative Scotland. I had to start 'engaging the public'; to start promoting my idea. That was the thing which made me most apprehensive about *The Glasgow Effect*, drawing attention to my own life and my 'lifestyle choices' in such a deliberate way.

Any artist using their own body or life experiences as 'art' is essentially making themselves into an 'object' to be criticised for aesthetic and/or other value. I'd hoped that after examining the negative consequences of such introspective behaviour in my 2009 book *Confessions of a Recovering Data Collector*, I would no longer self-inflict the anxiety this sort of work caused. But this time I was instrumentalising the methods of my former practice, creating a pastiche of those older works, in order to draw attention to the wider social and economic systems within which I was operating. But again the thoughts kept racing through my head: why am I putting myself through this?

I had known from the moment I started writing the funding application that I wasn't going to call the project *Think Global, Act Local!* As much as I believed in Patrick Geddes' principles (the great Scottish thinker who coined that famous phrase in 1915), I was always just using it as a benevolent sounding 'working title' to get through the selection process. In June 2015, I'd done a small sketch in my notebook mapping out my actual intentions. You could call it an alternative PR campaign for the city, drawing attention to the hypocrisy of the 'liberal elites' in power, who continue glossing over the cracks with 'culture' whilst presiding over a city and a country with such chronic inequalities. What I didn't realise at that time was the extent to which I'd be included in that number. But what I did know was that I had chosen a provocative combination of image, action and words.

I remember four conversations with very different people towards the end of 2015 in the run-up to the launch of *The Glasgow Effect*. They all had quite visceral responses. On 26 November, I met my ex-love Sarah-Jane in London for a catch-up. She was born in Ayr, but ended up moving 'down south' to escape what she had found to be an oppressive culture. I told her I'd been awarded the money. What a fluky chancer I think she thought. I sensed a mix of amusement and resentment. 'Money for nothing' it sounded like to her. Then there was my friend, curator Michelle Emery-Barker, who was actually born in New Zealand, but moved back to Glasgow with her family when she was 12 and now sounded as Glaswegian as they come (to me anyway). She pulled a disapproving face. 'Chips!' she said, 'Don't you think that's a bit of a stereotype?!' Maybe it was, but I had my mind set on using that image. I'd already taken the time to borrow an SLR camera from the university on 28 October 2015 and bought, photographed and eaten those infamous chips (pictured on the back cover). For me the image chimed so well with much of my previous work exploring food, over-consumption and obesity within our global capitalist system – from *Superfluous Consumption* in 2000, to *Eat 22* in 2001, to the *Vending Machine* in 2009, to *Anti-Capitalist Aerobics* and *The Other Forecast* in 2013.

Then there was Moira Jeffrey, an art writer who I had known since 2010 when she was forced by her boss at *The Scotsman* to interview me for an article as part of my *Press Release* project, which formed my MFA Degree Show at the CCA that June. She had initially been most sceptical of my motivations, writing, 'Is it a critique of the careerist structure of the art system or merely a symptom of it?',[2] but eventually she seemed to come round to some of my more provocative tactics. I bumped into her at a Creative Carbon Scotland event in Edinburgh on 30 November 2015 and showed her the promotional poster I had made with the image of the chips. I explained my motivations in relation to the demands of the university. She chuckled at the thought of the senior management's faces when they found out what I'd really done: 'That's brilliant, you're totally fucking them with their own agenda!' she said.[3]

And then finally, there was this kind Glaswegian man who I met during The Only Way is Ethics festival (founded by Glasgow's vegan virtuoso Craig Tannock),[4] which I took part in to launch the *Radical Renewable Art + Activism Fund* project in Glasgow in December 2015. 'You're not going to leave Glasgow for a year?' he said. 'There are so many people who were born here who never leave anyway!' And that was it. Although I didn't know quite how to articulate it at the time, he had just nailed the thing which I was most keen to highlight and explore through this action: the relationship between literal and social mobility, between class and carbon footprint (now visualised in my Carbon Graph on pages 138–9). It was the issue which I had begun to outline in my funding application: that it's the lifestyles of the most privileged people in our world that are doing by far the most damage to our climate, and that something big needs to change. I had written:

> *Think Global, Act Local!* is an essential critique of the 'internationalisation' agenda that permeates various levels of public policy and funding, which is the antithesis of what needs to be done to address climate change. In this 'internationalised' world, a circle of 'global elites' (particularly artworld professionals and academics) operate as though they are above the moral imperative to reduce our carbon emissions, believing it is 'OK' for them to travel because what they do and say is *so important*.[5]

I knew I had to put something online about the project, to provide a context for what I was doing, but I really didn't like the idea of writing a blog. I thought that would make me far too vulnerable. On 2 January 2016, I woke up anxious at 5am. I'd taken to writing in my notebook at those times as a good way to calm down. That morning, I scrawled: 'I don't want to blog relentlessly. I don't want to overproduce. I don't want to ego broadcast/pump out stuff without listening. CALL THE WHOLE THING OFF'.[6] I drifted back to sleep. By lunchtime, I'd eventually made my way over to the studio. The very least I should do was make a Facebook event page. I uploaded my beautiful image, carefully selected and enhanced from the 48 shots I'd taken of those chips back on 28 October 2015. I copied and pasted the text from the application blurb, switching 'the Strathclyde region' to 'Greater Glasgow'. I put the end date as 11.59pm on Saturday 31 December 2016. I set the 'venue', not to any specific address, but simply as the whole city of 'Glasgow, United Kingdom' and made the title *The Glasgow Effect*. That would do for now – I'd sort the rest out later in the week. This was a complex project, arising from so much thinking over so many years: the pressure had been slowly building the whole time I'd been living in Scotland. It was addressing so many anxieties about the way my life was turning out. I didn't really know how to describe it succinctly. I boiled it down to this simple question, which I pasted expectantly at the top of an empty Tumblr blog:

How would your career, social life, family ties, carbon footprint and mental health be affected if you could not leave the city where you live?[7]

Then I went home. I had a busy day on Monday. I was helping to coordinate two national days of action for the public ownership of our railways with protests at stations across the country. This was an annual event, scheduled to coincide with the New Year train fare increases – fares which had consistently risen above inflation since the privatisation of the railways in 1993, but which had seen sharp increases totalling 25 per cent since the Tories took power in 2010.[8] The actions were in England on Monday 4 January and in Scotland on Tuesday 5 January (because the Monday was a bank holiday up here). As a self-employed person, I never did bank holidays and only knew they were happening because there were less emails coming in as other people took the day off. It was calmer. I could spend less time maintaining my inbox and more time on the actual things I needed to get done. I made some more notes in my notebook, thinking about how to articulate what I was doing. I was reading the book, 24/7: *Late Capitalism & the Ends of Sleep* by Jonathan Crary, which is about how social media culture is accelerating all aspects of our life/work and demanding that we eke out more and more consumption and production of value from every moment of our waking lives.[9]

I like Glasgow and Glasgow likes me[10]

By the afternoon I was back on Facebook, this time sharing the photos coming in from the actions outside railway stations all around the country: Derby, Brighton, Carlisle, Coventry, Sheffield, Stoke-on-Trent and London King's Cross where Jeremy Corbyn – the new leader of the Labour party – had showed up for the photo op with then Shadow Transport Secretary Lilian Greenwood. I'd been helping to coordinate these annual New Year protests since 2012, in collaboration with the unions' Action for Rail campaign and We Own It (a group founded in 2013 to campaign for the public ownership of all public services). This was the first time that mainstream politicians had ever come along, which showed how much Bring Back British Rail, over the last six and a half years, had helped to shift the public discourse around renationalisation and socialist policy in general. My phone started to ring. It was BBC Radio 5 Live. They wanted me to do an interview about rail renationalisation on their *Drive* show that evening. I'd have to leave the studio just before 6pm to cycle down to the BBC studios at Pacific Quay. I still had a few more hours to work.

I carried on with what I was doing. But, then, started to see small notifications popping up in the bottom left corner of my screen: 'Jsh Div has commented on your event *The Glasgow Effect*', 'Rosanna Hall has shared your event *The Glasgow Effect*'. That's strange, I thought. It's obviously on people's radars. But I don't have time to get distracted, I've got to brush up all my arguments for my interview this evening. But the notifications kept coming. I glimpsed back at the Facebook event page... hmm, people seem to be pretty angry: 'This is completely ridiculous! Staying in large city for a year is not art. It's called life. Pretentious crap!' posted Louise Cameron at 3.05pm. I need to upload my original funding application to the blog to help explain more about the project, but that'll have to wait until tomorrow. 'Studying the effects of sitting about doing fuck all with my generous grant' said John Niblock at 3.19pm. 'Get tae fuck' wrote Dean Inglis at 3.24pm. Kate Inkster Macdonald had been on the Creative Scotland website and downloaded the list of grants they'd awarded. The £15,000 headline figure had been posted up next to 'Glasgow, Kelvindale'. I don't even live in 'Kelvindale' I thought, but there didn't seem much point quibbling over details or engaging in any petty social media spats with people I'd never met. This already looked as though it was too big for that. My heart started to race. I switched from my Bring Back British Rail inbox to my 'Ellie Harrison' one – it had started to fill up. There were messages from *The Herald*, *The National*, *Vice* magazine, *The Scottish Sun*, *The Scotsman*, the *Daily Mail*, *Dazed & Confused* amongst many others. The one from *The National* read:

> There's obviously been a lot of negative reaction to your project. You've been accused of being offensive,

patronising and indulging in taxpayer funded poverty porn. I think, perhaps, people assume you're effectively getting a grant to sit about in Glasgow and do nothing for a year. Do you think you could maybe try and explain what you hope to achieve with the project or even why you want to do this and why you think it necessary?

Oh god, this is getting serious I thought, but there's just not enough time to reply to all of this right now. Then I received an email notification from my OKCupid dating profile. Some smug looking woman had recognised me from Facebook and messaged to ask: 'Is this part of your project as well?' Lock down I thought. I quickly logged in and deleted that profile. It looked as though I'd have to put my love life on hold for a while! I had already set my personal Facebook profile to the highest privacy settings, but it was still too much of a risk with all those journalists snooping around. A few clicks and it was deactivated again. After that brief relapse of three months which had seen me through the worst of the winter and most of my recovery from the operation, I would have to go back to my trusty ghost profile again. The one with zero 'friends', which I could just use for updating the project and campaign pages for which I'm the 'admin', including *The Glasgow Effect* event which was hosted by the 'Ellie Harrison' artist's page. Phew, I feel a bit safer now. I dashed off a few short replies to the journalists, saying to look out for a blog post later in the week. I shut down my laptop, put it in my bag and jumped on my bike.

It was pitch black outside. I had to get across town to the BBC studios at Pacific Quay. We were going live after the sports news at 6.30pm. My heart was pounding as I pedalled. It felt as though anyone I passed on the street might just stop and attack me. How was I going to get through this? I signed in at the desk and was taken through to a small studio where I sat alone with a pair of headphones on, listening to the programme which was broadcasting live from Salford Quays. Deep breaths. I had a few moments to compose myself. Finally this double life I'd been living – this split personality of artist and activist – had come back to bite me. Was the former about to undermine the latter? The more water I drank, the dryer my mouth got. I thought I was going to be sick. But it was too late.

'Well let's speak to Ellie Harrison, volunteer and founder of the campaign Bring Back British Rail', said the chirpy radio presenter.

'Hi there', I said.[11]

I was up against an overpaid lobbyist from the free-market think tank the Institute of Economic Affairs. Founded by wealthy businessmen in 1955, it was one of that global network of think tanks inspired by the Mont Pelerin Society responsible for promoting neoliberal policy to our politicians over the last half century in order to strengthen their own power and to further accumulate wealth. I pulled myself together pretty quickly, using all my usual arguments to explain that privatisation was a totally inefficient way to run a public transport network, which was wasting around £1.2 billion of taxpayers' money every year,[12] and that fares had increased to such an extent that many people could no longer afford to use the train.[13] As is usual on the BBC, all my arguments were shut down by men with vested interests in the privatised system who are paid *not to care* about social justice or the environment. I came out feeling, as always, that I could have done better to refute their baseless claims; that there had just not been enough time, if only I'd been allowed that final right of reply! I looked out across the Clyde at the twinkling lights of the city centre on that quiet bank holiday evening. It was a beautiful view you got from that spot, so close to the Garden Festival site where I'd been with my family all those years before. It was hard to believe that in a parallel universe there was shitstorm now well underway. What the fuck was I going to do?

I called my friend Sarah. She was my oldest friend in Glasgow, who I'd known since 2002 when we studied together at Goldsmiths. We were the only two who left without applying to stay on for the Masters – me because I felt totally out of my depth, and Sarah because she skipped it to go straight on to a PhD! She was born in Stirling and had moved back up to Scotland a few years before to be closer to her mum. She'd seen what was hap-

A meme posted on *The Glasgow Effect* Facebook event page by Darren O'Connell on 4 January 2016, 5.39pm

pening with *The Glasgow Effect* event page. By that point it had become quite difficult for any regular social media users to miss. 'Why don't you come round here?' she said. I cycled up to her flat in Yorkhill, my whole body was shaking by that point. The adrenaline was too much. She calmed me down and cooked me dinner. I felt safe there. I didn't have a smartphone and my laptop's battery was totally dead, so I couldn't log on to see what was happening, even if I had wanted to. The irony of this social media shitstorm was that I wouldn't have had a clue about it, had I not been at its centre.

As we ate, Sarah scrolled through the comments on her iPhone. They were coming so thick and fast by that point it was almost impossible to focus on a coherent debate. 'I assume in the interests of fairness and transparency to the local communities and the "sustainable practice" of a "successful artist" you will be publishing a breakdown of how your £15,000 is spent?' demanded Lana Dalry at 3.31pm. 'If this is satire in an attempt to raise people's awareness of you regardless of consequences, then you are a genius. If not, then em, aye' said Charles Lindsay in Dalmarnock at 5.06pm. 'Congratulations on your career suicide! I'll be seeing you take my order in KFC at the 4 corners within a fortnight (if you get the job that is)... Dafty' wrote Stephen Honnan at 5.21pm. 'A creative way of saying she's goat a tag' added Peter Farley at 5.30pm. 'WE STILL HAVE PPL LIVING ON FOODBANKS FFS!' screamed Ewan Grant at 5.45pm. 'Am loving the comments tho! on yersel creative scotland! £15K well spent on this golden patter facebook thread' added Montague Ashley-Craig. I remember Sarah showing me a meme that someone had made of me in a sports car waving cash out of the window. That was pretty hilarious, if not to say ironic. If I went out and bought a sports car, what would I live off for the rest of the year, I thought? At least some people were just having fun.

Just after 7pm, the *Daily Record* published an article online – the first media outlet to add fuel to the fire. The headline read: 'London artist paid £15k public money to spend a year in Glasgow for research project branded "poverty safari"'. They'd asked Frank McAveety, the Labour leader of the Glasgow City Council, for his words of wisdom. He waded in, commenting:

> She aims to study what is termed the Glasgow Effect
> – the effect that staying in the city has on much of its
> population… If she contacts our Poverty Leadership
> Panel we can put her in touch with legions of single
> parent families living in poverty who can tell her in
> minutes what it's like to be poor in Glasgow and how
> that affects family health and prevents you from ever
> getting out of the city… That's called the Poverty Trap.
> It shouldn't take Ellie a year to discover it.[14]

Good one Frank, I thought. Maybe you and your cronies at the Council think 'market forces' are going to solve the 'poverty trap', while you just sit back and get plump on our council taxes? That's exactly why we're stuck with such persistent and worsening inequalities of wealth and health in our city after more than half a century of Labour rule. Hypocrisy kills. But the main quote in the *Daily Record* article came from Darren McGarvey (aka Loki the Scottish Rapper):

> It's horrendously crass to parachute someone in on a
> poverty safari while local authorities are cutting finance

to things like music tuition for Scotland's poorest kids. I don't know the artist personally but I think we'd all benefit more from an insight into what goes on in the minds of some of Scotland's middle class.[15]

Loki – I know him! He'd been pals with Oliver (my old flatmate) for a few years. We played his track 'Fancy Grammar' on our pilot radio show at Wunderbar festival in November 2011.[16] Oliver designed the cover for his 2014 album *Government Issue Music Protest* and had just booked him for his *Speaks & Loki: 2016 Scottish Mini-Tour*, funded by Creative Scotland, which was kicking off in March.[17] I'd even met Loki myself when he did a gig at Oliver's event at Transmission gallery on the night of the helicopter crash at the Clutha bar in December 2013. I liked his stuff. I like his politics. Obviously I'd not made much of an impression on him when we met in real life! I had been living here more than seven years; that would make the longest parachute drop in history! How long do you have to live in a place before you're permitted to criticise it? God, what the hell is his problem?

Sarah helped focus my mind. It was clear that I was going to have to make some sort of statement about all this. I knew I wouldn't need to go through any of the mainstream media. That's not how it worked any more. I could just post something up on Facebook and they would all be forced to scramble around trying to keep up. So much power lay in my hands – it felt as though the whole world was waiting to hear what I had to say. I was like The Sorcerer's Apprentice standing at the centre of the storm, attempting to conjure and control the events unfolding around me. I remember Sarah asking me: 'What message do you want to get across?' I didn't have the answers to the questions the journalists and everyone on Facebook had been demanding. Those would only start to emerge over the course of the year. All I could do at that point was publish my original funding application and explain the absurd circumstances which had forced me into applying for the money. I knew the university wasn't going to like it. I knew that Creative Scotland wasn't going to like it. I was going to have to stick my neck out and go it alone.

But that statement would have to wait until tomorrow. I was far too tired by then to put anything coherent together. I was also far too terrified to go back to my flat alone. What if there was an angry mob outside? Sarah said I could sleep on her sofa and got me some blankets. I switched my little Nokia phone off, as I do every night before bed, and I finally lay my head down to rest. I barely slept – words buzzing round and round my mind all night. By the time it was morning, I had scrawled out a first draft of my statement on a scrap of paper. Sarah had to go to work, so I left early and headed back home, sneaking through the front door with a quick glance around to check no one was watching. When I switched my phone back on there were messages from *The Scottish Sun*, BBC Radio Scotland, the university press office, Creative Scotland and BBC Radio Nottingham – whoa, it was news down there as well?! I didn't have a clue there'd been an hour-long 'phone-in' about *The Glasgow Effect* on the Kaye Adams Programme that morning.[18] I didn't really want to speak to anyone. Plus I now had this paralysing feeling that I needed to document all this somehow – that this was the project – and that, therefore, it all needed to be recorded for posterity. I'd have to work out some way of recording phone calls before I spoke to any of those journalists. I had to ensure I captured it all.

I plugged my laptop in, opened it up and stared at the screen. There were hundreds of messages flooding in, too many for one person to be able to reply to. I'd always been so diligent with my emails, devoting several hours each day to ensure everyone who contacted me on any of my four different email accounts got a response. That was just not going to be possible any more. I had to focus on getting my statement finished and publishing my original funding application online. Later that day, I crept out across the street to my local newsagents and bought a stack of newspapers. 'Artist's £15k... to stay in Glasgow' wrote the *Daily Mail*. 'What a load of old arts' wrote the *Daily Record*. '£15k for Artist that can't leave Glasgow for a Year' wrote Dundee's *The Courier*. 'Art of provoking a Backlash' it said on *The Herald's* front page. 'Artist's stay at home project has wrong effect' said *The Scotsman* and *The Scottish Sun* wrote '£15,000... it's Monet for nothing: artist paid to stay put'. I was simultaneously horrified, amazed and amused to see my face splashed across all of them, photographs pilfered from wherever they could find them on the internet. My friend Zoë texted me to say there was a picture of my face on the screens inside the Subway, not that I'd be able to go down there to check of course. This was totally insane.

But I couldn't allow myself to get totally distracted. It was the second day of action for the public ownership of our railways. I had been helping to organise a demo at Glasgow Central station at 5pm. I had all the Action for Rail flyers in my studio, which I needed to pick up and take down to meet the activists from the RMT (National Union of Rail, Maritime & Transport Workers) and TSSA (Transport Salaried Staffs' Association) unions at their offices on Hope Street. I had to make sure I got my statement out on Facebook and uploaded my original funding application to the blog before I set off. I remember the adrenaline building up until the point I was ready to hit publish at 4pm. I didn't have time to wait to see the response. I had to run out of the door and cycle across the city to pick up the flyers.

It was pissing with rain. I arrived in such a state that I skidded across the painted concrete floor. Lifting my left arm up to prevent it getting hit again, I crashed down on my hip. Ouch! I leapt back up. I didn't have time for another injury, so I limped upstairs as quickly as I could to grab the stuff I needed and then got back on my bike. I cycled down to Hope Street and met the others there: mainly a big group of burly blokes. We walked down to the station together, and I started to relax a little. I wasn't sure if they had seen the news and had put two and two together. I stayed there for an hour or so, handing out flyers and shouting 'Stop the rail rip-off! Renationalise our railways now!' at the top of my voice. It was quite a release. I met my friend Ivor afterwards and we went for some dinner in Stereo. I felt quite electric, like there was a huge neon sign above my head pointing downwards. Was everyone looking at me and muttering stuff behind my back?

When I got home, I tried to get some sleep, but I just couldn't – the fear of what was unfolding in that parallel universe was too much. I wanted to see it all with my own eyes, so I got up out of bed and sat down in front my laptop. That's when I got sucked into a Facebook wormhole – who were all these people and why were they so angry? I followed the links to examine their profiles to find out more about their lives. I must have visited hundreds of them…

Facebook wormhole

Below is a chronological selection from more than 8,800 comments and memes posted on *The Glasgow Effect* Facebook event page (quoted exactly as they appear).

> Anyone else think that this mass vitriolic reaction could be the Glasgow effect she is researching? We've played right into her fuckin hands.
> – Jethro Jones, 4 January 2016, 3.47pm

> Are all the angry artists on here just pissed off cos they never thought of it? Donno what the artistic outcome will be, but the comments on here are worth 15k in LOLS already.
> – Iain Mackie, Anniesland, 4 January 2016, 4.09pm

> If Creative Scotland are funding you to live in Glesga for a year I think I'm due 23 years fucking back payments.
> – Cllm McCln, 4 January 2016, 4.33pm

> I'm I the only one who is really excited by this artistic stroke of genius? Imagine the pain this artist will feel when she can't get on the number 9 to Braeheed, the McGills to the beautiful Paisley in Renfrewshire, how will she cope without being able to visit her much loved EK shopping centre, and no more trips to Bellshill either? Clydebank is outthe question, as is the majesty of Cumbernauld. Pure genius, I couldn't have put that 15G's to any better use. Bravo, Creative Scotland, Bravo!
> – Jim Barclay, 4 January 2016, 5.00pm

> Seriously though, had this been called anything other than 'the Glasgow Effect', no-one would have batted an eye-lid and people would know it's about the boring topic of Travelling for Work rather

than poverty. However, the artist's degree project was to, instead of producing a degree show, turn her work-space in to a pr office to get people to come along in the absence of art. So, mission accomplished on this second stage!
– David Hughes, 4 January 2016, 5.23pm

Thousands of artists articulate the 'poverty experience' and are marginalised culturally in favour of this kind of wishy washy shite
– Darren Loki McGarvey, 4 January 2016, 5.47pm

as an artist/academic you could have just stuck to your role and made this project about challenging this demand to travel and understanding of what it means to be a successful artist and just reflecting on the artistic process. good research must include a reflection on your position as a researcher, posing as your subject will not reveal anything valuable, especially as your experience is not going to be anything like the experience of people who are forced to stay in one place by the lack of local opportunities, and feelings of isolation
– Emzo GoedZo, 4 January 2016, 6.03pm

If gauging how people react to the news of your project is all part of the gig then please include the following comment: Go and piss thousands of people off on your own dime and put the begging bowl away.
– Darren Loki McGarvey, 4 January 2016, 6.15pm

should just give the 15k to amber leaf and stereo and cut out the middle man
– Louis Burns-Hart, 4 January 2016, 6.17pm

My grandparents worked all their lives, in industries like the shipyards and singer sewing factory. Tough industries, all their days. Now they struggle to pay their bills thanks to the cuts and limitations of our government. Heating their home, lighting their home, filling the fridge. Every penny accounted for. And they have not had the opportunity to travel out with greater Glasgow in far more than a year. Thank you so much for your, what must be, devastating sacrifice. Having to limit yourself to the boundaries of

one of the biggest, most bustling cities in the uk. Your 'art' is really making a a difference to this tragic world. What a saint you are. Wars will stop over this. The ice caps will most certainly grow back. My days it'll cure cancer! Have some perspective you pretentious twat.
- Mikey Irwin, Anniesland, 4 January 2016, 6.18pm

What an insult to proper glasgow-based artists, and a boggin waste of public money. On the flip side, rare occasion that folks are actually talking about arts funding! Wehey!
- Kendal Orr, 4 January 2016, 6.19pm

I'm enjoying that the event page itself actually has a giant chip on its shoulder.
- Joey Asbo, 4 January 2016, 6.27pm

Trident now looks like money well spent
- David Moses, 4 January 2016, 6.46pm

Sorry Ellie, but I know some folks who have been working on a very similar project over the past few years, though critics should note any funding is from the welfare budget rather than Arts funding. They've successfully maintained a low carbon footprint, thanks to having barely enough cash to afford a bus. If you'd like to see the results, they're currently exhibiting at a Job Centre Plus near you - places limited, so come early to avoid disappointment.
- John Bitmap, 4 January 2016, 7.04pm

Okay. I give up. I'm really confused by this. Am I missing something? Am I missing the 'art' part somewhere? Because all I've seen so far is someone who has a successful job being given £15,000 to attempt to live like someone who hasn't been so fortunate in life. Surely if you want to document life as someone who has no option and can't leave Glasgow City (even in the event of a family death), the more productive option would've been to put forward an application for funding for a film that includes real people who live in these real situations on a daily basis? Sure, it would still be poverty porn, but at least it would serve a purpose. I see no purpose in this. I personally know very creative people - Primarily writers and musicians - who

are genuinely talented, special people, who work tirelessly just to get their art noticed by a few people, who would even fit into the poverty guidelines this project, for whatever reason, is hoping to achieve (???) who could give you a fantastic perspective on what it's REALLY like to be an artist struggling to survive in Glasgow. On another note though, if anyone fancies giving me £15,000 to write my book over the next year from the comfort of my poverty-stricken area of Cumbernauld, please do let me know.
– Jess Ball, 4 January 2016, 7.13pm

I just created a big jobby that won't flush. Can I have £15k please?
– Andrew Marshall, Glasgow, 4 January 2016, 7.22pm

You're either Glasgow's best con-woman or Glasgow's best troll. Either way fair play to you.
– Blane McInarlin, 4 January 2016, 7.39pm

I feel sorry for this poor Lass who has done nothing apart from try to be creative ! I also think that she should though be very proud of the response from the people of Glasgow who have no gripe with her or her Art, but who will stand together in trying to make the collective better ! Some of the Replies have been more Art than Art itself ! #creativescotland
– John Soutar, Milton, 4 January 2016, 8.36pm

You can't choose the Glasgow life ... The Glasgow life chooses you ... Unless you get some money together and claim it is for a good cause
– David Bryans, Baillieston, 4 January 2016, 8.44pm

Didn't this person try to prevent the Great Western Road Tesco from opening up, because it would ruin the community or some hipster shit like that?
– Kerr Matheson, 4 January 2016, 8.47pm

Please tell me how this benefits the people of Glasgow or indeed our society? You have no right or real understanding of poverty in Glasgow, or indeed in any other part of the uk. You

insult thousands of hard-working Glaswegians who have pulled themselves up from poverty. Make a crass social statement, with no understanding of the influences or situation or circumstance in which people live. But expect to be able to make a case study by living 'free' of burden with 15k of tax payer money for a year? Hate to tell you this, but to experience poverty. You dont need to have a 15k bank account. You need to have lived it for many years, been born into it, or had EVERYTHING stripped away from you. That 15k should of been used to further the careers of street artists, sculptures, mural painters and other real art which would of brightened up our social spaces and brought real joy to others. Not as a free meal ticket to an idealistic bone idle, self aggrandising pseudo 'artist'. You insult the poor, you insult art and you insult the people of Glasgow.
– David Hope, 4 January 2016, 9.31pm

Welcome to Glasgow, love. Friendliest city in the uk. I'm sure you'll do fine!
– Andrew Nugent, 4 January 2016, 9.55pm

I feel sorry for the girl now, but how disconnected from reality do you have to be to have thought that this was a good idea, and maybe not have predicted how it might be perceived by some people – well everyone apart from hipsters and art students really. What sort of friends does she have that encouraged her to do this, what sort of bubble do these people live in? As for Creative Scotland they should be shut down for granting her £15,000 of taxpayers money to do....... nothing – this part really is beyond words.
– Oli Rattray, 4 January 2016, 10.06pm

'Art' gets people talking. You might not like it but the 15k has done it's job.
– If Råh Ïhseruq, 4 January 2016, 10.26pm

'The people are really friendly they said'
– Steveo Neill, Dundee, 4 January 2016, 10.27pm

This burds gonna have a wider man hunt than Osama had.
– Brian Lou Reilly, Glasgow, 4 January 2016, 10.54pm

I have an idea ! Why not use all the money wasted on arts funding to contribute towards providing a Universal Basic Income to everyone so they can fund whatever they want!
– Suzanne Patrick, 4 January 2016, 11.03pm

What if the true art of the project is the reaction of the public? I mean you can see quite a range of emotions in peoples' reactions to this. Maybe it's not even real. Maybe it's an elaborate social media form of meta-art. We're all different brush strokes decorating the digital canvas. It must be that 'cause £15k isn't enough to pay security companies to defend her from THE PURE RAGIN' GLASWEGIAN ARTIST POPULATION THAT ARE GONNAE TAKE HER A SQUARE GO. That's just my opinion though.
– Craig Porter, Glasgow, 4 January 2016, 11.26pm

'on yersel ellie hen! There has to be someone to push the boundaries. Sounds like a great way to test the waters to see if working solely in a Scottish city is actually a viable option for professional artists. After the tturner prize being held in Glasgow I'm not surprised that creative Scotland has funded this project– if successful you will prove that artists don't need to travel out of Scotland to be successful or make an impact. If unsuccessful you will be providing us with a powerful critique of the way the art world works in a globalised society. As it's only the fourth day of this 365 day project it's too early to expect a massive amount of content from you but it will be interesting to see what kind of work you choose to make in this time. Glasgow is already a vibrant art city, and you seem to have a lot of people hoping you make work that helps the existing and impoverished communities so you've definitely got a lot to work with there.. think the worst thing i've seen on this page is the 'poverty safari' comments–– hope you can turn that on its head. will check the blog in Feb– good luck! ps. loved your referendum piece
– Jen Ga, 5 January 2016, 12.10am

Have to say fair fucks for pulling off a con this simple and effective. We're all just gutted someone else thought of it first.
– Seàn McKinlay, 5 January 2016, 12.23am

This is not art, it is a chimera. Where are are the creative skills and ideas. More wind from the bagpipes! It has been done before, I for one did it in the 1940's and 50's. 'Limits of sustainable practice'? .... subsistence was our watchword, creativity was how you could keep a roof over your head and make your food last until the end of the week. 'Durational Performance'?...I could not leave my city for many years simply due to lack of funds and two weeks annual holiday. 'Carbon Footprint'.. I had a coal fire a when I could afford it, and a bike . No digital communications. Enforced, by lack of funds and a restrictive but respectful society where integrity and self respect was all, I had to resort to the simple act of taking up 'local opportunities' and EARNING A LIVING. The 'ART' in this project can only be in the creativity and ambiguity of the words used to justify the grant. Whoever made the decision to fund it has never been deprived or subsisted. Shame on you all for a disrespectful project being foisted upon the hardworking persons of Scotland.
– Sue Chadwick, 5 January 2016, 8.13am

Art pushes boundaries and challenges social convention by catching peoples attention, by posting/chating/debating ellie has caught your attention! Whether you agree or not is irrelevant.
– Rose Davidson, 5 January 2016, 9.36am

#prayforellie
– Laurie Macmillan, 5 January 2016, 10.40am

basically ur a stinkin cow now fuck aff back to ur art dungeon ya weirdo
– Gary McCulloch, Cumbernauld, 5 January 2016, 11.04am

Some perspective on £15,000 – The interest on our national debt costs 48 billion a year, mostly paid to banks who created the debt in the first place. That costs us uk taxpayers £131506849 per day or we could give away £15000 to 8767 projects per day all year, every day of the year. Cant see this 15k being the biggest waste of our money this year, can you? You don't need banks and interest to create money, we just 'choose' to do it that way.
– Iain Mackie, Anniesland, 5 January 2016, 11.50am

I'd like to point out that 'the Glasgow effect' that everyone
is already familiar with, may have virtually nothing to do with
adepravation. People die earlier in Glasgow regardless of their
income, right across the social scale. Its weird. A person earning
60k in Glasgow will typically live a shorter life than an person
earning 60k in London. Lets not make this about 'poorism' or
whatever, because its probably not.
– Cola Nuke, 5 January 2016, 12.05pm

The only reason anyone is interested in 'the glasgow effect' is that
the patter is on fire on the comments! We as a nation, will rip the
absolute shite outta ye hen, untill you piss back aff to whatever
entitled shire you hail from. If ah want to go on holiday, i have to
save up ma minimum wage (a cheeky wee 3 star all inclusive to
magaluf does the rest of us...) We think you should do the same...
– Jamie-Lee Love, Mansewood, 5 January 2016, 1.41pm

Poll: who is Ellie Harrison?
- a midden
- ANOTHER GAY MAN
- Darude
- Tyla Jackson
- Escaped Rolf Harris hiding from the law
- Benny Harvey
- The c**t punting aftershave in the arches toilet
- The real slim shady
- The wumman that turned the weans against us
- José Quitongo
- A talented artist
- A visionary!
- The white Pele
- Batman
- A Bigger, Blacker Dick
- She's Spartacus!
- A right numpty
- Gary Glitters hair
– Tomas Arr, 5 January 2016, 1.42pm

Hope you end up oot yer pus at a 3 day pairty in Easterhoose and end up taking it baw deep up the back pipe fae 40 wee needs oot their bin on Ket. All in the name of Art.
– Baz McCormack, 5 January 2016, 3.53pm

And let's add to that, an artist who makes work about public transport and climate change. These things do intersect – diet, lifestyle etc. I am not defending a project I haven't seen, but everything Ellie had made in the past is on the right side of this argument. I do now accept that chips are a taboo issue within 2 miles of The Dark Star.
– Alec Finlay, 5 January 2016, 3.57pm

I've glanced at her previous work and although not my cup of tea in terms of art, yes it's clearly coming from a place I would support. I think what this episode reveals is that political awareness of big picture systems (public transport, climate change) is different from sensitivity to intimate human-scale issues, and communities of place with fierce identities, where awareness of social class and ascriptive inequality are vital, particularly when your art is based around jolly japes... But what a brutal way to find that out.
– Luke Devlin, 5 January 2016, 3.57pm

I assumed that the chips were a reference back to her Eat 22 project where she documented everything she ate for a year. This project, as it lasts a year, and refers quite a bit to diet, seems similar. Chips and carbs generally featured heavily in that i.e. that she was eating them herself, not judging others for eating them. She was in my MA Art Theory course around this time, and I found her thoughtful and interesting.
– Bridget McKenzie, 5 January 2016, 4.00pm

Remember when this level of committal to local produce and the community used to be normal. Now it's the subject of some over budgeted vanity project masquerading as an experiment in the name of art.
– Johnny Lapsley, 5 January 2016, 5.50pm

I'll pay ye' £15k tae fuck right off.
- Lauren McGhee, 5 January 2016, 6.05pm

Just thought I'd point out that Ellie has been granted the £15k to
come up with a thought provoking and talked about art project,
not to sit about and do nothing for a year. Ironically by entering
into this highly volatile event, the people of Glasgow have all but
secured the funding for her. Creative Scotland will be overjoyed
at the exposure and debate that this has caused. And as an artist
Ellie will be pleased with the result no matter what.
- Androo Faulkner, 5 January 2016, 6.22pm

Just what Glasgow needs.... Another ugly freeloading chick
- Craig Fowler, 5 January 2016, 6.44pm

i'm gonna dive in front and take a bullet for ellie harrison now:
glasgow is a bland city awash with bad food, all those street
performers doing the ed sheeran thing years after the fact, bad
dogs, a return fare that randomly changes depending on the
mood of the driver and the weird money box thing that i don't
think exists anywhere else in the world or at any other time period..
you know, you put your money in and you can see it and then it
disappears, and there's no change. why is that a thing? there's a
wee collection dish i can put the money on, why does it go into the
pegless pachinko machine-looking thing? why are you always at
least ten minutes late arriving to renfield street when it isn't even
rush hour. first group i know you're reading this.
- Kain Schankula, Glasgow, 5 January 2016, 7.24pm

Trolling your employers and your funding body at once, and - it
seems from all the comments here - annoying/bewildering a large
percentage of the Glasgow artistic community and angering
so many of the good folk of Glasgow in general..... It will be
interesting to see the fallout from all this. It feels like you have also
humiliated the people who were defending your project/right to
funding before your eventual response. I don't know whether to
wish you good luck or suggest building a fallout shelter.
- Justine Ross, 5 January 2016, 7.54pm

This isnae. Naw. H'od oan. This isnae wan e' those durational performances, is it?
– Paul Cannon, 5 January 2016, 9.05pm

Let's be honest hen, yer dain this fir the banter
– Allan Hollinsworth, 5 January 2016, 10.28pm

The Indian philosopher Amartya Sen notes that there is a difference between fasting and starving. Although they are very similar types of functioning, fasting is choosing to go without food even when one has other options. It is the exercise of choice. The ability to choose to fast, as opposed to starving without choice, means having what Sen calls effective freedom, or capability. For those who don't have that effective freedom, who are starving, the one who is fasting can seem contemptuous, even if that is not the intention of the one who fasts. This is part of the reason why we are all still talking about an art project.
– Luke Devlin, 6 January 2016, 2.11pm

15 grand sort oot yer teeth n that mick jagger haircut yer a lassie no a wee boy fae the 70s
– Franky Dyer, Thailand, 6 January 2016, 7.40pm

'The artist must prophesy not in the sense that he foretells things to come, but in the sense that he tells his audience, at the risk of their displeasure, the secrets of their own hearts'. I thought you might like this quote from Collingwood (1938, Principles of Art)
– Murdo Macdonald, 11 January 2016, 12.43am

Adrian McCallum, 5 January 2016, 12.10am

Stevie Jukes, 4 January 2016, 6.10pm

Joe Crogan, 4 January 2016, 5.23pm

Craig Peters, 4 January 2016, 8.12pm

Ibiza Sandy, 5 January 2016, 4.34pm

Andy Blip, 7 January 2016, 1.58pm

I suddenly noticed it was 4am. I was sat alone in a freezing cold flat wearing my pyjamas, the old dressing gown which my dad bought me for Christmas when I was 12 with a thick red woolly jumper, which my Welsh granny knitted me in the '90s, pulled down over the top. I was desperately tapping away at my keyboard, trying in vain to reply to all the messages that I'd received. I was starting to lose the plot. Eventually I crept back into bed. The next day, I didn't want to get up. I knew I'd have to face the music with the university and Creative Scotland now that the agreement I had negotiated to 'donate' them the grant had been exposed as against the funding body's guidelines.[19] So, this is what it feels like to be at the eye of a storm? I was no longer The Sorcerer's Apprentice, more the subject of a witch-hunt. The paranoia was kicking in.

I remember a phone call and other correspondence with the university, then the doorbell rang. I nearly jumped out of my skin. I crept tentatively to the intercom.

'Hello?'

'It's *The Courier*', a man's voice said.

Fuck! How did they get my address? I froze and fell silent, then quietly put the handset back on the hook and retreated into my bedroom. I felt under siege. A couple of hours later, I crept down the stairs towards the front door. There was a massive box in the hall addressed to one of my neighbours. Oh god, I thought, it was an actual courier. Ha! That shows me not to be so suspicious. But I spoke too soon, the next day there was a picture of my house in Dundee's *The Courier*. It showed the whole block to make it look as if I owned a four storey mansion, even though I just live in the attic flat. I was so angry. I might have pissed off half of Scotland, but was it really a crime to put a picture of some chips that I ate myself on the internet? I'd been doing that for decades. How dare they put a picture of my house in the paper! I called their offices and demanded that they take the photo down from their website. It was too late for the print edition. It felt so wrong that they would do that to a woman living on her own. Now anyone could be showing up at my front door with pitchforks, letter bombs or whatever else. These journalists were so shallow: determined to personalise everything and not look at the bigger systemic failings

which had led to this situation. But I guess that was inevitable –
in this act of self-destruction, I'd inadvertently used one of Saul
Alinsky's *Rules for Radicals* upon myself: 'Pick the target, freeze
it, personalise it, and polarise it'.[20]

Before I knew it, it was after 4pm. I hadn't eaten anything or
turned on my heating. I suddenly noticed that my whole body
was shaking and my fingers were blue. It was a disaster for my
chilblains. It was very unlike me to forget to eat as well. I don't
think I'd ever done that before. There was just so much going on
inside my head. My heart raced continually. Leonie Bell, Director
of Arts & Engagement at Creative Scotland, came round to my
flat. She was really kind, given everything I'd done – she gave me
a hug. I made her a cup of tea. I remember how embarrassed I
was that the only milk I had to offer was 'almond milk'. What a
middle class cliché I was. I never normally even bought that stuff!
It had just been on special offer at the Co-op and I'd felt like a
wee treat to cheer myself up in those bleak first few days of the
New Year. Everyone was keen to offer me 'PR advice', as though
there was some easy way of smoothing over the huge cracks in
our society, which had been exposed. What I did ask Creative
Scotland for was access to their Press Data system, which would
automatically retrieve every single article in which *The Glasgow
Effect* was mentioned. They agreed to send me print outs of them
all.

When Leonie left, I felt that I should perhaps say a bit more
about the project than I'd been able to cover in my statement on
Facebook the previous day. I had been invited to go on *Channel
4 News* and on *Newsnight* but said no to both – I knew how bad
that would look to do a live link to the heart of London's media
and quite unnecessary too. I realised that it was within my power
to pick the media outlet that best reflected my own values and the
one which I most wanted to support by giving them an 'exclusive'
that was bound to solicit a lot of traffic. So I chose CommonSpace,
the news service and social media hub of Common Weal. Founded
at the time of the Referendum on Scottish Independence in
2014, Common Weal is like a mini New Economics Foundation
for Scotland – part of the new wave of progressive think tanks
fighting back against the neoliberals in the battle of ideas. They

sent me some questions and I spent the rest of the evening writing my answers to provide a bit more context for what I was doing. I wrote:

> I see it as the role of the artist to stick their neck out in order to raise important social and political issues. And, although I've had to deal with a barrage of personal attacks: making myself into a middle class punchbag, I don't regret the decision to use the title *The Glasgow Effect*. I was aware of the issues around class this would throw up, but I wanted to expose the 'tale of two cities' which is highlighted by two similar sounding phrases: 'The Glasgow Miracle' – commonly used in the artworld to refer to Glasgow's post-industrial renaissance as a global centre for culture, and 'The Glasgow Effect', which, as we all know, is its antithesis in PR terms.[21]

The article went live the following morning, and was again requoted by the mainstream media who continued to run stories about *The Glasgow Effect* every day for a week, when fortunately they finally started to lose interest. David Bowie died, so they switched attentions to that. When the Press Data cuttings finally arrived in the post from Creative Scotland, the document was nearly an inch thick.

I began to realise why these journalists were all so angry. I had consciously exposed my own privilege in order to draw attention to conduct across the 'creative class'. There was one man at *The Herald* who jumped on the bandwagon with everyone else, advising that 'Ellie Harrison should wrap up the Glasgow Effect and move on', whilst his own Twitter profile bragged about his love of sports cars. As long as they could keep the bile directed at me, they would deflect attention from unethical and unsustainable lifestyles they, and their wealthy paymasters, were living themselves. None of them wanted an angry mob on the scale of the one that had descended on *The Glasgow Effect* event page turning up on their doorstep.

Most people in the artworld disowned the project too for the same reasons. This was 'public engagement' on a scale that terrified these twee middle class arts organisations and academia. It was engagement with the 'wrong' sort of public. One local journalist, Brian Beadie got what I was doing from the start, writing on the Kiltr website: 'This work is addressing the middle classes, and taking the piss out of them if anyone, not the working classes'.[22] There was also a man called Darren Chadwick who I spotted engaging quite a bit in the Facebook debate. Remaining open-minded, he allowed his own views to evolve slowly as I released more information about my motivations via my statement on Facebook, in the original funding application and then finally in my interview on CommonSpace. It had become an addictive puzzle for some, enthusiastically researching and piecing together the clues bit-by-bit. On 5 January 2016 at 5.31pm he posted: 'I am actually coming around to the idea… Part of it is to show that the system is fucked…' Exactly Darren, that's what I thought. And if an artist can't throw a spanner in the works to make that apparent, then who can?

The Divide

When Darren McGarvey composed himself, he wrote an interesting essay on the Bella Caledonia website, where he clearly articulated the crux of the backlash against *The Glasgow Effect*:

> If only they [Scotland's art establishment] could grasp
> the fact people are not actually annoyed at Ellie or even
> conceptual art – infuriating as it is at times. If only they
> could grapple with the thorny reality that people are
> actually annoyed at the big floppy-haired elephant in the
> green room: they are annoyed at rising social inequality
> and how this expresses itself culturally.[23]

We have an unequal and unrepresentative artworld, because it reflects our increasingly unequal society as a whole. On 4 March 2016 at the *Declaration* human rights festival at the CCA, I watched the film *The Divide*, which also, coincidentally, stars Darren McGarvey.[24] It is a documentary based on the book, *The Spirit Level* by Richard Wilkinson and Kate Pickett, which shows 'how almost everything – from life expectancy to mental illness, violence to illiteracy – is affected not by how wealthy a society is, but how equal it is'.[25]

The Divide cuts between different parts of the world to offer 'fly on the wall' accounts of different people's lives within our global economic system. From Darren knocking about the streets of Pollok talking about all the mental health issues and problems with addiction he has faced as a result of growing up in poverty, to a man earning a six figure salary in America. Both are equally miserable: Darren because of the deep sense of injustice he feels in comparison to others, and the man in America because his high-consumption lifestyle – a huge house with a massive mortgage, cars, school fees and much more – mean he is locked into a soulless job in banking with a long commute to the city, just to keep everything afloat. He never has any time to see his family. Both of them live in fear of attack from others on the streets. As Wilkinson and Pickett show, 'the quality of social relations deteriorates in less equal societies'.[26] Anger and anxiety abound. After coming into contact with many more middle class people through her campaigning, Glaswegian activist and anti-poverty campaigner Cathy McCormack came to realise 'their problems are different but they certainly have them, and they're damaged people'.[27] One of her heroes, Nelson Mandela (1918–2013), the former President of South Africa, captured the problem succinctly when he famously said that in an unequal world 'all of us, rich and poor, are robbed of our humanity'.[28]

Small is Beautiful

It is these increased social inequalities coupled with catastrophic environmental destruction and frequent financial instability and uncertainty that the New Economics Foundation have shown are the three key consequences of neoliberal policy. These are the most pertinent issues of our time and *The Glasgow Effect* sought to make visible the connections between them. Someone else who got what I was trying to do was Roanne Dods. I first met her before my bike crash the previous September, when I took part in *Imagination* – the 'festival of ideas' she coordinated with

Glasgow-based writer Gerry Hassan at Govanhill Baths. Roanne had worked in the arts for several decades but had become disenchanted with the establishment and so started producing projects aimed more directly at social change. She founded her own progressive think tank Mission Models Money and set up the *Small is Beautiful* festival of local social enterprise. She was inspired by the influential book *Small is Beautiful: A Study of Economics as if People Mattered* by economist EF Schumacher (1911–77). Written in 1973, it was this book which had also inspired the New Economics Foundation's original motto 'economics as if people and the planet mattered'.

When I found the *Small is Beautiful* book on my mum's bookshelf in October 2018 and read it myself, I came to realise the significance of Schumacher's ideas to the 'low-carbon lifestyle of the future' I was attempting to live during *The Glasgow Effect* (described in Chapter 7).[29] Much like Patrick Geddes' famous slogan 'Think Global, Act Local', *Small is Beautiful* demonstrates that we must address the world's problems – namely inequality and the depletion of natural resources – at a local level; that we can only address our global environmental problems by solving the social and economic ones close to home – by creating a city, society and economic system where we can access 'those things we all need to live happily and well' without the need to travel. Schumacher describes a whole new approach to economics inspired by Buddhist principles, which enable us to know when enough is enough. So that we can all learn to enjoy living within our means:

> ...production from local resources for local needs is the most rational and economic way of life, while dependence on imports from afar and the consequent need to produce for export to unknown and distant peoples is highly uneconomic and justifiable only in exceptional cases and on a small scale. Just as the modern economist would admit that a high rate of consumption of transport services between man's home and his place of work signifies a misfortune and not a high standard of life, so the Buddhist economist would

hold that to satisfy human wants from faraway sources rather than from sources nearby, signifies a failure rather than a success.[30]

Roanne's knowledge of Schumacher meant she was well aware of the significance of what I was trying to achieve with *The Glasgow Effect*, even if I hadn't quite pieced it all together yet. She was keen to support me. She was angry at the way the mainstream artworld had just jumped on the bandwagon, embarrassed at the attention I'd drawn to their privileged lives. She called me several times in those first few weeks of January to check I was doing okay. I went round to her house for dinner where I met her friend Ruth Little, an Australian writer. I remember Ruth saying: 'It doesn't matter what you say, it's become toxic. You're as untouchable as nuclear waste right now'. I agreed. There was no point me saying anything else through social media or the mainstream media for that matter. I could not be disassociated from the caricature they had made me out to be. 'Ridicule is man's most potent weapon' – it seemed I'd unleashed another of Saul Alinsky's *Rules for Radicals* upon myself. Anything I now said would inevitably be twisted, misread and attacked. Instead I would have to be judged on my action; on 'deeds not words' as that famous Suffragette slogan from the early 20th century went.[31] It would be action that would not be swayed by their short-termist demands. As I said in the statement I published on Facebook on 5 January 2016, by the fifth day of the project I'd already 'ticked Creative Scotland's "Public Engagement" box and fulfilled the university's "Impact" agenda and so can get on with the real work'.[32] That meant work entirely focused on long-term benefits, unfolding over the decades to come. And it wasn't as if I had been silenced online altogether, I could continue expressing my views anonymously through the many projects and campaigns I was (and would soon be) running.

Roanne started talking about organising a public discussion event about *The Glasgow Effect* 'stooshie' (that was a new Scots word to add to my ever-expanding vocabulary). She wanted to get me together with Darren McGarvey and Katie Gallogly-Swan, another young working class activist from Coatbridge in North Lanarkshire who'd written a brilliant comeback to Darren's piece on Bella Caledonia. She argued that class divisions – Darren's 'two different worlds'[33] – were not as black and white as he claimed.[34] Because of my newfound 'toxicity', I was resistant to taking part. I felt it was far too early for me to be able to articulate the complexity of what I was doing. I needed the whole year to clarify exactly why I had taken this action, and to start to produce some results. But Roanne wouldn't let it go. She really wanted to create a forum where people could discuss these important issues face-to-face, without the ability to hurl abuse and then hide behind the computer screen. She said the others had all agreed to take part and that the date was set for 3 February 2016. I couldn't be the only one not there – that would look terrible – so I reluctantly agreed.

I had so many brilliant and intense conversations with Roanne around that time. A week or so before the event she told me she had oesophageal cancer, which could not be cured. The doctors had only given her a year to live. Roanne said that she was certain her cancer had been caused by continual and ongoing stress resulting from traumatic childhood experiences. She was only 50. Despite her diagnosis, her drive to keep working only intensified. She chaired *The Glasgow Effect: A Discussion* event at the Glad Café in the Southside on 3 February 2016. It turned out to be another electrifying evening which Darren recounted in 'The Changeling' chapter of his book, particularly his terror at meeting me![35] Listening back to the discussion, you can see that the seeds of what he went on to write in *Poverty Safari* were sown right there in the debate that ensued.[36]

Roanne had been concerned about public health for a while, organising many events within *Imagination* festival aimed at finding holistic solutions to Scotland's many problems and demanding systemic change so as to stop them recurring. She was keen to point out to the audience at the Glad Café the findings of the Glasgow Centre for Population Health (GCPH) research to date, that 'excess mortality' in Glasgow is experienced 'across virtually the whole population: all ages... both males and females, in deprived and non-deprived neighbourhoods'.[37] At that time, the mystery of 'the Glasgow Effect' – why people die younger in Glasgow than in similar post-industrial cities in northern England – was still unsolved. Little did any of us know that the GCPH were just putting the finishing touches to their epic report, *History, Politics & Vulnerability: Explaining Excess Mortality in Scotland & Glasgow*,[38] which was due for release that May. I only found that out when I received several aggressive emails from GCPH following the discussion event. One was 1,676 words long.[39] How dare an artist – an 'outsider', an English-sounding person – with no experience in public health weigh into the debate? What could I possibly offer that all these public health 'professionals' had failed to find?

Could there be a worse insult?

Had I known everything that I know now about 'the Glasgow effect' then perhaps I wouldn't have had the guts to do what I did. But that would have been a great shame. Instinct and naivety can create powerful art. Even after seven and a half years of living in Glasgow, I didn't *really* know the city. I didn't *really* understand Scottish culture. That was the point. As one journalist wrote: 'If one of her strategic aims was to gain a better understanding of the city then she's learned a valuable lesson already'.[40] Those first few months of the project were a steep learning curve and some striking characteristics started to emerge. There appeared to be something oppressive about Glaswegian culture. People were scared of standing out and being seen as different, hence the mad rush to join the social media pile-on. Roanne Dods called it 'tall poppy syndrome':

> The idea... that if you stick your head above the parapet, people knock you down. And that can be isolating; terrifying; difficult. So conformity, and therefore lack of diversity, is what happens... And we have this very strong media in Scotland... We have a lot of newspapers where you're kind of kicked if you fail and you're kicked if you succeed... so that already creates a culture of conformity.[41]

Carol Craig also highlights this 'syndrome' in her book, *The Tears that Made the Clyde*. She traces it throughout the west of Scotland's entire culture, particularly its humour, which she says derives almost entirely from 'putting folk down'. Glaswegian patter is 'particularly good for emotional abuse'.[42] In terms of the impact on health, 'the endemic, low-level abuse in west of Scotland culture' is profound.[43] When I wrote in my CommonSpace interview published on 7 January 2016 that 'I see it as the role of the artist to stick their neck out in order to raise important social and political issues',[44] not only did some sadistic smartarse upload a photo of a guillotine to the Facebook event page, this statement was the antithesis of Glasgow's oppressive culture of conformity, which is preventing people fulfilling one of their key 'psychosocial' needs to express their 'individuality'.

And it is this culture which Carol Craig claims has held Glasgow back. It was easier for those who had the time, education and resources necessary to help improve the living conditions for everyone to simply leave instead. She writes that

> those who were disgruntled were more likely to emigrate than put up a political fight. Indeed emigration is often seen as a 'safety valve', relieving pressure for radical reforms.[45]

Emigration certainly has been a massive 'brain drain' on the city over the last two centuries:

> ...this environment undermined Scotland as a whole. Between 1821 and 1945 over two million Scots migrated. Over half of these migrants went to the USA, Canada, Australia and New Zealand and were intent on finding a better life. In the 1960s, when Archie Hind wrote his novel [*The Dear Green Place*], as many as 40,000 Scots a year left their homeland. Many were from Glasgow and the west of Scotland.[46]

I guess you could call this 'social mobility' too – leaving your community behind for the sake of your own interests. The paradox of Glasgow's oppressive culture of 'putting folk down' was that if it was initially intended to ensure people didn't get 'above their stations' and, therefore, leave their communities behind, it often had the opposite effect: of forcing them away. That was the most interesting element of *The Glasgow Effect* project as a 'psychological experiment' – after the shitstorm and all the 'emotional abuse' that was aimed at me, leaving was the one thing I couldn't do. I was forced to invest all my time, education and resources in Glasgow, whether it wanted me or not.

But the paradox goes even further, as this oppressive culture of 'putting folk down' is simultaneously twinned with what *Evening Times* columnist Gary Lamont refers to as Glasgow's 'fierce patriotism'[47] – a blind allegiance to a place, which shuts down criticism when we need it the most. It was the complete opposite of what Glaswegian activist Tom Anderson – founder of Glasgow's first Socialist Sunday School – had advised 100 years before. The second of his 'Ten Proletarian Maxims' from 1918 read: 'Thou shalt not be a patriot... Your duty to yourself and your class demands that you be a citizen of the world'.[48]

Gary Lamont's 'fierce patriotism' was a completely alien concept to me. I could never take offence (to the extent that I would have to publicly declare it online) if someone criticised the place where I grew up or lived.[49] We need a 'ruthless criticism of everything existing',[50] as Marx himself demanded – that's how we affect positive social change. Many people in Glasgow were dismissive of Carol Craig's book when it came out in 2010, because it was brave enough to do just that – to realise artist and writer Alasdair Gray's simple mantra 'Let Glasgow Flourish by Telling the Truth'.[51]

But what became clear during *The Glasgow Effect* shitstorm was that what people were most angry at was who was perceived to be doing the criticising. Just as, perhaps, you can only get away with telling a 'gay joke' so long as you're gay yourself, the so-called 'Scottish Cringe' (described as 'a persistent sense of inferiority, self-pity and embarrassment at overt expressions of Scottishness')[52] is only okay, as long as you have a Scottish accent. It's a phenomenon best summed up in a bit of unheeded advice once offered by my mum in one of her more conservative (with a small 'c') moments: 'Never insult a woman's husband', she said. 'It's fine for her to moan on about him all day long, but as soon as you join in or launch an attack on him yourself, she'll become defensive'.

The fact that I appeared to be English was one of the things that aggravated people the most. I was the 'colonist' coming up here and taking their money. Money which many people thought should have gone to 'proper Glasgow-based artists' (as Kendal Orr described them) – an authentic Glaswegian with a Glaswegian accent. A friend at my studios overheard an artist there who'd known me for over six years saying: 'It should have gone to a Scottish artist'. The Glasgow Effect vividly exposed to me the dark underbelly of nationalism: fear and hatred of 'the other'.[53]

People were also aggravated by the fact that I'd described the project as 'extreme'. I wrote in my funding application that 'I believe it is the role of the artist to take the extreme lifestyle decisions, which would not be possible for anyone in a less privileged position, and, in doing so, expose the contradictions in all our lives'.[54] But nobody made the connection between these two key points of contention. It was clear that many people born and raised in Glasgow never left for socio-economic reasons. The point was that this decision was only 'extreme' for someone in the 'creative class', whose career has led them into an absurd, transient and carbon-intensive existence such as my own. It was extreme, quite simply, because I wasn't 'from here' to start with.

I was an outsider forcing myself to stay – it was a self-imposed exile from my 'native place' (where I was born) or what the Palestinian philosopher Edward Said would call one's 'true home'. In his 1984 essay, *Reflections on Exile*, he also explores the curious conflicting relationship between 'nationalism and its essential association with exile'.[55] Whereas nationalism has historically resulted from a 'condition of estrangement' – a feeling of being oppressed within one's own home (by English landowners, by American oil tycoons or by other Scots – as the play *The Cheviot, the Stag & the Black, Black Oil* brilliantly illustrates), one of the outcomes of nationalism, and indeed patriotism, is to turn against and further isolate, those who don't appear to fit in. Edward Said writes:

Nationalisms are about groups, but in a very acute sense exile is a solitude experienced outside the group: the deprivations felt at not being with others in the communal habitation. How, then, does one surmount the loneliness of exile without falling into the encompassing and thumping language of national pride, collective sentiments, group passions? What is there worth saving and holding on to between the extremes of exile on the one hand, and the often bloody-minded affirmations of nationalism on the other? Do nationalism and exile have any intrinsic attributes? Are they simply two conflicting varieties of paranoia?[56]

The exile therefore finds themselves caught in a permanent limbo, never granted permission to be critical of a place no matter how long they may live there and never truly able to fit in. As Edward Said shows, exiles therefore exist in a continual state of 'resentment' for non-exiles, constantly thinking:

They belong in their surroundings, you feel, whereas an exile is always out of place. What is it like to be born in a place, to stay and live there, to know that you are of it, more or less forever?[57]

That I will never now know.

CHAPTER 6

Creative Destruction

But is it art?

NOW THAT THE dust has settled, this chapter takes a detour to think about *The Glasgow Effect* as an artwork – what it might mean and how it might be interpreted – and to start to explore some of the issues it raises about class, capitalism, art, education and much more. Defining 'conceptual art' in his book about the subject, Tony Godfrey writes that it

> is not about forms or materials, but about ideas and meanings. It cannot be defined in terms of any medium or style, but rather by the way it questions what art is... Because the work does not take a traditional form it demands a more active response from the viewer, indeed it could be argued that the conceptual work of art only truly exists in the viewer's mental participation.[1]

And participate they did. People from different backgrounds came together on the levelling platform of social media. *The Glasgow Effect* event page became a space where anyone could have their say. In January 2016 it appeared in the 'newsfeed' of more than one million people, 248,000 of whom viewed the page with more than 8,800 actively engaging by posting comments and making memes. On 5 January 2016, it was 'trending' nationally. It certainly 'touched a lot of people', as Limmy's art school character might say.[2] In those first few weeks of 2016, it provoked an explosion of outrage; of anger and frustration; of humour and ideas.

Central to the debate was the question of whether or not what I was doing was 'art', and what the purpose of 'art' is or should be in our contemporary society. Everyone who engaged was challenged to consider this question on some level and to come to their own conclusion. For people like Sue Chadwick it was clear: 'This is not art, it is a chimera. Where... are the creative skills and ideas? More wind from the bagpipes!' In her book, *This is Not Art: Activism & Other 'Not-Art'*, artist and writer Alana Jelinek offers a definition of 'art' which chimes with the clear distinctions I'd previously drawn between my art and my activism:

> Art is not political action. Art is not education. Art does not exist to make the world a better place. Art disrupts and resists the status quo, and if it fails in this prime objective it serves only to deaden a disenfranchised society further.[3]

There's no doubt in my mind that those chips, that social media shitstorm, the 'funding controversy' and my proposed individualistic localised action amidst accelerating global social, environmental and economic crises, were all part of an inseparable whole. *The Glasgow Effect* was an artwork which illustrated both the power and the necessity of art in a deeply troubled society such as our own. It was an artwork that both 'disrupted' and 'resisted' the status quo as well as creating an inclusive platform for discussion. It was an artwork which held up a mirror to an angry and unethical world, completely divided by economic meltdown and six years of austerity – a signal of the direction things were heading at the start of 2016.

Some of the most powerful artworks use provocative tactics to wake people up to what's really happening in the world around them. On the evening of 4 January 2016, when my friend Sarah asked me 'What message do you want to get across?', I made the decision to go it alone against the institutions – Creative Scotland and the university – and to lure them both into the fray. When I wrote in my statement 'The fact that this university, like most others in the UK, now requires its lecturing staff to be fundraisers and is willing to pay them to be absent from teaching as a result, should be the focus of this debate',[4] it was a provocation not dissimilar to the tactics used by the Futurist artists in 1913. In the heated political climate just before the First World War, they organised theatre shows deliberately selling the same tickets to several people to consciously cause a 'traffic jam, bickering, wrangling' as people tried to take their seats.[5] At what point did it dawn on them that there was no show? That maybe they were it?

There's also the German artist Hans Haacke who, in the '80s and '90s, did much to challenge so-called 'philanthropic' funding streams in the arts (specifically the sponsorship of exhibitions by tobacco giants like Philip Morris – a practice now illegal), which compromise artists' ideas – using their work to legitimise unethical business models. He aimed to make visible invisible injustices for all to see. In his 1972 work *Rhine Water Purification Plant*, he wanted to draw attention to the pollution of the river Rhine from the sewage works nearby. So he brought some of its water into a nearby gallery to fill a giant fish tank which he then made home to some fish. Suddenly, now that people could see the state of the water, they were angry at the artist for his cruelty – how dare he place those fish into such filthy water? When all he was doing was drawing attention to injustices which were unfolding beyond the gallery on a vast scale. They were 'shooting the messenger' and in doing so, were neglecting to see that also – in designing a filtration system – he was actually proposing a solution.

Glaswegian artist and writer Alasdair Gray emphasises the important role artists play in our society: 'People need artists: either to look up to or to look down upon'.[6] *The Glasgow Effect* enabled me to take on that dual role. Firstly as the model citizen boldly pioneering the 'low-carbon lifestyle of the future' (described in Chapter 7) and secondly as a two-dimensional object of ridicule on social media – a symbol of the detached 'liberal elite' I too had come to despise. French philosopher Gilles Deleuze (1925–95) describes the main role of the philosopher within society as being to 'play the fool'.[7] Like Diogenes the Cynic living in his barrel in Athens more than two millennia ago – he was the only one who had pushed past ridicule and so could think the unthinkable and say the unsayable. Or, in the words of the Situationist International movement in Paris in 1968, to 'Be realistic, demand the impossible!' By sacrificing myself to the social media trolls, I was able to raise questions so few artists are prepared to ask. Questions about persistent and worsening inequalities in Glasgow and Scotland and questions about conduct across the 'creative class'.

I took a big risk making *The Glasgow Effect*. The risk of ridicule, the risk of failure and of personal and financial loss. At a time of accelerating global social, environmental and economic crises, we need more artists who are prepared to do the same. Yet the opposite is happening – austerity has helped create a culture of fear and conformity in the artworld. In his series of articles in 2016–7, artist and writer Morgan Quaintance poses the question 'Why is there not more politically-engaged art in these turbulent times?'[8] One of his answers (echoing Darren McGarvey's sentiments), is that in an increasingly unequal society, the UK's artworld has become dominated by people from privileged backgrounds who are increasingly removed from what's going on in the 'real world'.[9] He refers to what's been happening as 'The New Conservatism'. But he also argues that:

> I suspect silence, resignation or apathy are fuelled by something far more basic, comfort. Put simply, people are adverse to personal risk and lifestyle change.[10]

In his article exploring 'Complicity and the UK Art World's Performance of Progression',[11] Morgan Quaintance shows the extent to which contemporary art has been corrupted through its relationship with private finance (or the so-called 'philanthropists'). This is something Alana Jelinek also illustrates in her book, *This is Not Art: Activism and Other 'Not-Art'*:

> Despite the fact that many in the artworld claim a Leftist politics, and some are even politically motivated to critique neoliberalism and undermine its structures, most in the artworld most of the time are actually replicating neoliberal structures and embodying and perpetuating its values.[12]

The spell of neoliberalism has created an artworld full of hypocrites. As well as the fear and conformity caused by austerity, many artists are so in awe of an artworld with its glamorous 'high-visibility, international activity',[13] that they simply don't want to rock the boat. The promise of 'popularity', 'image' and 'financial success' is just too tempting, even though it may be a mirage.[14] The real tragedy of this corruption of our values is that it is the thrifty lifestyle to which many artists are accustomed – where autonomy and creativity are paramount – that can and should offer the path to happier, healthier and more sustainable lives for us all. And it is only a creative education, offering that eclectic mix of essential life skills, which makes this way of living possible. These skills are:

- Critical thinking – enables you to see through the poisonous rhetoric of the right-wing media, determine fact from fiction and reject the omnipresent advertising which demands 'we waste money we don't have on things we don't need' or else we'll be seen as a 'failure'.[15]
- Practical skills – enable you to 'waste not, want not' and 'make do and mend'; to build and make your own things, love and repair the objects that you do own and keep using them for as long as possible; to develop 'a healthier relationship with "stuff"'.[16]
- Confidence – enables you to 'Be realistic, demand the impossible!' Even if it means standing out from the crowd; to 'speak truth to power' and to know how to get your voice heard.
- Self-motivation – ensures you're driven by so many other things – curiosity, the need for answers, simply 'for the love of it',[17] the desire to change the world and much more – all ahead of money.[18]

Money can't buy you love

It is often said that we live in a world that 'knows the price of everything and the value of nothing'. This sad state of being has resulted from the overbearing presence of our financial system in all our day-to-day lives, encouraged by neoliberal policy – where 'transactions' have replaced 'conversations' – and, more broadly, from the systems our society uses to measure its 'success'. In the '50s, the most omnipotent of all these systems was created by the British economist John Maynard Keynes (1883–1946) when he devised the United Nations System of National Accounts. If your country didn't comply with this system, then you couldn't be a member of the United Nations club. The system defines how to measure a country's Gross Domestic Product (GDP) – a single figure which captures the total 'monetary value of all the goods and services it produces' – suggesting that an increase in GDP is a reflection of that country's 'success'.[19]

In her film, *Who's Counting?* (developed from her 1988 book, *If Women Counted: A New Feminist Economics*),[20] Marilyn Waring shows how damaging this GDP measure has become:

> The system recognises no value other than money,
> regardless of how that money is made.

GDP counts petrol sales, cigarette sales, alcohol sales, arms sales and so on. If your house burns down, GDP goes up, because you have to pay to get it rebuilt. But the most damaging aspect of the system is its flipside, that GDP completely ignores all the activity which does improve our well-being and make us happy; the activity which is 'priceless' – that is the entire 'core economy' of family, neighbourhood and community on which all else in our society is built. That is all the vital caring work, which has been predominately performed by women. As Marilyn Waring points out, under this system: 'If there isn't money involved, if there isn't a price, you don't measure it'.[21]

The Glasgow Effect did have a price tag attached, a whopping 15 grand and therefore it could be measured. Despite the fact that many argued that the money was extravagant and unnecessary, the irony was that without it, there would have been no artwork; my action would never have been noticed; its value never assessed. There would have been no debate. The funding for *The Glasgow Effect* was a means to an end – it caused the 'controversy' which provided a platform for people to participate. As Barry Hale posted on Facebook at 2.40pm on 18 February 2016, after

I launched my public survey asking 'Does more Money = better Art?', '...money can mean unheard voices can be given a forum through which to be heard'. I just stood back and let everyone have their say.

Much of this debate focused on what else 15 grand could have been spent on: '18,987 copies of Darude Sandstorm' or '10,791 double cheeseburgers from McDonald's' apparently.[22] In her book, *The Tears that Made the Clyde*, Carol Craig shows that Glaswegians have a particular obsession with money because the history of 'poverty and insecurity of employment in Glasgow may have predisposed its citizens to pay particular attention to materialistic goals'.[23] It plays out in a culture of 'instant gratification' – mindless consumption or hedonism – 'drinking a substantial amount of your wages' on payday,[24] which is clearly not good for long-term mental, physical or financial health. You only need glance at some of the memes posted on *The Glasgow Effect* event page to see these 'extrinsic' values on display. People were fantasising about what they would do with a similar 'windfall' – throw all the money in the air, buy a sports car, a big gold chain or loads of weed? Or go on holiday to Magaluf, or Tahiti even? I kept wondering, but then what would they have lived off for the rest of the year? Plus obviously the holiday was out of the question.

In the eyes of many, I was that artworld hypocrite, the 'aloof academic' bragging about the value of all the grants they've been awarded, as though they are badges of 'success' (behaviour which is the norm on staff profiles on most university websites). It's true I did not need to apply for that money to take this action. I wanted to draw attention to a dysfunctional higher education system which is also hypocritical – claiming to prioritise 'intrinsic' values as a PR strategy, whilst actually being obsessed with the complete opposite – 'popularity', 'image' and 'financial success'.

But more broadly, *The Glasgow Effect* was able to highlight one of the fundamental flaws evident across so many aspects of our society which only serves to entrench inequalities. It is captured in a phrase used by EF Schumacher in *Small is Beautiful*: 'Nothing succeeds like success and, equally, nothing fails like failure'.[25] That is, the more privileged you are to begin with, the more likely you are to be rewarded. It's what the right-wing Canadian 'public intellectual' Jordan Peterson calls the 'Pareto Principle' – that 'success' and financial rewards gravitate towards those who are already successful and wealthy and therefore accumulate disproportionately with them. Jordan Peterson describes the 'Pareto Principle' as:

> ...when one good thing happens to you, it makes you a little more powerful and attractive and so that fractionally increases the possibility that another good thing will happen to you and then that spirals out of control.[26]

This became most clear to me, firstly (from the bottom of the pile), when I was temporarily disabled after my bike accident and dependent on bus routes which had been axed by the profiteering bus company First because they weren't seen as 'commercially viable'. The powerless are less able to fight back. Then, secondly (from the top of the pile), when I started working in academia – being paid a good salary and yet supported and incentivised to go out and get yet more money. It was also the first time I'd had a proper pension (with the Universities Superannuation Scheme). Disregarding the wholly unethical practice of the scheme itself – the fact that USS boss Bill Galvin pays himself a salary of £566,000 and that they have 113 staff earning more than £100,000,[27] and that they are investing our remaining funds in arms and fossil fuels (despite an ongoing 'divestment' campaign),[28] how is it fair that the more privileged you are to start with – a tenured job in academia – the more you can save into a pension, therefore the wealthier you will be as a pensioner? Surely it should be the other way around?

These dysfunctional systems across society just ensure that privileges and inequalities are entrenched with age. Because the 'Pareto Principle' phenomenon is described in the Bible – 'To those who have everything, more will be given, from those who have nothing, everything will be taken'[29] – Jordan Peterson believes that it is 'natural' and therefore cannot be counteracted. I believe it is something which can and must be challenged across the whole of our society through all the systems we use to distribute our resources, and the kind of values that these encourage. The National Health Service was founded in 1948 on Marx's equalising principle of 'From each according to his ability, to each according to his needs!'[30] Yet our increasingly marketised healthcare system, operating in an increasingly unequal society, now delivers an 'inverse care model' – where those most in need of good healthcare are least likely to have access to it. Wealthy patients get 25 per cent more time with their GPs, because there are fewer people requiring appointments in wealthy areas.[31] Health inequalities are further entrenched.

Around the time Creative Scotland was set up in 2010 (as a result of the merger of the Scottish Arts Council and Scottish Screen), there was a lot of concern about how the way our new national arts funding body was run and the values it espoused would affect the sort of culture which would be produced and who would get to produce it. Roanne Dods was one of the many people who worked hard to highlight Creative Scotland's 'corporate ethos that seems designed to set artist against artist and company against company in the search for resources' – encouraging a sort or entrepreneurial attitude focused, above all, on making money.[32] But there were also concerns about its closeness to the Scottish Government's political agenda. The Creative Scotland campaign set up in 2009 by Glasgow-based *Variant* magazine (which 'bit the hand that fed it' too much and sadly eventually had its own funding cut) claimed Creative Scotland

> remains a confusing and self-contradictory set of propos-
> als overwhelmingly seeking to make artists instruments of
> government policy – in the words of the Bill, artists are to
> 'support the government's overarching purpose'.[33]

The 'arms-length principle' – first defined when the Arts Council of Great Britain was founded in 1946 (coincidentally also by John Maynard Keynes) – was designed to ensure that distance was maintained between the government and the funding body. The 'arms-length principle' sought to ensure that art could remain politically autonomous, so as to preserve its critical edge. Under these new plans hatched by the Scottish Government, was all Scotland's culture destined to become advertising for 'brand Scotland'? Everything I wrote in my *Think Global, Act Local!* funding application was true except for just one line: 'All the while reflecting positively on the original site of its making: Glasgow, Scotland, as a centre for cultural activity'.[34] My intention in switching the name to *The Glasgow Effect* was to draw attention to gross inequalities in our country in the hope of addressing them, not to gloss over them.

It's not that I don't think public funding plays a vital role in supporting the arts sector (in its current form under our present global economic system). It clearly does. But we must ensure it remains politically autonomous so that it can 'speak truth to power', and that it prioritises those who need it the most. On 1 March 2016, I was invited by the Scottish Artists Union to speak via Skype to the Cross-Party Group on Culture at the Scottish Parliament in Edinburgh. I began by screening the track 'The Glasgow Effect' by Glaswegian rapper Wee D,[35] which had just been released in response to the project on 18 February 2016. I analysed Wee D's criticisms one-by-one (many of which also fixate on what else 15 grand could have been spent on) and then concluded my talk by saying that:

> I think what we both want to see is arts funding, increased arts funding let's say, that is distributed more equitably. And, arts funding that is available and accessible to people from all sorts of backgrounds, and, most importantly, I think we both want to see art that is for everyone and that is connected to the reality of most people's lives.[36]

Since the end of 'the "golden age" of the arts' and 'the age of austerity' that has followed, arts budgets have been slashed. Cuts have been perpetrated by the Tories and SNP alike – forcing our local authorities, as Darren McGarvey describes, to cut finance 'to things like music tuition for Scotland's poorest kids'.[37] 'Efficiency savings' mean public funding for the arts is now almost entirely centralised, with remote 'gatekeepers of culture' deciding where the money is spent. Austerity has purposefully made arts funding into a scarce resource, which now has to be fought over – creating conflict amongst artists,[38] and discouraging risk-taking.

As Morgan Quaintance has shown, the 'price is an artworld in which politics, dissent and non-conformism are increasingly marginalised or else passed over in silence'.[39] Private sponsorship – the alternative to public funding that is the norm in America and which is increasingly being promoted in the UK under austerity – only makes things worse. You will always be compromised by the agenda of the wealthy individual (the so-called 'philanthropist') or corporation that is awarding the money.[40] They will always seek to legitimise their unethical activities by associating themselves with your art. In the case of the multinational oil companies – Shell and BP – who have sponsored the British Museum, the National Portrait Gallery, the Royal Opera House and the Tate amongst others,[41] they are using art and culture to erase the 'unsightly environmental destruction' they're responsible for.[42] Just as Hans Haacke took on the tobacco giants in the '80s and '90s, activist organisations such as Platform London have shown this 'artwashing' by the oil companies who are causing our climate crisis, as well as blocking international efforts to cut carbon emissions, to be the key fight of our times.

These numerous issues with arts funding go some way towards explaining why Darren McGarvey was so keen to stress during *The Glasgow Effect: A Discussion* event at the Glad Café on 3 February 2016, that all his creative production his '17 albums' and his 'youth project' has all been funded off his 'own back':

> I could never take funding from a funding body, a government funding body and then turn around a say 'I'm doing this project that's going to investigate such and such'. Because I know that if I really want to 'investigate such and such', I'm going right to the fucking source. It's the fucking government. It's the structures of power. And everywhere that they are. At local level, national level and international level. And you simply cannot do that, when your whole thing is funded from the government.[43]

Unless, of course, you have no immediate plans to reapply or, like *Variant* magazine, your critical integrity trumps all else. It's for the same reason as Darren McGarvey that I have funded the year it's taken to write this book completely off my 'own back' (by maintaining my thrifty lifestyle and continuing to teach part-time). German artist Gustav Metzger (1926–2017) was one of the first to highlight these issues with arts funding, but he decided that non-participation was the best way to address them. He called the first ever collective Art Strike in 1977 to try to bring down the 'art system',[44] which he saw as corrupted by the commercial values of the art market and by publicly-funded artists who were being used as puppets for the state. It is as clear now, as it was back then, that an alternative is necessary.

I started thinking about the need for a new ethical and autonomous alternative arts funding stream when I was on the Campaign Lab 'economic justice' campaigning course in 2014. These initial thoughts evolved into the *Radical Renewable Art + Activism Fund* project which I began work on the following year. The aim was to challenge the inherent compromises demanded by existing funding streams, by raising revenue from investing in a renewable energy co-operative instead – something which has social and environmental goods in its own right – reducing carbon emissions and dependency on fossil fuels, as well as increasing community participation in the democratic running of the co-operative. The key difference between a co-operative business model and a capitalist one is that it operates on a 'one member, one vote' basis where everyone has equal power, regardless of how much money they may have invested (shares in capitalist companies operate as 'one pound, one vote' and therefore the 'Pareto Principle' is perpetuated). The idea was that any surplus generated would be used as an uncompromised source of funds for more politicised art and activism which can 'disrupt and resist the status quo'.[45] One of RRAAF's mottos is 'where the money comes from is as important as where it goes'. An alternative funding stream like RRAAF is all the more relevant in an increasingly precarious environment where living costs continue to rise and therefore time to invest in creative and political activity, which doesn't offer any financial return, is scarce. In our neoliberal age of privatisation, it's time and education that have become the real luxuries.

Every human being is an artist

But it has to be said that to be obsessed with funding systems at all may be to have to succumb to a certain neoliberal way of thinking, where 'money talks' (sorry, I am one of 'Thatcher's Children' after all!). The whole point is that creative and political activity is not expensive – anyone can and should be able to get involved. As Schumacher points out in *Small is Beautiful*: 'philosophy, the arts and religion cost very, very little money'.[46] And, in fact, in order to ensure they remain part of the realm of 'intrinsic' values, it is essential that they remain within the 'priceless' economy, which although not counted by GDP, are the things which actually give meaning to our lives. Inclusive creative and political activity is what can offer everyone the chance to fulfil our dual 'psychosocial' needs – to express our 'individuality' as well as increase our sense of 'belonging'. It is participating in this activity which offers the healthy, sustainable and low-cost alternative to what Common Weal describe as 'debt fuelled, wasteful, unsatisfying consumerism used to build and sustain our identities via what we buy'.[47]

And very often the bigger the budget the more meaningless the artwork. The perfect example being millionaire artist Anish Kapoor's hideous blot on London's skyline erected for the Olympics in 2012.[48] The *ArcelorMittal Orbit* cost £22.9 million – £3 million of which was from taxpayers (200 times more public funding than *The Glasgow Effect*) with £19 million coming from the multinational steel and mining company ArcelorMittal, which it is named after. So, not only is the artwork essentially just a big advert for a company which has assisted in putting many privatised steelworks in Britain out of business, and a gross waste of the natural resources used to build it, but punters also have to pay £15 to visit. Residents of London borough of Newham where it's situated said: 'It doesn't reflect the area', 'It's not sophisticated... it's nothing. It's just a nothing' and 'In this area, Newham, there's a lot of room for improvement on a practical level, before you can get to, you know, things like art'.[49] These huge amounts of money have resulted in something which is totally disconnected from the human scale or from any specific locality.[50]

When I got the results of the public survey which I ran on *The Glasgow Effect* Facebook event page from 15 February to 26 April 2016, asking 'Does more Money = better Art?', 87 per cent of the 639 people who took part agreed with me, that 'No' it does not. German artist Joseph Beuys, who was most well-known for coining the phrase 'every human being is an artist' in 1973,[51] said, 'If a person is an artist he can use the most primitive of instruments: a broken knife is enough'.[52] His core message was that creativity is such a fundamental part of what it means to be a human, that we must all be allowed the time and the education to be able to manifest it in positive ways. Because, it is creativity – in its simplest terms 'the *power* to create something from nothing'[53] – which, if liberated, is the driving force for social change. As Beuys said, it's only 'Through the idea of art, democracy will one day be a reality'.[54] But, as writer Oli Mould argues in his 2018 book, *Against Creativity*, the problem is that the concept of 'creativity' has been hijacked by neoliberalism. Creativity has been made into an 'industry', artists have become increasingly 'professionalised' and being an 'artist' has been made into something that by its very definition (as shown in the NS-SEC in Chapter 1) is a 'middle class job'. Art is, and has come to be seen as, the pursuit of the privileged few.

'That's robbery!' said one of the audience members at *The Glasgow Effect: A Discussion* at the Glad Café. One of the memes posted on the Facebook event page by Jim Be at 6.41am on 5 January 2016 read:

The earth without 'art' is just 'eh'.

Echoing this sentiment, Schumacher claims that we can all live without some scientific education (especially seeing most technological innovation is completely unnecessary and causes many more new problems than it solves).[55] But if we are forced to live without a creative education, he says: 'Unless I get my understanding from another source, I simply miss my life'.[56] The

Does more Money = better Art?

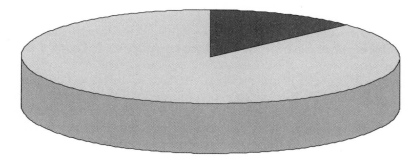

Public survey conducted on *The Glasgow Effect* Facebook event
page from 15 February to 26 April 2016 (639 participants)

■ Yes
□ No

arts exist to help us to create meaning, and it is a 'sense of purpose
and meaning in life' which Harry Burns, Scotland's former Chief
Medical Officer, says is one of the six core things necessary to
create wellness.[57] But as Carol Craig documents in her book, *The
Tears that Made the Clyde*, the problem is that there is a long-held
resistance in Glaswegian culture to art or any expression of emotion
or creativity. In *Poverty Safari*, Darren McGarvey recounts being
chastised by his classmates at school for simply using the word
'beautiful'.[58] An 'aggressive philistinism'[59] prevails which is seen
as defending Glaswegian values 'against the middle classes'. Carol
Craig writes:

> The difficulty with this is that these values (*it's no for
> the likes of us and it's crap anyway*) keep people in a
> restricted world and encourages time and money to
> be spent on drinking, watching telly, bingo, gambling,
> football or consumer goods.[60]

Activities which are clearly not good for mental, physical and
financial health, nor very sustainable (unless, of course, you're
playing football instead of merely watching). If we do not provide
people with the tools necessary to channel their alienation, anger

and frustration in positive creative directions and remove the stigma of doing so, then it can only manifest in far more (self-) destructive ways.

I'm as mad as hell, and I'm not going to take this anymore![61]

Anger and frustration were two of the overriding emotions provoked by *The Glasgow Effect*. People have every right to be angry in a world where the richest 1 per cent own two-thirds of global wealth and the richest 10 per cent cause half of global carbon emissions[62] (which actually includes nearly all of us here in Scotland where we're using three times our fair share of the world's resources every year in order to fuel our carbon-intensive lifestyles), when all the while the 'people who are poor and powerless bear the brunt'.[63] People have every right to be angry at those in power who are doing nothing to remedy this situation – putting their own selfish interests above all else. It was alienation, anger and frustration that pushed me into creating *The Glasgow Effect* in the first place. It was a scream, which was then amplified by the city. Enough is enough, we cannot go on like this![64] At the very start of 2016, the vitriolic response to *The Glasgow Effect* became a signal of the direction our world was heading towards.

People are angry at an 'extractive, inaccessible economy' which has left them powerless. As writer Aditya Chakrabortty describes:

> Spread across Britain today are people and places united by a common condition: they are largely powerless. Their economies have been emptied out, their services cut to the bone, their incomes under threat. The market discards them; the media ignores them; the state disregards them.[65]

This has been exacerbated by economic meltdown and the years of austerity since 2010, but it is a situation which Marilyn Waring would argue has been with us ever since the invention and implementation of the United Nations System of National

Accounts. Crucially, the Gross Domestic Product (GDP) measure does not take into account how a country's wealth is distributed, its 'spatial polarisation',[66] nor the way it is centralised in certain places (and in the hands of the few), leaving other places threadbare:

> ...it doesn't matter what they say... GDP might grow up, but GDP is utterly unrelated to the well-being of the community. It tells you nothing about levels of poverty, it tells you nothing about the distribution of poverty, it tells you nothing about primary health care, educational standards, environmental cleanliness and folks have realised that this uni-dimensional economic fabrication, just doesn't bear any relationship to their lives.[67]

I described *The Glasgow Effect* as 'one woman's protest against globalisation' – a tongue-in-cheek way of questioning the value of attempting such a 'David and Goliath' struggle. But it was a protest against the same 'extractive, inaccessible economy' that fuelled the Brexit vote and the election of Donald Trump as President of America later that same year. I was initially concerned about the possibility of the UK leaving the European Union from a human rights and environmental protection perspective, but in my dazed and sleepy state on the morning of 24 June 2016, I immediately began to see the opportunity Brexit presented for 'taking back control' of all our essential public services and infrastructure which have been privatised.[68] I didn't blame the people who voted to 'Leave'. There is a reason they feel they have 'no control' over their lives, because they are receiving such huge bills for privatised utilities and services, which are largely owned by foreign companies purely motivated by profit. I saw the public ownership of our essential services as the positive message to inspire the disaffected – the solution (the alternative to the bigoted right-wing rhetoric) which would actually reduce living costs and therefore improve their quality of life.

What has angered me the most since the Brexit vote has been all the talk in the mainstream media from the Brexiteers about 'striking new trade deals' which prioritise these foreign investors over the interests of local populations. Most international trade deals include the most evil of evil Investor-State Dispute Settlement (ISDS) clauses which allow private companies to sue our governments as and when we decide to take back what is ours.[69] Whereas Brexit could be an opportunity to localise our economy and supply chains to a more human and accessible scale – thereby significantly increasing our self-sustainability and reducing our carbon emissions – these Tory Brexiteers' idiotic attitude is for us to reduce trade with the European Union (our nearest neighbours) and to increase trade with Australia instead![70] The people voted against globalisation and what are they being offered instead? Globalisation on steroids.

Donald Trump's approach is different. Despite claiming that 'climate change is a hoax' and withdrawing America from the Paris Agreement – the international pact agreed at United Nations Climate Change Conference in Paris in December 2015 (COP21), which aims to limit global temperature increases to 1.5°C above 1990 levels – the irony is that through his 'trade wars' with China and elsewhere he is actually localising the American economy, reducing carbon emissions from imports and exports and creating jobs for the local population. Without wanting to defend Trump, there's clearly a *very fine line* to be trodden between the need to localise our economies and the need to preserve human rights and not veer into nationalism and fascism – a very fine line which was crossed in the tone of much of *The Glasgow Effect* debate.

Fascists use 'divide and rule' tactics to encourage populations to fight amongst themselves in order to deflect attention from the real injustices being perpetrated by those in power. I'm as angry as anyone at the indulgent and exclusive globalised artworld which wastes public money shipping slick artworks (and high-profile artists) around the world, rather than engaging and including those on their doorstep. That was what motivated me to do *The Glasgow Effect* in the first place.[71] And that was why the project had been premised on the exact opposite – investing all my time, education and skills in Glasgow instead. However, we must be cautious that 'artworld bashing' (which I am certainly guilty of) does not just become a distraction from the more serious problems that we face. An indulgent and exclusive globalised artworld is a symptom of a grossly unequal world, and not the cause. As German artist Hito Steyerl writes in her 2017 book, *Duty Free Art: Art in the Age of Planetary Civil War*, 'in relation to other industries the art sector' is 'a blip'. It merely functions as a useful scapegoat:

> Contemporary art is just a hash for all that's opaque, unintelligible, and unfair, for top-down class war and all-out inequality. It's the tip of an iceberg acting as a spear... Predictably, this leads to resentment and outright anger. Art is increasingly labelled as a decadent, rootless, out-of-touch, cosmopolitan urban elite activity. In one sense, this is a perfectly honest a partly pertinent description...
>
> On the other hand, talk of 'rootless cosmopolitans' is clearly reminiscent of both Nazi and Stalinist propaganda, who relished in branding dissenting intellectuals as 'parasites' within 'healthy national bodies'. In both regimes this kind of jargon was used to get rid of minority intelligentsias, formal experiments, and progressive agendas; not to improve access for locals or improve or broaden the appeal of art. The 'anti-elitist' discourse in culture is at present mainly deployed by conservative elites, who hope to deflect attention from their own economic privileges by relaunching stereotypes of 'degenerate art'.[72]

There's a reason why all the right-wing media want to dismiss and ridicule art – it is part of an anti-intellectualisation strategy aimed at keeping people in their place or – in the case of Glasgow – forcing them to escape to somewhere less oppressive. It is because, as artist Bob & Roberta Smith says, 'Art makes people powerful'. Examining and questioning the world around you is the first step towards beginning to change it. It's a creative education which enables us to perceive, interpret and therefore to change our existence for the better. The critical thinking, practical skills, self-motivation and confidence developed through this creative education make people more likely to 'speak truth to power' and to know how to get their voices heard. As Schumacher states in

Small is Beautiful, it is education that is our 'greatest resource'.[73]

> Development does not start with goods; it starts with
> people and their education, organisation, and discipline.
> Without these three, all resources remain latent,
> untapped potential.[74]

It takes education, as well as 'organisation, and discipline' to
enable people to channel their alienation, anger and frustration
in ways that are creative, rather than (self-)destructive. Creating
something takes time and effort and often provides 'delayed
gratification' (joy or 'eudaimonic' well-being) rather than a
quick hit. Whereas destroying something is always the easy
option. Destruction can be perversely enjoyable, providing
'instant gratification' ('hedonic' pleasure),[75] but it is clearly not
good for long-term mental, physical and financial health, nor
very sustainable. Writing in the '30s, in a time not dissimilar to
our own, following the Wall Street Crash of 1929 and the rise of
nationalism and fascism in the decade that followed, the German
philosopher Walter Benjamin said that mankind's 'alienation has
reached such a degree that it can experience its own destruction
as an aesthetic pleasure of the first order'.[76]

In her book, *Artificial Hells: Participatory Art & the Politics of
Spectatorship*, art critic Claire Bishop quotes German philosopher
Boris Groys who provocatively notes that it is only destruction that
is 'truly participatory', something in which everyone can equally
participate:

> ...it is clear that there is an intimate relationship between
> destruction and participative art. When a Futurist action
> destroys art in its traditional form, it also invites all specta-
> tors to participate in this act of destruction, because it does
> not require any specific artistic skills. In this sense fascism
> is more democratic than communism, of course. It is the
> only thing we can all participate in.[77]

Career suicide

The Glasgow Effect provoked an explosion of 'creative destruction' – a term used by economist Joseph Schumpeter to describe the way in which capitalism evolves: creating the new by destroying the old.[78] Artist Hannah Black says 'Contemporary art is only interesting when it's trying to abolish itself',[79] and so *The Glasgow Effect* became anti-art artwork. It was an attempt to kill off the sort of over-budgeted and individualistic art that is celebrated under our present global economic system that results from angry and unethical, 'manipulative and competitive' world obsessed with 'extrinsic' goals,[80] in order to make way for something much more inclusive and optimistic to emerge.

The point is that we must democratise creativity so it that it becomes part of everyone's lives and is not just the preserve of 'professional artists' who are 'specialists' in their field (and therefore clueless about everything else). I probably did study art for too long – six years of higher education to the point where my interests have become esoteric (to the point where I even use the word 'esoteric'!). When I wrote my *Think Global, Act Local!* funding application in 2015, I used quote marks around certain words like 'success' to question their meaning and interpretation in our society and others like 'action research' to parody the jargon used in academic circles. But that nuance was lost in the stampede and I found myself breaking another one of Saul Alinsky's *Rules for Radicals*: 'Never go outside the expertise of your people'. Or in the words of another meme posted up on the Facebook event page:

> If it's inaccessible to the poor it's neither radical nor revolutionary.

But what made *The Glasgow Effect* personally so painful to endure – what I described as 'having to live and channel so many contradictions through one human body for such a long time'[81] – was that as well as being a 'protest against globalisation', it was also, essentially, a protest against myself. It was the person I'd found myself becoming – the 'professional artist' or the 'aloof academic' who'd been pushed into applying for a grant to fund a self-indulgent artwork vs the person I wanted to be – the one I knew could contribute more of my time, education and skills into improving our local community.

Committing 'career suicide' was the only way to effectively 'address the contradictions and compromises' which had been building in my own lifestyle 'as my career as an artist/academic has progressed' – one of the project's key stated aims.[82] The more successful we become, the more temptations there are to live unethical lives. It was committing 'career suicide' that helped me address the 'Pareto Principle' head on – enabling artists from less privileged backgrounds to use the project as a springboard for furthering their own careers. As Schumacher says:

> The true problems of living – in politics, economics,
> education, marriage, etc – are always problems of
> overcoming or reconciling opposites.[83]

Killing off the 'professional artist' or the 'aloof academic' character I'd found myself becoming was what enabled me to (temporarily) 'overcome and reconcile the opposites', between:

> the individual/the collective
> ego/anonymity
> artist/activist.

As I wrote in the newsletter I sent out to my mailing list on 11 March 2016:

> Of course, the irony of all of this public attention is
> that: to do meaningful, successful 'community' work
> of any sort, you must put aside your ego and work in
> a collaborative, low-key way. For that reason, I will be
> keeping a low profile while I get on with the work for the
> rest of the year.

It was this painful process that enabled me to evolve into a new role, where my 'education, organisation and discipline' could be put to better use. The problem with the 'colonists' and the problems with Glasgow School of Art, which Darren McGarvey and Limmy both highlight, relate to the way education has also come to be seen as commodity in our neoliberal world. It has come to be seen as the private property of the person who receives it, a 'passport to privilege' as Schumacher describes.[84] Something you can acquire for your own personal gain and then take with you when you leave on the next step of your career. Schumacher says this illustrates the 'ingrained selfishness on the part of the people who are quite prepared to receive and not prepared to give'.[85] It is the opposite of what Glaswegian trade unionist Jimmy Reid saw as the value of education – an essential tool for creating a 'sense of purpose and meaning in life' and then contributing your knowledge back to your community. It is the opposite of what we need to do to make Glasgow a more equal, sustainable and connected city. As Schumacher says:

> Unless virtually all educated people see themselves as servants of their country – and that means after all as servants of the common people – there cannot possibly be enough leadership and enough communication of know-how to solve this problem of unemployment or unproductive employment...[86]

Where art and politics become one

> Art won't save the world. Go volunteer at a soup kitchen you pretentious fuck.

That was the demand of another meme posted on *The Glasgow Effect* event page by Chris Irish at 1.56am on 5 January 2016. Curiously, not only did I agree with much of the sentiment, but that's also essentially what I was planning to do. In framing the artwork as a year-long 'durational performance', all I needed to do to create it was to stay in Glasgow for one year and not leave. I could have just 'sat about and done nothing for a year' as many people assumed I was, and the artwork as a 'symbolic act of resistance' (slashing my own carbon footprint for transport to zero) would have been the same.[87] That is the problem with most art in its present form: that it can only ever be 'symbolic' – a tool for thinking, enabling insights and provoking debates. But it's clear that what's needed to actually 'save the world' in this time of accelerating global social, environmental and economic crises, is direct political action instead.

In his book, *Tell Them I Said No*, writer Martin Herbert profiles artists who have, through various means and for various reasons, rejected the mainstream artworld. Writing about the German artist Charlotte Posenenske, who in 1968 at a time of heightened global unrest, gave up making art in order to become a social worker, Martin Herbert says her tactic was to:

> First perform the limits of art vis-à-vis social change;
> second do something that really contributes to it.[88]

The Glasgow Effect aimed to do the same. I didn't 'volunteer at a soup kitchen', but I did devote large amounts of my year to activism aimed at investigating, exposing and challenging the root causes of the injustice in our city and beyond, as opposed to just 'firefighting' the symptoms. It was activism that required 'big picture' thinking and a long-term view, which very few people have the luxury of the time and education necessary to do. People often can't 'make Glasgow' (as the slogan goes) as they're either far too busy working/travelling for work or they're far too disempowered to know where to start. And yet, as we'll see in more detail in the next chapter, it is community activism which is part of the solution to 'the Glasgow effect'. In their analysis of Glasgow Centre for Population Health's 2016 report, Common Weal highlight the dual role that activism can play in improving health and the extent to which community participation in Glasgow is far lower than in Liverpool and Manchester:

> Political activism, as a quite intense form of mental
> stimulation, can therefore play a double role in terms
> of public health: it can stave off the damaging effects
> of social isolation and exclusion by establishing deep
> ties and bonds within communities, while at the same

time having a tangible impact on government, and therefore improving public health via more socially just legislation.[89]

This was a connection that Glaswegian activist and anti-poverty campaigner Cathy McCormack was well aware of. Writing about her experience of living in damp and poorly insulated housing in Easterhouse in the '80s and '90s, she said:

When I was suffering from depression I asked my doctor if he could give me a prescription for a warm dry home but all he could offer me was anti-depressants. I realised my doctor was employed to treat the symptoms of our health problems in the same way that the officials were employed to treat the symptoms of our dampness. I refused his kind of medicine and joined my community's fight for justice instead.[90]

Cathy McCormack also understood the power of creativity for improving individual well-being and in fighting for social change. When her community felt shut out of the sanctioned City of Culture activities in Glasgow in 1990, she again took matters into her own hands. With a group of local activists they set up the Easthall Theatre Group and wrote and performed their own play. *Dampbusters* used humour to draw attention to the reality of the living conditions for many of the city's residents, completely glossed over by the 'creative city' narrative. Cathy and her friend Marie Knox dressed up as Aspagilum and Penicillin – the two main moulds that were growing in their houses as a result of damp and which were causing chronic respiratory disease. On the cover of their programme they subverted the official festival slogan, 'There's a lot Glasgowing on', to 'There's a lot of culture growing on' to reflect the type of 'culture' which was more ubiquitous in Easterhouse at that time.[91]

This is the sort of inclusive creative and political activity we need more of in the future if we want to improve well-being in the multiple ways Common Weal identify above. Although Alasdair Gray's idea about why the world needs artists – 'to look up to or to look down upon' – is useful for understanding the role I performed in *The Glasgow Effect*, it is not as useful moving forward. It sets up a hierarchy (either looking up or looking down) and only serves to separate artists off as 'special'. Artists aren't the special ones, they're just the lucky ones. The ones who have discovered the secret – how to create a 'sense of purpose and meaning in life' without needing to resort to 'debt fuelled, wasteful, unsatisfying consumerism'.[92]

The real task for us all is to create the social conditions where 'every human being [actually can be] an artist' as Joseph Beuys proclaimed; where there are no longer any remote 'gatekeepers of culture' who determine what is and what isn't 'art' and who's allowed to take part. We must create the social conditions where the distinctions between artist and non-artist dissolve and where art and politics become one.

Public property

Before I get too sidetracked by building utopia (that's what Part 3 is all about), I need to come back to the reality of living *The Glasgow Effect* – a project which was unfolding in circumstances that were clearly far from ideal. As much as I may have wanted to become the anonymous, altruistic activist, the irony was that once I'd accepted public money to undertake this individualistic action, I became 'public property'. Thousands of people all over Scotland were expecting a return on their 'investment'. They believed it was their right to tell me what I should do, like I was actually now the 'servant of the common people' that Schumacher suggests we need.[93]

There was never any question that I was going to 'sit about and do nothing for a year'. There was just so much I knew I wanted to try to sort out – stuff that had been concerning and frustrating me since I moved to the city seven and a half years before. I felt such a huge weight of responsibility as a result of accepting that money – I had to try to make a difference for the disempowered silent majority being most disregarded by those in power. The UK Government defines 'The Seven Principles of Public Life' as 'selflessness', 'integrity', 'objectivity', 'accountability', 'openness', 'honesty' and 'leadership'.[94] So, in my newfound role as public servant, I did my best to try to live by them. If only our politicians also did, democracy might not be in such a crisis.

Once the 'funding controversy' had died down, my first task was to carefully plan how to spend the £15,000. Over the first three months of the year, I looked back at all my living expenses from previous years in order to work out the bare minimum I would need to survive. £8,400 would work out at £700 per month. Once all my housing costs were covered, this left me with about £80 per week for food and entertainment. My thrifty lifestyle (plus the fact that my travel costs were zero) meant I could maximise the amount of money I could invest in the projects and campaigns themselves (£6,600). I kept meticulous records and receipts for everything so that I could 'submit myself to the scrutiny necessary' to ensure my 'accountability'.[95]

These receipts were eventually handed over to a 'researcher' at Glasgow Caledonian University as part of their FinWell study into the finances of 50 people across the city living on 'low incomes'.[96] FinWell (which stands for Financial Well-being) was a 'research project', which I found out about at a Glasgow Centre for Population Health seminar called 'The hidden financial lives of low-income households' on 30 June 2016. Being part of the study was an interesting experience. As well as offering me an insider's perspective on the industry that is population health (described in Chapter 7), it was also what enabled me to truly understand the intimate connection between mental, physical and financial health in our neoliberal world – crisis in one of these areas will likely cause the other two to quickly deteriorate. And that what's actually most important for well-being is not a high income, but a secure and stable one.[97] There was so much public scrutiny over what I spent the money on, but I always thought it was more interesting to focus on what I *didn't* spend it on. Zero went towards directly supporting the oil industry, zero went to the privatised public transport companies and zero went to Tesco.

I did my best to try to settle into my year in the city, but my life was now under the microscope. I had to try to live by example, to 'be the change I wanted to see in the world' as Mahatma Gandhi (1869–1948), leader of the Indian independence movement against British colonial rule, once wisely advised. Or in the wise words of Glaswegian trade unionist Jimmy Reid, who famously said in his speech to the Upper Clyde Shipbuilders in 1971:

> There will be no hooliganism, there will be no vandalism, there will be no bevvying, because the world is watching...[98]

CHAPTER 7

Lived Reality

Low-carbon lifestyle of the future

AGAINST ALL ODDS, my year in Glasgow in 2016 was an attempt to live what I see as the 'low-carbon lifestyle of the future'. It was an attempt to live in the way I believe we all must – to improve our individual well-being and that of our society, whilst also making dramatic reductions to our carbon emissions (as shown in my Carbon Graph on pages 138–9). We all need to learn to enjoy living within our means. We need to start 'equalising assets and education' across society,[1] so that wealth and resources are more equally shared around the world, and therefore, eventually, there is less need for migration and the 'dislocation' and (self-)destructive behaviours it creates.

The 'low-carbon lifestyle of the future' is one where we are no longer trapped in the destructive cycle of 'living to work, working to earn, earning to consume',[2] and so lack the 'time to live sustainably, to care for each other, and simply enjoy life'.[3] It is a lifestyle which is enabled and encouraged by a city which is sensitively designed, as Patrick Geddes suggested, around a balance of 'place' (geography), 'work' (economy) and 'folk' (anthropology),[4] so that we can fulfil all our 'intrinsic' goals – community, affiliation, self-acceptance, physical health and safety – within a short distance of our homes. In this happy and healthy new world, it's the very fact that we do not travel far from home and have so many friends, family and colleagues nearby that makes these goals easier to achieve. It is a city where education is inclusive and accessible and seen as a life-long process; where we all have the time, the curiosity and the means to find out about our surroundings, about how our city works – who has responsibility and power over what – so that we can locate where the levers of social change lie.

Patrick Geddes proposed that every city should have its

own live 'exhibition', where plans for the city's development would be updated and presented in real time. It would be a welcoming and accessible space where every citizen could go to 'learn about the place they live' and take on the responsibility of studying and critically interrogating it to help improve the situation for all of us. Geddes saw this 'city exhibition' as a 'precondition for parliamentary democracy', arguing that it was impossible to know how to vote, if you do not understand the place where you live and its place in the world.[5] Empowered with this knowledge, we can all become 'active citizens' and not just 'passive consumers'. Instead of mindless consumption, we can invest our time and energy in holding our politicians to account. This 'low-carbon lifestyle of the future' is both what will enable and what will sustain a real participatory democracy in which we can all be involved.

The 'low-carbon lifestyle of the future' is enabled by:

- a city where we all have access to healthy, affordable and sustainable food; where we have the knowledge, skills and the time necessary to be able to cook ourselves and others healthy meals. (Food)
- a city where everywhere we need to go is in walking or cycling distance; where we have the time necessary to travel slowly under our own steam; or there is luxurious, seamless and accessible free local public transport to get us anywhere else in the region we need to go. (Transport)
- a city where the 'costs of living' – our housing, energy and everything else – are minimal, so that money is no longer the motivating force for everything we do. (Housing)
- a city where there are no adverts spreading disinformation and 'false consciousness'; where creative education is inclusive and accessible to all, happening in our communities instead of elite establishments as an ongoing part of all our lives; where we can all acquire the 'organisation, and discipline'[6] necessary to channel our energy into something creative or productive instead

of mindless consumption and other (self-)destructive behaviours. (Education)

- a city where we all have meaningful work, where we can all contribute to delivering the essential goods, services and infrastructure which enable our city to function and so we can all see the fruits of our own labour in front of our very eyes. (Meaningful work)
- a city which is 'fearless',[7] of hope and not hate, not governed by the politics of 'divide and rule', where we are all free to try out new things, express our 'individuality' and speak up and have our voices heard without fear of attack; where 'every human being [actually can be] an artist' and an activist; where these distinctions dissolve and where art and politics become one. (Self-expression)
- a city where all the powers necessary to deliver the essential goods, services and infrastructure our citizens need lie with the local or regional authority and not in the hands of a distant administration (in Edinburgh, London or Brussels). (Local democracy)
- a city where we all have the 'nurturing family' and the 'supportive network of people' nearby which Harry Burns, Scotland's former Chief Medical Officer, says are key elements of well-being. (Love and emotional support)
- a city that is just so pleasant to be in that people actually don't want to leave. (Travelling without moving)

This city is not Glasgow. Despite being confined here, I was determined to try to live this way regardless. And so the aims of *The Glasgow Effect* project became two-fold – to identify and explore the huge barriers preventing us all from adopting this lifestyle and, through my action, working to start to remove as many of them as possible. These barriers are economic (caused by poverty and a broken global economic system fuelled by consumption and obsessed with 'growth'), these barriers are physical (caused by completely insensitive town planning) and, finally, these barriers are social and cultural (fuelled by class prejudice and the (self-)destructive values and behaviours that prevail). These barriers are now so deeply ingrained and intertwined that it will take a long time and some radical ideas to unpick them so we can finally enable everyone in our city to release their 'latent, untapped potential'[8] and find a 'sense of purpose and meaning in life'.[9]

Citizen's Basic Income

Some people during *The Glasgow Effect* debate were able to see the way I was proposing to live as a model for how we should start to transform our economy in the future – to enable everybody to have more autonomy and control over their lives, key things which Harry Burns says are necessary to improve public health.[10] Writing in *The National* on 9 January 2016, musician and activist Pat Kane said:

> What I think Ellie Harrison exposes is a coming crisis that most of us will have to face – in Glasgow, Scotland and across the developed world: What will it mean to do 'good work' in the future?
>
> Ellie Harrison's project – her 'extreme lifestyle experiment' – is an anticipation of what's to come. We will all have to relax our standards on what we regard as a legitimate or respectable 'job', as the new pieces of our socio-economic future settle... Harrison's mix of occupational skills, community activism, education and self-expression – and her enthusiastic interest in how all these elements fit together – is going to be more and more the mainstream experience of 'work' in our societies. We should learn from her, and from the new wave of socially-engaged artists like her...
>
> In a world where certain kinds of employment will not be returning to heal a broken ex-working class, we must try to think unconventionally about the elements of a good, satisfying, purposeful life... A basic or citizen's income is the kind of topic that a country like Scotland –

which can think about its collective direction as easily as breathing – should pursue, from street-level to elite-level.[11]

Pat Kane is specifically talking about what some have referred to as 'the coming automation' – the fact that, as well as the triple crises of increasing social inequalities, catastrophic environmental destruction and frequent financial instability and uncertainty, we're on the cusp of another massive deindustrialisation of our economy, which may well make many of these things worse. Many more of our routine jobs are becoming automated. Forty per cent of jobs in Europe and America are predicted to be lost in the next 20 years.[12] Pat Kane is promoting one possible way of mitigating this crisis, by redistributing some of the wealth generated by the producers and the owners of the technology that is causing such widespread redundancy – the big global tech giants – in the form of a Citizen's Basic Income (or Universal Basic Income as it is sometimes known). This would be a modest regular salary for every single citizen to enable them to live a decent quality of life, especially in the absence of full-time work. The idea is that a regular, reliable income would relieve the continual state of stress caused by poverty. Without the pressure to 'work to earn' and to 'earn to consume', we could instead invest time in the important activities previously ignored by our mainstream economy – that is, the 'core economy' of family, neighbourhood and community, of care and of love, volunteering or creative and political activity. Indeed, poverty is typically experienced in 'households with a preponderance of very old or very young members' who need to be looked after, precisely because the mainstream economy ignores most care work.[13]

Citizen's Basic Income (CBI) is an idea that has been promoted in Scotland since the early '90s by feminist economist Ailsa McKay. She was influenced by Marilyn Waring and sadly died in 2014.[14] The CBI concept builds on the 'Wages for Housework' campaigns during the Women's Liberation movement of the '70s, which demanded pay for all the unseen labour which women did at home – cooking, cleaning, washing, bringing up the kids and more. This was the work which enabled their husbands to go out to work – to participate in the mainstream economy – and then come home and expect to be fed. Although they never won their wages as such, their action made visible this hidden labour and created one of the most important conversations about equality since the Suffragettes in the early 20th century. From a feminist perspective, a Citizen's Basic Income offers to give women financial independence and autonomy, 'reducing unequal power relationships' and offering 'more choice over life decisions'.[15] If your husband is abusive or isn't pulling his weight, with a guaranteed income, it becomes easier to leave.

Pat Kane suggests that the funding I received from Creative Scotland to live in Glasgow for a year could be seen as a pilot for what Citizen's Basic Income could achieve. Indeed the way I chose to distribute the money in regular, reliable monthly payments was very like a CBI. As a single person living alone, I didn't have anyone else (in Glasgow) to look after or care for, which meant all my time and energy could be channelled into 'volunteering' or creative and political activity instead – all powered by the skill of self-motivation developed through the many years of creative education I've been privileged to have. Given that 32 per cent of adults in Glasgow still have no educational qualifications at all (compared with 29 per cent in Liverpool and 23 per cent in Manchester),[16] we would need significant improvements to our education system to ensure that a Citizen's Basic Income enables positive social change. It is also is a very individualistic 'solution' (something I'll come back to in Chapter 11).

Biographic solutions to systemic contradictions

Public health 'specialists' have referred to our present time (from 2000 onwards) as the 'fifth wave' of public health. This is preceded by four previous 'waves' of broad improvements in health dating back to the 1830s, when calls were first made for comprehensive 'sanitation' – sewer networks and clean water, which everyone could access. The 'fifth wave' is categorised by what they call the three 'complex challenges' of 'obesity, inequality and loss of well-being' together with broader ecological problems caused by 'exponential growth in population, money creation and energy usage'.[17] In 2006, the number of obese and overweight people in the world overtook, for the first time, 'the numbers who are malnourished and underweight'.[18] Scotland has the highest levels of obesity in the UK, particularly amongst its middle classes,[19] which is increasing rates of diabetes and cancer and costing NHS Scotland £600 million a year.[20] 62 per cent of men and women in Glasgow are considered 'overweight',[21] as well as 20 per cent of Glasgow's children, 8 per cent of whom are obese.[22] If all my years of dieting and all that 'research' attending Weight Watchers meetings at Maryhill Central Hall has taught me anything, it's that gaining weight is caused by only one thing: consuming more energy than you use. You're either eating too much or not exercising enough (and usually it's a combination of the two).

The chips touched a nerve in Glasgow for two reasons. The first may have been summed up in the quote Murdo Macdonald shared during *The Glasgow Effect* debate (on page 168): they drew attention to an unhealthy reality. But secondly, the chips touched a nerve because they were seen to be *personalising the issue of poor health*. That image (pictured on the back cover) – in all its vivid yellow greasiness – claimed that there are what the critic of consumer culture Zygmunt Bauman has described as, 'biographic solutions to systemic contradictions'.[23] It's worth digesting that quote for a minute (pardon the pun) because it sums up everything that's wrong with our contemporary culture – under the spell of neoliberalism – that is causing a mental health crisis to the extent that 'depression is now the condition most treated by the NHS'.[24]

Those chips conveyed the wrong message. It was the opposite of my *Anti-Capitalist Diet Plan* (described on page 100) – that we have an obesity 'epidemic', not because of the flaws of individuals ('biographic solutions') but because of the flaws in the profit-making systems which we're using to produce and supply our food (and indeed all the other resources which affect our health). The same symptoms would not be recurring all over the world if there was not a systemic cause ('systemic contradictions'). To talk of 'lifestyle choices' at all is to deny that some people have the time, education and resources necessary to make better choices. The structure of our city, society and economic system does not make these choices any easier to make. We can only solve the challenges of the 'fifth wave' of public health by taking a holistic approach which takes into account ecological and 'psychosocial' problems,[25] and by creating a culture where 'healthy choices and behaviours become the norm'[26] and are not just the preserve of the privileged few.

Again writing about her experience of living in damp social housing in Easterhouse in the '80s and '90s, Cathy McCormack was infuriated by the 'Good Hearted Glasgow' government health campaign focused on personal solutions to Scotland's health crisis.[27] Launched by the English TV show host Robert Kilroy Silk and supported by Glasgow District Council, the Scottish Office, Greater Glasgow Health Board and the UK Government, the message was: 'Let's all go jogging, stop smoking and eat brown bread'. She said:

> It dawned on me then that health promotion like everything else had become individualised... either the gap between rich and poor had become much wider than anyone had ever imagined and these people had not a clue about our reality or else their campaigns were a deliberate conspiracy to cover up the real cause of our appalling health record.[28]

The real cause being the poor quality of the housing stock which led to condensation, dampness and outbreaks of mould that caused chronic respiratory disease. Conditions only improved after Cathy and her community in Easthall in Easterhouse led their sustained and successful anti-dampness campaign, which took them to Westminster, the European Commission and the United Nations to raise the money necessary to upgrade the housing and install a ground-breaking solar heating project. In 1990 she made the documentary, *Your Health is Your Wealth* to

> restore realism to the health debate by 'acknowledging that many ordinary people see a healthy way of life as a luxury'.[29]

Fortunately, it has been one of the main aims of the Glasgow Centre for Population Health (GCPH) since it was founded in

2004 (by NHS Greater Glasgow & Clyde, Glasgow City Council and the University of Glasgow with funding from the Scottish Government), to illustrate the extent to which structural problems are the main causes of our city's comparatively poor health.

History, politics and vulnerability

'Economic policies matter for population health'.[30] Those were the words which struck me the most as I sat in my studio reading GCPH's epic 353-page report published in May 2016, when I was five months into my year. *History, Politics & Vulnerability: Explaining Excess Mortality in Scotland & Glasgow* shows that premature deaths in Glasgow are 30 per cent higher than in Liverpool or Manchester, and 20 per cent higher in Scotland compared to England and Wales as a whole.[31] The report synthesises research into 40 possible causes for this 'excess mortality', and purports to have solved the mystery of the so-called 'Glasgow effect'. There is no simple answer – like chips or alcohol or nasty weather (although these are all taken into consideration). Instead they show that poverty and deprivation – the main causes of poor health in any society are 'extremely complex constructs',[32] and argue that despite Glasgow being, in fact, no less 'deprived' than those similar post-industrial cities in northern England, our population has been made more 'vulnerable' to poor health as a result of 'a series of historical processes which have cumulatively impacted on the city' since the Second World War.

Overcrowding

The first of these factors is described as the 'lagged effects of high historical levels of deprivation'. Before the 'slum clearances' began in the 20th century (during the second and third 'waves' of public health), Glasgow had some of the worst overcrowding and most squalid housing conditions in Europe, which contributed to poor mental and physical health. This is something clearly evidenced in Carol Craig's book, *The Tears that Made the Clyde*: 'even up to the 1960s overcrowding was a serious problem'.[33] Glasgow did not catch up with Manchester and Liverpool with comparable household sizes until the 2011 Census.[34] The knock-on effects of poor housing conditions are therefore thought to be one of the things contributing to 'the Glasgow effect'.

Rip it up and start again

It is therefore a cruel irony that it was the 'big solutions' to the health problems caused by slum housing, which have caused huge problems of their own. 'The nature and scale of urban change in the post-war period (1945–80)' is described as the next key factor contributing to 'the Glasgow effect'. During this period so much city centre accommodation was totally flattened. On the recommendation of Robert Bruce, the city's Chief Engineer who produced the infamous 'Bruce Report' in 1945, tens of thousands of people were moved out to new peripheral housing estates – Easterhouse, Pollok, Castlemilk and Drumchapel were the biggest. They were built very quickly to poor standards and with few local amenities. Glasgow had a much greater emphasis on high-rises than Liverpool or Manchester, but the two worst aspects of these redevelopments were, firstly, the fact that existing communities which went back generations – with strong social ties (community and affiliation) – were completely smashed up when people were relocated, causing the 'dismantling of social networks'.[35] And secondly, because most people were moved several miles outside the city centre, they essentially became out of sight and out of mind of those in power. There was, therefore, much less per capita investment in housing repairs and maintenance causing the quality of housing to further deteriorate very quickly, as Cathy McCormack's experience in Easterhouse clearly shows.

Brain drain

Worse still was the deliberate depopulation of the city implemented by the Scottish Office from the late '50s. Glasgow's population plummeted from its peak at over one million people in 1939[36] to less than half of that by the '90s when the city began to rebrand itself with the 'creative city' image promoting the narrative of 'the Glasgow miracle'. This dramatic decline was the result of a short-sighted policy of deliberate urban sprawl, which was the brainchild of a privileged English town planner called Patrick Abercrombie (1879–1957). The 'Abercrombie Plan' of 1946 (known as the Clyde Valley Regional Plan) aimed to spread Glasgow's population out across the Central Belt to newly built 'New Towns'. Some of these were nearby: East Kilbride in South Lanarkshire and Cumbernauld in North Lanarkshire, but some were much further afield: Livingston in West Lothian, Irvine in North Ayrshire and Glenrothes in Fife. From the late '50s, as part of the Scottish Office's wider regional 'modernisation' agenda, the New Towns encouraged 'the selective removal of the city's population on a mass scale'. It was reverse social cleansing, where younger, skilled workers with families were relocated out of the city, 'skimming the cream of Glasgow',[37] and leaving the city with 'a seriously unbalanced population with a very high proportion of the old, the very poor and the almost unemployable'.[38] It was a massive 'brain drain'.

Hypocrisy kills

Glasgow's communities, left totally disorientated and disempowered by all this upheaval, were then less able to unite to fight back against a hypocritical Labour council, which continued to betray their interests in favour of enriching themselves. In the '80s, Glasgow's Labour establishment embraced neoliberalism. They continued to 'prioritise inner-city gentrification and commercial development', which only exacerbated inequalities and the damaging impacts of Margaret Thatcher's policies on Glasgow's 'vulnerable' population. The so-called Labour council was 'guided by the maxim... that "what's good for business is good for Glasgow"'. As GCPH show:

> These were seen as quite 'astonishing' developments in such a 'solidly Labour city', and were soon to lead to the identification of Glasgow as a so-called 'dual city' with 'dual urban policy': on the one hand high budget, high profile retail and property development in the city

centre led by (what has been referred to as) a 'growth coalition' in which the city council and the SDA [Scottish Development Agency] played lead roles; but on the other hand much lower resourced and very limited mitigation and management of poverty, and an intensifying social crisis in the city's poorer areas, principally in the peripheral estates.[39]

History, Politics & Vulnerability makes the stark contrast with the response of Liverpool's Labour council at a similar time. Upholding their socialist principles, the city united and fought back – refusing to put many of the Thatcher government's policies into action. This 'fostered widespread participation and politicisation',[40] and created a real sense of solidarity across Liverpool's population. As a result, Liverpool now has higher levels of 'trust, reciprocity, volunteering',[41] which has helped to build the social networks that the New Economics Foundation show are the 'very immune system of society'[42] – that all important 'core economy', which makes all of us less 'vulnerable' to poor health.

The World Health Organisation's study *Closing the Gap in a Generation: Health Equity through Action on the Social Determinants of Health*, which first kicked off discussion about 'the Glasgow effect' in 2008, said the 'toxic combination of bad policies, economics, and politics is, in large measure responsible for the fact that a majority of people in the world do not enjoy the good health that is biologically possible'.[43] They used Glasgow as a prime example of this 'toxicity' – illustrating the stark inequalities between the average life expectancy of a boy born in Calton (54 years) and the 82 years a boy born just 12km away in Lenzie in East Dunbartonshire could expect to live.[44] 'Social injustice is killing people on a grand scale' said Michael Marmot, the study's lead author. It is clueless hypocrites in power who are causing these problems and, unfortunately, in Glasgow's Labour establishment these characters were still rife.

On 24 May 2016, I had the pleasure of meeting another Labour councillor. He had been a former member of the Strathclyde Partnership for Transport board, so I naively assumed he might be a good person to discuss our terrible public transport with, and my ideas for making it better. I quickly discovered that (like most other people in positions of power in this city) he was not a man who actually used or relied on public transport himself. When I asked him if he had a car, he took the question as an insult: 'Of course I've got a car!' A BMW in fact. Later in our conversation he bragged about all the flats he owned in Bridgeton and how much their prices had increased since all that public money had been spent regenerating Bridgeton Cross (including the Olympia Building where Glasgow Centre for Population Health is now based) and moving Glasgow Women's Library into the area.

And it wasn't just these 'wee hard men' (as historian Christopher Harvie describes them)[45] that were to blame. There was also the woman who was former Executive Member for Transport, Environment & Sustainability. She completely ignored all three of the emails I sent her in 2016 to try to arrange a meeting to discuss my ideas for making Glasgow a 'more equal, sustainable and connected city'. I did not see or hear from her at all until 2018, when I was cycling through Dennistoun and she pulled out of a side road in her little red car and nearly knocked me over. 'It's that fucking Labour councillor!' I exclaimed when I spotted her face through the windscreen. But by then she'd whizzed off down the road leaving me in a cloud of fumes.

Other causes of 'the Glasgow effect'

Glasgow Centre for Population Health conclude their report with a few of the other factors which may have made Glasgow's population more 'vulnerable' to poor health. These include, interestingly, *less ethnic diversity* in Glasgow than in Liverpool and Manchester and the 'healthy migrant effect' that results from healthier peoples moving in.[46] Though this 'effect' may improve the overall average life expectancy of the city, it actually only exacerbates inequalities between the migrants and those 'from here'. They also cite *a more negative physical environment*, where 60 per cent of Glasgow's population live within 500m of vacant and derelict land – no doubt a result of such massive depopulation and deindustrialisation which has left many large wastelands.[47] The lower levels of *educational attainment* mentioned previously also contribute to poor health. And finally, they cite the *inadequate measurement of poverty and deprivation* as a possible reason why Glasgow's health record appears so bad (something I'll come back to in Chapter 10) – admitting their research uses metrics, which fail to capture 'the complex, multi-dimensional, "lived reality" of deprivation and poverty' in Glasgow.[48]

Diseases of despair

Although Glasgow Centre for Population Health show that 'excess mortality' is observed among all social classes in Scotland, it is in the poorest neighbourhoods where it is most linked to high rates of death as a result of 'psychosocial' problems and the so-called 'diseases of despair': alcohol, drugs, suicide and violence.[49] They write:

> Thus, the effects of poverty and deprivation, other negative impacts of deindustrialisation (eg de-skilling, role redefinition), the psychosocial impacts of marginalisation and social exclusion – all factors which are common to many populations in a Britain that has been characterised by significantly widening inequality over the past 35 years – have been made worse in Glasgow (and Scotland) by existing vulnerabilities (brought about by a series of historical factors), feelings of powerlessness, and other 'modifying' factors [described above]. This has led to relatively greater stress, worse mental and physical health, compensated for – in some cases – by greater reliance on alcohol and drugs related 'coping mechanisms', resulting in yet worse health outcomes.[50]

Glasgow's massive post-war redevelopments clobbered our city's working class twice. In terms of the theory of 'dislocation', people were now less able to fulfil either of their inherent 'psychosocial' needs; neither the ability to be seen as individuals, nor the ability to identify with others and develop a sense of 'belonging' – something which had previously been so strong. Working class solidarity was smashed to smithereens and amongst the individualism promoted by neoliberal policy of the '80s, 'pain was privatised'.[51]

The structural transformations of the city have created what Schumacher describes in *Small is Beautiful* as 'metaphysical disease', in that it is poor mental health more than anything, which is driving our 'excess mortality'. Schumacher makes clear that if the disease is metaphysical, 'the cure must therefore be metaphysical'. The cure is the 'education, organisation, and discipline' which he prescribes. But this must be a creative education that is not focused on learning facts, but instead on understanding ourselves, understanding the world around us and our place within it and helping us to develop our ethical codes:

> Education which fails to clarify our central convictions
> is mere training or indulgence. For it is our central
> convictions that are in disorder, and, as long as the
> present anti-metaphysical temper persists, the disorder
> will grow worse.[52]

The elephant in the room

There was one glaring omission from Glasgow Centre for Population Health's report, which I noticed on my very first visit to look round the art school on 17 January 2008. Why was there a massive six-lane motorway going right through the city centre? That was just so weird. What a hostile environment – the noise and air pollution – it has created for everyone who isn't in a car. Building the M8 motorway was another key recommendation of the infamous 'Bruce Report' of 1945 that was taken forward (fortunately not everything Bruce recommended was realised as he wanted to demolish and rebuild the entire city!).[53] The motorway has severed the city in two, disconnecting so many communities in the north: Possil, Milton, Sighthill, Springburn, Royston and many more. And yet nowhere on the 353 pages of *History, Politics & Vulnerability* are the words 'M8' or 'motorway' even mentioned.

It was only during my 'extreme lifestyle experiment' in 2016 – where I refused to leave Glasgow's city limits, or use any vehicles except my bike – that I truly began to appreciate the significance of Glasgow's car-centric infrastructure for all of our mental, physical and financial health. I made two key discoveries. Firstly, that despite the fact that Glasgow gives far more of its space over to roads – 25 per cent of the city compared to just 12 per cent in Edinburgh[54] – it has 'one of the lowest levels of car ownership in Britain'.[55] There is, in fact, a silent majority of people in this city who don't have access to cars, which includes many of our most vulnerable populations: children and elderly and disabled people. And despite the assumptions of *The Glasgow Effect* meme-makers that I must own a sports car or have gone out and bought one straight away with my £15k, I was, and have always been, one of them. Glasgow's roads, built in the post-war period by those arrogant and short-sighted town planners, only served to privatise 'mobility' – to entrench inequalities and create division. It is Glasgow's motorways that have created the 'two different worlds' which Darren McGarvey describes,[56] where more than half of the population are made to feel excluded. When I began to crystallise these ideas in November 2016, I wrote:

> Glasgow is the city with the lowest car ownership in
> Scotland (49 per cent of households compared to 86
> per cent in Aberdeen), yet its cityscape is completely
> dominated by the sight, noise and smell of motorways.
> This car-centric infrastructure has created a divided city
> of 'haves' and 'have nots' – those who own cars and
> can glide over the epic flyovers and experience their
> spectacular views and those who have to negotiate the
> underworld of underpasses and endure the noise and air
> pollution which filters down from above.

My second key discovery came from the 'lived reality' of only walking and using my bike. My whole perception of distance shifted. I could now see quite starkly how isolated I was – why did my friends all live in other parts of the city and not next door? The ridiculousness of my solitary existence was made all the more apparent. If I wanted to visit a friend in the East End or in the Southside, I had to factor that distance into my day. I couldn't really stay late at people's houses, or drink too much as I knew I had to be awake enough to cycle home afterwards. It was particularly bleak in the winter.

But when I began to analyse the data being collected on my Trackimo GPS tracking device, I could see quite clearly the places I was frequenting and the sort of distances I was cycling as I went about my weekly routine. I was mainly just travelling about the West End, the city centre, Dennistoun and parts of the Southside (see my Annual Heatmap in Chapter 8). It was then that I realised the extent to which all of the big peripheral housing estates – Easterhouse, Pollok, Castlemilk and Drumchapel – were well beyond my reach. They were all more than 5km out of the city – the distance Cycling UK (the national charity supporting cyclists and promoting bicycle use) say is a reasonable distance people could be expected to travel by bike in a daily commute.

Suddenly I understood why cycling in Glasgow was seen as an 'irritating trope of middle class life'.[57] Of course 'posh people and hipsters' living in the West End, Dennistoun and swanky parts of the Southside were more likely to cycle into town, because it just isn't so fucking far! Meanwhile people living in the big peripheral housing estates have been left stranded at the mercy of private bus companies. The post-war redevelopments were the opposite of what we need to do to build a healthy and sustainable city – that is to have a densely populated city centre where everyone can easily walk or cycle to where they need to go,[58] and where everyone lives together rather than being segregated by social class. Glasgow's inner-city tenements should never have been flattened and replaced by roads, they should have refurbished or rebuilt where they were.

In 1960, before the worst of the damage was done, architectural critic Ian Nairn described Glasgow as 'the most friendly of Britain's big cities, and probably the most dignified and cohesive as well'.[59] That certainly had never been my experience, even before the 'chips hit the fan'! But it was during *The Glasgow Effect* that I began to realise the extent of our city's spatial segregation. Glasgow's plan for repopulation – developing its 'knowledge economy', celebrating its 'creative industries' and promoting the narrative of 'the Glasgow miracle' is only exacerbating these inequalities. All it is doing is luring more affluent, and therefore more 'mobile' middle class people to capitalise on new opportunities the city has to offer. We have a growing population of 'professionals' (artists and others) operating in an inner sphere, meanwhile the majority of the city's residents live in places far off the beaten arts track. It was no wonder I didn't discover 'the Glasgow effect' even existed until I'd been living here for five years. Big arts festivals like Glasgow International are more concerned with having an esoteric conversation with a global artworld elite rather than including, communicating with or representing people in our city's peripheral areas.

In her book, *Estates*, Lynsey Hanley maintains that 'part of the problem with estates was their estate-ness' – the fact that they were built for a specific group, often on the periphery, rather than evolving organically.[60] This is something Cathy McCormack came to describe quite starkly as an 'economic apartheid':

> this isn't just about having no job or less money to spend;
> it's about a kind of social and economic apartheid in
> this city... [Journalist Alf Young describes this as being]
> about people who feel physically and spiritually shut
> off from life in the city mainstream, hidden away from
> the Glasgow of new jobs, designer shopping, waterfront
> apartments, festivals, boutique hotels and café culture.[61]

When Cathy visited South Africa herself in the late '90s she realised that Glasgow's apartheid could in some respects be seen as worse than the racial apartheid which Nelson Mandela had helped to overthrow. She claimed this was because it was *invisible* and committed by whites on whites. Unlike in South Africa where black people at least felt solidarity with one another and so could unite to fight back, in Easterhouse 'the abuse was camouflaged and the pain privatised'.[62] Chronic social isolation and fear meant that people had to face their problems alone. To add insult to injury, when Easterhouse finally got its own shopping centre in 1971 (having to make do with barely any local amenities for nearly 20 years before then)[63] it was named the 'Township Centre' – referencing the term used in South Africa to 'refer to the often underdeveloped segregated urban areas' where black people were forced to live. She came to describe the oppression of the poor in Glasgow as a 'War Without Bullets', an *invisible* war 'waged... with briefcases instead of guns' where people in places like Easterhouse 'have become the human fag ends of international capitalism'.[64]

This 'economic apartheid' extends to and is facilitated by Glasgow's terrible transport system. 'A developed country is not a place where the poor have cars. It's where the rich use public transportation'. Those are the wise words of the former mayor of Bogotá in Colombia, Gustavo Petro. His point is that we better serve the needs of the poor and create a more equal and better integrated society if we provide a luxurious, efficient and affordable public transport network, which everyone wants to use, so that it is no longer seen as a 'hardship' to have to take the bus. It's the opposite of what Margaret Thatcher had in mind when she supposedly said anyone still travelling on a bus at the age of 26 should count themselves a 'failure', but then, as we know, all her economic policy was totally geared towards creating 'haves' and 'have nots' as a way of materialistically measuring 'success'. In Edinburgh, publicly-owned Lothian Buses offer regular and reliable services which currently cost £1.70 for a single (compared to the £2.50 on First Glasgow) and people from all walks of life use the buses on their daily commute. As someone wrote on one of our campaign Facebook pages recently: 'I saw Lord Steel on a bus in Edinburgh once, that would never happen in Glasgow!'

If people in the peripheral areas are disconnected from the social, economic and cultural life of the city, then rather than smashing everything up and moving them all closer again, we must reconnect them with some radical new transport links. In Bogotá they have just built a new cable car linking the impoverished neighbourhood of Ciudad Bolivar to the city's main public transport network, in order to 'increase opportunities and reduce crime'.[65] That's how you counteract the 'Pareto Principle' – by investing the most in the communities most in need. If only we had such forward-thinking and less self-serving leadership, Glasgow could now have a Subway extension to Castlemilk, the East End and many other places and the Strathclyde Tram joining Easterhouse and Drumchapel, which have all been proposed over the years.[66] By failing to address the dire state our city's public transport network, inequalities have been entrenched since our motorways were built and our buses were deregulated in 1986.

Remunicipalisation!

Another important, and perhaps inevitable, outcome of my 'extreme lifestyle experiment' was the localisation of my activism. *The Glasgow Effect* was all about 'mobility', or lack thereof, so it made sense for transport to become the core focus.[67] As well as its role in entrenching inequalities in our city, transport is also 'the elephant in the room' in terms of carbon emissions. Transport is the biggest contributor to carbon and air pollution of all sectors in Scotland's economy,[68] and the only one which has increased its emissions since our 'world leading' Climate Change (Scotland) Act 2009. It's the one which, until now, has proved too inconvenient for our car-loving politicians to properly address. But another profound moment of realisation came in August 2016, when I read the report, *Taken for a Ride: How UK Public Transport Subsidies Entrench Inequality*, written by The Equality Trust – the progressive think tank founded by Richard Wilkinson and Kate Pickett in 2015, following the stark findings of their book, *The Spirit Level*. The report says:

> Public transport is a significant and escalating cost
> for many people. But while transport may be a drain
> on the finances of some, for others the cost is more
> debilitating... This matters, as it means the poorest in
> society are unable to travel as far or as often, limiting
> their ability to compete with the better off for jobs
> and decent pay... Our transport system is a driver
> of inequality, and societies which have high levels of
> economic inequality have worse health, more crime, less
> social mobility and lower levels of trust.[69]

Putting cars aside for a moment, the *Taken for a Ride* report shows that even within our public transport network, public money is disproportionately being used to subsidise the elements which are predominantly used by the most privileged people (ie the national rail network or Glasgow's Subway) rather than our bus network, which far more people actually rely on (bus trips make up three quarters of all public transport journeys in Scotland).[70] What struck home to me then was the extent to which those travelling longer distances on our national rail network were generally the wealthier people. They were people, like me, who were receiving the larger pay packets that seem to make longer commutes feel 'worthwhile', despite the fact that they cause stress and erode well-being.[71] This inequitable investment is being perpetuated by the fact that privileged people have more power and 'louder voices' and are therefore able to fight for and to win greater investment for the infrastructure that they use. It's no wonder public transport is so much better in Edinburgh and London (where it is nearly all still publicly-owned and/or publicly-controlled), because that's where all the politicians live and work! And worse still, because the public transport seems fine in those cities, the people in power never realise quite how bad it is elsewhere or see any urgency in acting to improve it. That's the 'Pareto Principle' in action once again.

I suddenly realised that for the last seven years working on the Bring Back British Rail campaign, I had been one of these privileged people exacerbating this situation. Now, finally, I had the chance to turn my attention, my power and my privilege, to sorting out the failing privatised bus network much closer to home. My whole perspective on renationalisation started to shift. Although the public ownership of the 'means of production' – the essentials we all need to live happily and well (energy, water, transport, housing, health and education) – had helped to provide a better quality of life for many people in the post-war period, nationalisation also stripped power away from local authorities making our economy and political systems far more centralised and less accessible to citizens outside London. Nationalisation had been disempowering. What we desperately need is a return to local ownership and control. 'Before

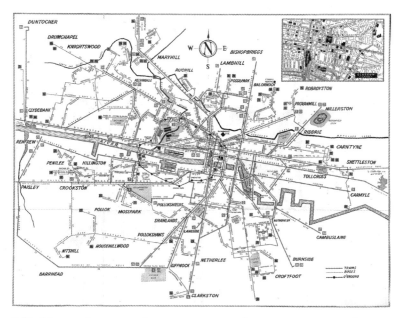

1940s Glasgow Corporation Transport Route Map showing Trams, Buses and Subway. (Glasgow City Council, formerly Glasgow Corporation)

nationalisation, there was municipalisation' writes Tom Crewe in his essay on the rise and fall of local government in Britain. The late 19th century saw the development of all-powerful municipal authorities such as the Glasgow Corporation, responsible for serving the needs of their citizens directly:

> Councils of differing political complexions in every part of the country bought out gas, water, electricity and tramway companies, on practical rather than ide-ological grounds. It made sense for a local authority to deliver essential services to its residents, safely, fairly, accountably and at reasonable prices, reinvesting the considerable income in further improvements. By 1901, one reformer looked 'forward to the time when joy will be considered as much a necessity in a city as anything else. In that time the citizens, well convinced

that all the prime necessities of life must be municipal-
ised, will not fail to demand that this great necessity
of joy, or... the means of joy, be also supplied by their
local councils'.[72]

Not something many people would associate with Glasgow City
Council now!

Municipalism was a system which gave cities freedom and
autonomy from the nation-states in which they were located.[73]
The Glasgow Corporation had all the powers necessary to raise
revenue and deliver essential goods, services and infrastructure to
our residents directly. As a result, at the end of 19th century, when
Glasgow's Subway (completed in 1896) and Tramway network
were first opened, our city's public transport network was seen as
one of the best in the world – 100 miles of tram lines (pictured
opposite), with 1,000 pollution-free electric trams serving all parts
of the city. But in the post-war period, the short-sighted and self-
serving decisions of those in power saw that nearly all of it was
destroyed. One of the saddest things I learnt about in 2016 was
in Glasgow's Transport Museum, where they tell the story of the
crowds that congregated in the city centre to watch the procession
of the 'last tram' on 4 September 1962 before they were all ripped
up. Two hundred and fifty thousand people (nearly a quarter of the
city's population) came out to watch that last tram rolling through
Glasgow. How were they meant to get to work the next day
I thought?

It was this same misguided thinking which pushed Richard
Beeching under instruction from Tory Transport Secretary Ernest
Marples to recommend closing half of Britain's 4,000-odd stations
and 6,000 miles of our railways (one third of our network) in
1963.[74] The 'Beeching Cuts' coupled with the 'Bruce Report'
saw important stations lost in Glasgow at Buchanan Street and
St Enoch (both of which neatly integrated with the Subway) and
at Glasgow Cross – still there in ghost form behind the Mercat
building (waiting to be brought back to life as part of Glasgow's
long-awaited Crossrail scheme). In removing one third of our
network, they were the ones that made our railway network into
something which only served the privileged few, a 'rich man's toy'

as it has recently been called.[75] In his article about the 'Beeching Cuts' journalist Tony Gosling highlights the extent to which the wealthy oil lobby and the road and car lobby were responsible for these devastating decisions. At a time when North Sea oil had just been discovered:

> The agenda was threefold: shift travel away from unionised public transport, increase oil consumption – and therefore fuel and vehicle tax revenue – and finally to shift power, forever, into the hands of giant private oil companies like Exxon, Shell and BP... This was a brave new fossil fuel led world where a dependency on energy would drive economic growth like never before. Governments and people alike would have to get used to the car. Public transport was way too fuel-efficient.[76]

Whereas it used to be possible to get to most places in the UK on public transport, it was these drastic decisions the '60s which privatised 'mobility'. Those who could afford to own cars were fine, everyone else was left stranded. But the saddest thing I discovered in 2016 was the story of what happened to Strathclyde's Buses. Margaret Thatcher's plan to deregulate the buses in Scotland, Wales and England (not Northern Ireland) was particularly vindictive because London remained exempt. Under her deregulated model, which came into effect on 26 October 1986 (a date known as 'D-Day'), anyone could start a bus company and operate wherever they thought they could make most money. The network was no longer planned to serve the people that needed access to public transport, but a total free-for-all instead. Writer Kenny Barclay says of those first few years:

> Competition for passengers was fierce with existing operators suddenly facing new rival operators; congestion and bitter battles took place across the country. In order to survive companies had to work hard to win new passengers as well as keep their existing passengers. New liveries, marketing campaigns and new vehicles both big and small arrived.[77]

At the time of deregulation, the majority of the routes in Strathclyde which had been publicly-owned and operated by the Strathclyde Passenger Transport Executive (an ancestor of SPT, 1975–96) were transferred to an arms-length public company called Strathclyde's Buses.[78] Although it was still wholly owned by the council, it was no longer permitted to receive any public subsidy and now faced this 'fierce competition'. Similar arms-length public bus companies were set up to serve other cities in Scotland: Lothian Buses in Edinburgh, Tayside Bus Company in Dundee and Grampian in Aberdeen (which later became FirstGroup). Only one of these, Lothian, still remains publicly-owned.

The story in Glasgow is a tragic one. Strathclyde's Buses merged with the smaller Kelvin Central Bus Company and the council sold its holdings to the management for £30 million in a 'workers' takeover' in 1993. Unfortunately, these workers no longer saw the strategic long-term significance of the public asset they were now guardians of and were more interested in making a quick buck. After a failed takeover attempt by Stagecoach in 1994 (stopped by the Competitions & Markets Authority), FirstGroup bought the company in 1996 – the £110 million sum they offered divided up between all 2,000 employees who owned shares. A newspaper article at the time said:

> 'I'm obviously delighted for our members', said Des Divers, a convenor for the Transport & General Workers' Union in Scotland and himself one of the lucky employee shareholders… '£35,000 less tax isn't going to last forever, but I dare say many of the staff will be taking a well-earned holiday'.[79]

Our futures sold off for a few holidays! The fact that a union representative (the T&G is now Unite) could be celebrating this terrible event shows the complete lack of understanding at that time of the long-term consequences of privatisation, and the way in which neoliberal policies had created a selfish culture of 'carpetbagging' focused on individual short-term gain. The decision was no doubt driven by the history of 'poverty and insecurity of employment in Glasgow', which Carol Craig suggests 'may have predisposed its citizens to pay particular attention to materialistic goals'.[80]

Whatever the case, we're all living with the consequences now. The balance of power over our public transport now lies firmly with the private bus companies. When the Strathclyde Passenger Transport Executive attempted to fight back by proposing to start rebuilding part of our tram network with the new Strathclyde Tram linking Drumchapel to Easterhouse, it was the private bus companies that blocked it. They saw it as a threat to their profits and challenged the proposals at a Westminster commission in 1996, who eventually ruled against the scheme, which would have been open by the year 2000.[81] FirstGroup now have a near monopoly on all Glasgow's bus routes, meanwhile Lothian Buses have gone from strength to strength in Edinburgh. The facts say it all. It's time to take back control of our buses and remunicipalise all our public transport now!

Reflection and action

Why do these people make such terrible decisions? Why was Patrick Geddes such a successful 'community sensitive' town planner,[82] whilst Abercrombie and Bruce both proposed ripping the city apart? As well as the blatant vested interests of people like Richard Beeching and Ernest Marples (who actually owned the road building company Marples Ridgeway), Des Divers and the other bus drivers, and the billionaire bosses of First and Stagecoach, it was because none of them followed the simple mantra 'Think Global, Act Local'.[83]

Geddes was what was known as a 'generalist'. He refused to specialise in any one thing, but instead developed knowledge of many different disciplines, working as 'an ecologist, a botanist, a theorist of cities, an advocate of the arts, a community activist, a publisher, a town planner and an educator'.[84] This was what enabled him to see how our society, economy and environment are so intimately interconnected. Geddes' definition of the 'arts', for example, included within it: art, architecture, social reform, transport, accounting, insurance, banking, stock raising, labour and more. His work was based on the idea that 'By Creating We Think',[85] and 'By

Living We Learn'.[86] He described his own work in the planning of cities and societies as 'applied sociology', in other words, social thinking applied by activists who used any or all of these 'arts' to build social groups.[87]

The main purpose of the city for Geddes was to help build and nurture these social groups, not to rip them apart. 'Groups of people *in the first place* and buildings after' he noted to stress the hierarchy a sensitive planner should use to help us fulfil our 'intrinsic' needs for community and affiliation. It's the planning 'specialists' like Abercrombie and Bruce that lacked this overview – the ability to predict the long-term consequences of their arrogant ideas. They fell into the modernist 'progress trap' which Schumacher warned against – the naïve assumption that: 'What is good for the rich must also be good for the poor',[88] without considering that maybe it was the rich that had got it wrong. They relied on technology (in this case the private motor car) to clean up the mess their ill-thought through plans had made. That's why the 'whack-a-mole' phenomenon has been so evident in Glasgow – you bash down one problem (overcrowding) and it resurfaces somewhere elsewhere with a vengeance (total alienation). That's what Abercrombie and Bruce did, that's what Beeching and Marples did, and that's why we're still dealing with the consequences now.

With all this evidence showing how structural problems way beyond our control are the cause of our 'health behaviours', it begs the question: is there any point in individual action at all? This is the big question at the heart of Darren McGarvey's book *Poverty Safari*: 'Can a left-wing structural critique be married to an ethics of personal responsibility?'[89] And it's also the central question of *The Glasgow Effect*. I think it's not a question of 'can', it's a question of 'it must'. It's not good enough to sit back and critique the world, 'the point', as Marx said, 'is to change it'.[90] You have to move from thinking into action, from seeing that 'the system is fucked', to doing something to 'unfuck it' (as Robin McAlpine from Common Weal says).[91] This necessitates that you take on different roles. You have to be the artist and the activist, the thinker and the doer, to reflect and then to act. That's what the 'Think Global, Act Local' mantra is all about.

During my year in Glasgow, I came to see the process in three parts: education, action and reflection. Firstly, I was looking, listening, absorbing, reading and thinking about everything that had happened, what I described as 'researching and learning about the world'. Then I was 'taking action to address the problems that I could see', then I was reflecting and adapting my behaviour and actions in response, then starting all over again at the beginning.[92] That, I discovered, is what 'action research' is all about.

Another of Cathy McCormack's heroes is Paulo Freire – the Brazilian thinker, educator and activist who founded the field of 'popular education' in the '60s. Popular education is education that happens not within the confines of elitist establishments, but out in communities as a tool for empowerment. His aim was to enable oppressed peoples (starting with the 'peasants' of Latin America) to analyse and understand the causes of their oppression, so that they could then come together to fight for their liberation. In the '90s and 2000s, Cathy worked on many popular education initiatives in Easterhouse and around Glasgow aimed at this 'consciousness raising'. By allowing people the time and space to 'communicate their feelings'[93] and 'think for themselves',[94] they enabled 'people... to explore the root causes of poverty'.[95] She writes:

> Our aim was to promote an education that was based on the clear analysis of the nature of inequality, exploitation and oppression and committed to progressive social and political change.[96]

When I found Paulo Freire's book *Pedagogy of the Oppressed* on my mum's bookshelf in December 2017 and read it myself, I was struck by his concept of 'praxis'. Freire defines 'praxis' as reflection plus action: 'reflection and action upon the world in order to transform it'.[97] And moreover, that it is this reflection upon the world which compels a person into action. He writes that

> discovery cannot be purely intellectual but must involve action; nor can it be limited to mere activism, but must include serious reflection: only then will it be a praxis.[98]

I discovered that Paulo Freire was articulating the working practice I'd been trying to develop since I moved to Glasgow in 2008, which had come to fruition during *The Glasgow Effect*. Praxis is a virtuous circle in that any action taken must then 'become the object of critical reflection'.[99] It's only on these terms that I can justify the self-indulgent activity of writing a book, if it helps me to analyse and make sense of all that has happened so that I know what action is most necessary next. The reason why so little changes in this neoliberal world is because people are 'specialised', 'professionalised', segregated and pigeon-holed and so are prevented from living the revolutionary praxis of thinking and acting.

We need to stop 'researching' and start fighting!

The Glasgow Centre for Population Health report makes policy recommendations to improve the city's health, that include:

> Addressing the costs of living: reducing costs which impact most on the poorest groups (including childcare, housing, heating, transport and food) relative to income is an important component in a strategy to reduce poverty and inequality.[100]

And so, when we were about to launch our local public transport campaign in autumn 2016 – calling for Glasgow's profiteering bus companies to be brought back into public ownership so that our transport network could be properly regulated and planned to serve everyone in our city – I wrote to GCPH to ask them to get involved. On 23 November 2016 at 3.46pm they replied to tell me, 'While... we would agree on much of your campaign aims, as an organisation we do not normally sign up or endorse specific campaigns'. Not one to take no for an answer, I replied to them on 27 November 2016 at 9.34pm to say:

> Thanks for getting back to me. Given that wealth and health inequalities are continuing to worsen, GCPH might consider changing tactics slightly to get more involved in campaigning? I suggest you look at the New Economics

Foundation's 'rebrand' which followed Brexit, reflecting the state of urgency we find ourselves in... I was at Glasgow's State of the City Economy conference on Friday and it appears the council's economic strategy for the next five years relies on increasing tourism from 2million to 3million tourists a year. A hell of a lot more carbon and air pollution and I'm not sure that's going to do anything to reduce inequalities, do you?

By this time in 2016, New Economics Foundation – which had already begun to describe itself as a think tank 'for ideas and new solutions that thinks – and then acts' and had launched NEON (New Economy Organisers Network) as its organising arm – was stepping up action at all levels. They aimed to provide positive solutions to all those who expressed their anger and sense of powerlessness through the Brexit vote. This was symbolised in the change of their motto from 'economics as if people and the planet mattered' (inspired by EF Schumacher) to 'building a new economy where people really take control'. NEF were actually offering real ways of creating a more accessible and human-scaled economy, and then campaigning to make them happen. On 28 November 2016 at 12.27pm, GCPH replied to say 'as a publicly funded body' it is 'not appropriate for us to be involved in campaigning'.

It's the same 'whoever pays the piper calls the tune' problem that both Darren McGarvey and I had identified. To reiterate the point he made during *The Glasgow Effect: A Discussion* event, 'It's the fucking government. It's the structures of power. And everywhere that they are. At local level, national level and international level. And you simply cannot [challenge them], when your whole thing is funded from the government'.[101] The reason why we have such awful public transport is because the existing campaign groups and organisations meant to be advocating for something better all have vested interests. Transform Scotland and its 'sister organisation', the Campaign for Better Transport in England, are both funded by the private bus and train companies and so won't publicly support public ownership. The completely inaccurately named 'Bus Users Scotland' is actually run by industry insiders. Transport Focus claims to be the 'transport user voice' but both these organisations are funded – yes, you've guessed it – by 'the fucking government'! This is why we need autonomous individuals to unite to fight back. In the book, *The Revolution Will Not Be Funded: Beyond the Non-Profit Industrial Complex*, the Incite! women's collective in America make it clear that only free labour is the true 'labour of liberation'.[102] It's only through volunteering our time – working off our 'own back' – that we can remain uncompromised by hidden agendas – something which could be enabled by *actually* reducing the 'costs of living'. The point is that we now know all too well what's causing inequalities in wealth and health in Glasgow. As Michael Marmot made clear back in 2008:

Social injustice is killing people on a grand scale.[103]

We need to stop 'researching' and start fighting! The very fact that GCPH won't get involved in campaigning for the actual solutions that their research advocates shows how our funding systems have been set up to prevent people from completing the virtuous circle of reflection *and* action, which is the path to meaningful social change. The American writer and activist Alice Walker famously said 'activism is my rent for living on the planet'. She is stressing that not only do we all have a duty to get involved in fighting for social, environmental and economic justice, but if we are denied the opportunity to do so then we cannot be whole human beings. This wholeness is the difference between being a 'specialist' or a 'professional' who Schumacher would say produces 'cleverness' and being a 'generalist' like Geddes who produces 'wisdom'. These specialists are always doomed to

try and cure a disease by intensifying its causes. The disease having been caused by allowing cleverness to displace wisdom, *no amount of clever research is likely to produce a cure.*[104]

Wisdom on the other hand can only be found 'inside oneself', by living our values and reaffirming our ethical codes by allowing time for a continual cycle of education, action and reflection:

To be able to find [wisdom], one has first to liberate oneself from such masters as greed and envy. The stillness following liberation – even if only momentary – produces the insights of wisdom which are obtainable in no other way.[105]

We have allowed 'cleverness to displace wisdom' and the way our 'research' is funded and carried out simply perpetuates the status quo. It's what Cathy McCormack and Darren McGarvey both refer to as the 'poverty industry' – where you have whole swathes of 'specialists' studying poverty for a full-time job. Cathy writes:

The trouble is that my 'field' – poverty – has become an industry. And when people work in the industry, they want to get to the top. They become competitive. They become protective. Can you imagine there are people out there who actually want to preserve what I've spent decades trying to eradicate. They've got jobs dealing with different aspects of poverty – and if they fix it, they're out of a job. You aren't going to win in the popularity stakes if you're trying to make a difference in people's lives.[106]

One of the earliest critiques of this 'poverty industry' was offered by British filmmaker Ken Loach in his 1969 documentary *Save the Children*. He 'bit the hand that fed him' so hard that the film was banned by the charity which funded it for 42 years. Loach documented Save the Children's colonial exploits in Africa and its operations in deprived parts of the UK, describing them as 'a mere sticking plaster or temporary measure'. 'Until these questions are addressed politically, then these fire-brigade rescue jobs cannot possibly solve it!'[107] The message is that 'charities are only necessary when societies fail' and, ultimately, that:

Socialism is the only answer, all the rest are propaganda![108]

Practising what we preach, preaching what we practise

This brings me back to that central question, 'Can a left-wing structural critique be married to an ethics of personal responsibility?' And, indeed, what the purpose is of any individualistic localised action amidst accelerating global social, environmental and economic crises. In many ways, my action during *The Glasgow Effect* – as 'one woman's protest against globalisation' – was doomed to fail from the beginning. One woman is never going to bring down the whole global capitalist machine, nor solve climate change, no matter how hard she might try. It was always only going to be a 'symbolic act'. But, nevertheless, it was an essential one.

It was only by taking this action that I was able to make a stand against the forces of globalisation which are hurtling us towards total climate breakdown. It was only by taking this action that I was able to attempt to live in a 'prefigurative' way[109] (in the way I hope we will all be able to live in the future) and, in doing so, to reaffirm my values on a daily basis, every time I hopped on my bike. It is only by living your values that you are able to act instinctively in line with them in times of stress or pressure. Unless we all attempt to 'actively address the contradictions and compromises' in our own lives and to challenge the status quo, then we'll never be able to act with integrity and are doomed to become hypocrites. This is why most people in power in our country make terrible decisions, because the very fact that they are in power means they're no longer living their values (if they ever did). This is one reason why we must reform our political systems to create a situation where power is shared far more equitably and truly loaned by people from the bottom-up (something I'll come back to in Chapter 10); we need a system which allows everyone the time and space for education, action and reflection.

Darren McGarvey came to realise the need for 'personal responsibility' when he made the connection between his drinking and drug-taking and his deteriorating mental health. He made the tough decision to swim against the tide and quit and his life began to improve.[110] Carol Craig's main criticism of the Glasgow Centre for Population Health's work is that it does not take into account the extent to which 'structural problems' create 'cultures', which 'once established… develop a life of their own and are able to reproduce themselves', even if the original structures have long-since changed.[111]

The 'dislocation' experienced in Glasgow as a result of widespread immigration from the Highlands, Ireland and elsewhere (coupled with squalid living conditions) may have been the initial causes of alcoholism across the city, but this now lives on as a supposedly inalienable 'culture' which journalist Lesley Riddoch says 'doesn't do emotion (without a large skelp of drink)'.[112] Carol Tannahill, former GCPH director, says 'we need to change the attitude in Scotland that you can't have a good time unless you have a drink in you'.[113] If the inspiring history of Glasgow's temperance movement shows us anything, it's that this is a culture that – with 'education, organisation and discipline' – can change. But in contemporary Glasgow on so many levels – fuelled by our manipulative consumer driven economy and the class prejudice that results – many people simply aren't doing themselves 'any favours' in terms improving their health.[114]

Although people in parts of Glasgow, including Milton and Castlemilk, find it difficult to access affordable fresh food due, in part, 'to poor transport links' (which it must be our number one priority to address), leading these places to be labelled 'food deserts',[115] there are many places in the city where healthy food is just being passed over. At a recent CommonSpace Forum event, a local food activist described being forced to throw away 'hundreds of courgettes' in Dalmarnock because people weren't interested in eating them, despite (or perhaps because of) the fact they were locally grown.[116] During *The Glasgow Effect: A Discussion* event, Katie Gallogly-Swan captured this issue succinctly, warning of the danger of working class people coming to build their identities around foods and behaviours which are actually killing them, because this was an important way of differentiating themselves from the 'middle class tubes'.[117] She said:

But then I think the problem is when it becomes that 'punching up' turns into a serious thing. I mean I don't mind my sister making fun of me for being a vegetarian and ordering my veggie box. I couldn't give two shits. But the problem is when that's used as a marker of difference. And then my sister uses her... buying something that is patently unhealthy and bad, because she sees that as a part of her identity... when we get to that stage we have to realise that that's not... 'punching up', it's not doing us any favours.[118]

In her book, Cathy McCormack continually describes her need to 'pop out for a fag' as the thing that differentiates her from the more privileged people she worked alongside in her campaigns and popular education initiatives, which clearly wasn't doing her 'any favours' either. When he finally confronts his own class prejudice in 'The Changeling' chapter of his book, Darren McGarvey comes to realise that:

Those irritating tropes of middle class life, around veganism, cycling and healthy eating, really served a practical function and were not necessarily as pretentious as I thought. As well as being markedly cheaper, many of these seemingly indulgent lifestyle choices were about living in accordance with the needs of the wider community, the environment and devising a sustainable lifestyle that integrated those needs.[119]

If working class people have been rejecting these things as a way of disassociating themselves from 'middle class life', then this is what Marxists would term 'false consciousness'. Illustrated quite neatly in the text which accompanies Steve Bell's 'Tower of Babel' cartoon (in Chapter 1), 'false consciousness' is 'a state of mind that prevents a person from recognising the injustice of their current situation' and which therefore causes them to aspire to goals that are not in their interests.[120] The consumption 'choices' that they have been making instead are actually serving to make a global economic elite even wealthier, at the expense of their mental, physical and financial health. For example, Scotland has been duped into believing it has an inalienable 'sweet tooth'[121] that is fuelling our obesity and diabetes epidemic, which actually only results from a capitalist ploy during the slave trade to 'create a market' for Caribbean sugar. People are making a lot of money from selling us this unhealthy crap.

When I first went vegan in 2009 and discovered I could buy a bag of lentils from my local shop for about £2 and that if I took the time to soak them overnight, they could feed me for two weeks, I found that knowledge truly liberating. Not only had I found the 'win-win-win' sweetspot of being simultaneously the cheapest, healthiest and most environmentally friendly food, but also the less money I spent on the 'costs of living', the less I had to work, the more free time I had to spend on the things which might actually create joy (or 'eudaimonic' well-being) and give a sense of purpose and meaning to my life: my art and my activism. The 'win-win-win' sweetspot of cycling as a mode of transport is captured perfectly in a sticker I saw stuck to the toilet door in Bike for Good in Finnieston in 2016:[122]

🚗 – this one runs on money and makes you fat
🚲 – this one runs on fat and saves you money

Anything that you have to keep taking to the petrol station and pumping full of money is the opposite of 'freedom' – it's a crippling financial burden. This highlights again the deep injustice of moving Glasgow's working class so far away from the city centre so that 'mobility' now always has a cost attached. And explains why cycling in Glasgow is still 'dominated by the most affluent of the population: those in the least deprived decile are nearly three times more likely to cycle than those in the most deprived decile'.[123] Those who commute by bike are shown to be half as 'stressed' and to have much better mental and physical health,[124] all of which further exacerbates inequalities in wealth and health.

The point is that Glasgow hasn't always been like this, it can and it must change its (self-)destructive 'cultures'.[125] It has its reputation for 'radical socialism' for a reason. In 1891, the Govan Clarion Cycling Club was founded to encourage people living in the Glasgow's slums to take up the healthy hobby and get out of the city on their bikes.[126] It was these same socialists who were behind the temperance movement and many other inspiring initiatives around that time (late 19th and early 20th centuries), which I'll come back to in Chapter 12. But as the story of Ernest Marples and Richard Beeching shows, there is a more sinister and manipulative agenda to car-centric culture, which is propping up a broken economic model based on consumerism and the insidious pressure to spend, at the expense of our own well-being, and that of our environment.

Thrift radiates happiness

'Thrift Radiates Happiness'. Those are the wise words inscribed upon the ceiling of the Birmingham Municipal Bank building opened on the city's Broad Street in 1933 in the midst of the Great Depression. Owned by the Birmingham Corporation on behalf of the city's people, the bank encouraged local people to save in bonds which were then used for local investment.[127] The bank no longer exists, the building is no longer open to the public and in the post-war period this simple fact of life – that 'thrift radiates happiness' – has been purposefully obscured from the masses.

At the start of Industrial Revolution, Glasgow's favourite liberal economist Adam Smith (1723–90) was himself very judgemental of people who spent their money on 'ornaments of dress, trinkets, fine furniture or jewellery' (things they might 'want', but don't actually 'need'), saying that this was a display of 'a base and selfish disposition'.[128] A century and a half later after much rampant capitalist 'growth', German sociologist Max Weber (1864–1920) came to realise in the chapter 'Asceticism and the Spirit of Capitalism' of his 1905 book, *The Protestant Ethic & the Spirit of Capitalism*, that, in fact, it was these ascetic values, which Adam Smith and many others maintained, which had been the driving force of capitalist expansion.[129]

The capitalists were often the ones that didn't consume and therefore accumulated wealth and power more quickly (like Scrooge himself). Instead they relied on those further down the pecking order to participate in mindless consumption, despite the fact that, as Common Weal describe, 'this habit encourages debt, is hugely wasteful and ultimately unsatisfying'.[130] This isn't about defining our identities by what we eat, drink, smoke, wear or own, it is the mass manipulation by a global economic elite who need everyone else to keep on consuming, just to keep the whole stupid system afloat.

Steve Jobs (1955–2011), the billionaire founder of global tech giant Apple was clearly one of these capitalists. He wore the same clothes every day as he didn't want to waste time making pointless decisions every morning about what to wear – being swayed by so-called 'fashion'. He just wanted to get on with the task of flogging more iPhones and piling up the money, making you feel that without the latest model your life would be incomplete and you'd be seen as a 'failure'.

Economist John Maynard Keynes (the man who brought us the United Nations System of National Accounts *and* the Arts Council of Great Britain) is also partly responsible for creating and perpetuating the (self-)destructive consumption-driven economy which we've had since the Second World War. He described what he called the 'paradox of thrift' (another 'fallacy of composition') – the fact that if we all behaved in a thrifty way (as artists often do) then the whole consumption-driven economy would fall apart. Therefore, he said, we must supress 'traditional wisdom' about the cheap, healthy and sustainable ways we should be living in order to prioritise economic 'growth' (as measured in GDP). As Schumacher describes:

> Economic progress, [Keynes] counselled, is obtainable only
> if we employ those powerful human drives of selfishness,
> which religion and traditional wisdom universally call
> upon us to resist.[131]

Glasgow is a city whose leaders promote the fact that it's the 'UK's second biggest shopping centre outside London'. Our whole city centre is geared towards employing 'those powerful human drives of selfishness' to keep our economy afloat. My friend Anna (who helped with the editing of this book) heard one of the Glasgow Centre for Population Health researchers joke at an early seminar that a more honest marketing slogan for the city – an alternative to 'Scotland with Style' (launched in 2004) – might be 'Go Shopping, Get Pissed'.[132] It's no coincidence that the vast majority of our remaining bus routes take you directly to the shopping centres (and not to the places where your friends and family might live). It takes 'education, organisation, and discipline' as well as courage in this city to go against the grain. If you don't 'Go Shopping, Get Pissed', what do you do?

This is where we come back to the arts. If art and cultural participation are also seen as 'middle class things', then that really is 'robbery' from the working classes, as these things can and should offer the antidote to consumer culture – a forum for free ideas and discussion away from the marketplace. It's using your brain for something challenging like writing or reading, or making or viewing art, which enables you to start to see it as your most important asset, which makes it easier to turn away from drugs and alcohol and to protect it at all costs. 'Your Health Is Your Wealth' as Cathy McCormack so wisely said, mentally and physically. And as I discovered in 2015 after breaking both my arms, your health is also your 'mobility'.

In the dark days of *The Glasgow Effect*, when I was cycling through the pissing rain and cursing out loud, or getting stuck in (self-)destructive Facebook wormholes, or waking up in the night with extreme anxiety and 'home sickness', I found it hard to remember any good things about this city. I forced myself to think hard. Back to the Scottish Government's protection of the NHS in comparison to the Tories in England (which had come to my rescue on several occasions and no doubt will do again), to the fact that (unlike England) we still have publicly-owned water,[133] to the free higher education, to the relatively cheap living costs (compared to London), but it was also the huge amount of free culture that I kept coming back to as the one reason why I had stayed here so long.

I kept meticulous records of every meeting, cultural, political or social event that I attended in 2016. There were 571 in total, most of which were free (funded with public money), meaning my annual bill for 'entertainment' came to only £129.50, about

£2.50 a week. We must create a culture where all these brilliant events are seen as being there for everyone, they must be accessible and inclusive and inviting so anyone can go along. As well as that ingrained class prejudice, there's a real structural barrier in the form of a shambolic and overpriced privatised public transport system, which means that for many people it's simply impossible to get into the town, let alone home again, in the evenings. Not only have we created 'food deserts' through our failure to address our public transport crisis, we have wilfully created 'cultural deserts' as well. It's no wonder our educational attainment is so poor. Both Carol Craig and Cathy McCormack argue that most poverty in Glasgow is 'more psychological and spiritual than material'.[134] And because 'Inequality in arts participation is most closely associated with education',[135] another vicious circle has emerged.

All of these behaviours – eating unhealthy, expensive processed foods, smoking, drinking, other 'mindless consumption', not being able to cycle and rejecting art and cultural participation because 'it's no for the likes of us and it's crap anyway'[136] – serve to entrench inequalities and reduce well-being amongst the most deprived people in the city. And this 'aggressive philistinism'[137] is also potentially shutting people out of a fast-changing 'labour market' too, as the skills acquired through a creative education become those least at risk of 'automation' and are therefore the ones most in demand in our new 'knowledge economy'. Likewise, Darren McGarvey and his followers' resistance to applying for public arts funding for their work is also exacerbating inequalities. One of Loki's many pieces about middle class people includes the line 'they can't be that creative, they're all subsidised with public funding'.[138] As well as his stated political reasons for not applying for public funding, it is clear his resistance is also about pride and the stigma of accepting any money from the state. It shows the extent to which the poisonous rhetoric of the right-wing media from the '80s onwards (like the 'NHS' insult that I remember from the playground) has created deep shame in accepting welfare payments or other public funding. Neither of these things are a 'begging bowl' and if working class people just stand aside and let the city's growing middle class population cream off all the public money, then they're actually just helping to maintain the status quo.

Hostile environments

I want to end this chapter, as I did the previous one, by returning to the reality of living *The Glasgow Effect*, which was intensely stressful and exhausting. It really all came back to my original fear in undertaking the project before it began – that 'any artist using their own body or life experiences as 'art' is essentially making themselves into an 'object' to be criticised for aesthetic and/or other value'. It was the dual 'objectification' that I put myself through – firstly as that two-dimensional object of ridicule on social media (to be 'looked down upon') and secondly as the model citizen boldly pioneering a new sustainable lifestyle for us all (an object to be 'looked up to'), which created the pressure.

The only way to survive was to try to distance myself from the character(s) I was playing.[139] But although I could see the funny side of a lot of the comments and memes posted on Facebook (see Chapter 5), there was still a huge amount of anxiety caused from being at the eye of the storm – another key reason why my 'extreme lifestyle experiment' was doomed to fail. There is a 'law' named after the economist Charles Goodhart aimed at showing how the object of any experiment will be affected by the very fact that it is being experimented on, which makes it near impossible to truly capture 'real life'. Goodhart's Law states that:

Any observed statistical regularity will tend to collapse once pressure is placed upon it for control purposes.[140]

The pressure placed upon me was immense. As one of the audience members said at the end of *The Glasgow Effect: A Discussion* event, 'The thing that you've done, Ellie, is that you've poked Glasgow. And to say you're a true Glaswegian, you've gotta be attacked!' I realised that in some weird way – in framing the project as I had – I'd managed to simulate the continual state of stress caused by poverty and deprivation (as a result of 'marginalisation and social exclusion' and financial instability and uncertainty) which Darren McGarvey describes so vividly in *Poverty Safari*. I had created a 'hostile environment' much more akin to the reality many people in Glasgow experience (those beyond the bubble of 'the Glasgow miracle') living in a continual state of fear and anxiety. And what made it worse was that I had to endure much of it completely alone. The social isolation and the lack of 'local love' I'd felt ever since I arrived in the city in 2008 only intensified.

It wasn't as if I expected much sympathy, I had brought all this upon myself. 'Part psychological experiment, part protest, part strike' – I'd tried to cram too much into one project and then to frame the whole goddam thing as 'art'. The framework of the 'durational performance' was useful in that it created a motivational structure – I wanted to work flat-out all year to get some important new projects and campaigns up and running in the short time that I had – but it was also what isolated me even further. It felt unethical to get too close to any other humans for fear of co-opting them as part of the 'art'. Although I did make new friends and start to build some community and affiliation as a result of the projects and campaigns I started work on in 2016, I couldn't really rebuild my 'core economy' in any meaningful way. My family were hundreds of miles away and I had to put my love life on hold.

The fear and anxiety took its toll. I had so much anger to process, I couldn't help but internalise some of the bigotry which was aimed at me. I looked around at all my friends – why were so many of them English? I felt disgusted. Are immigrants doomed to gravitate together, rather than integrate? I didn't want to be seen with them. I looked around at people in the artworld and I was still appalled at their decadent behaviour. I didn't want to be associated with that either. The more I alienated myself, the less I could be bundled in with any particular nationality or class. I was 'dislocating' myself deliberately, because I didn't want affiliation with any of those groups.

When my parents came to visit, I remember telling my mum to keep her voice down in public places so that people wouldn't hear our English accents. I internalised all that Anglophobia, alongside all the internalised homophobia that had been there for decades. *The Glasgow Effect* had forced me into the cliché of the 'middle class socialist' that writer George Orwell described, the 'sanctimonious hypocrite' who hated the poor as much as they hated the rich.[141] I hated the Scottish as much as I hated the English. I became a lonely misanthrope. It's taken the writing of the book to help heal myself altogether. But at the time, I just sat with all that anger, walked away from the computer screen and tried to channel it in more positive directions.

The outsiders

I also found myself gravitating to other 'outsiders'. On 23 January 2016 I attended a workshop at Transmission gallery with artist Diane Torr. We then started to hang out more often. She had a really interesting history – born in Canada 1948, she grew up near Aberdeen with an abusive father. Her older brother Donald was gay and ran away from home, so Diane quickly followed when she was just 16. She went to London and then to Totnes to study performance art, before moving to New York in the late '70s. There she pioneered radical feminist gender-bending performances and became famous as a 'drag king', holding workshops to help women try out being a 'man for a day'. Her whole life had been about transcending class, gender and sexuality barriers and resisting conforming to social norms. She moved back to Scotland in 2002, so that she had the protection of the NHS and would no longer have to live in fear of bankruptcy if she fell ill, as so many Americans do. In America, she'd had to learn how to hustle to get by. She told me the story of how she managed to land herself a council house in Hyndland when she had returned.

That summer when she was visiting her daughter Martina in New York, she collapsed. After ten days in hospital, she was diagnosed with a brain tumour and told to return to Scotland for a course of radiotherapy. She needed people in Glasgow to rally round and got together a rota of friends to support her during in trips to Gartnavel Hospital. I went along with her and in autumn 2016 we spent lots of evenings together where she told me stories about her life and her frustrations at the Scottish artworld, which she felt wasn't interested in her work. She always had to travel to Germany, elsewhere in Europe or back to America for work. She liked what I was doing. I think she saw a bit of herself in me, because, as she said in her soft Scottish-American accent, 'no one else is doing that sort of stuff'. She was going through an intense and disorientating treatment, yet she was always good fun and a calming influence for me to be around.

CHAPTER 8

Aftershock

The end is the beginning

ON THURSDAY 22 December 2016, nine days before *The Glasgow Effect* was due to end, I was still working flat-out in my studio. One of the big projects I'd been developing since May was coming to fruition and I needed to capture as much as I could before attempting to relax over Christmas. I'd managed to lose my little Nokia mobile phone earlier in the week and hadn't yet got myself a replacement, so was communicating with my family via email. My mum, my dad and my uncle were all on the train heading up from London Euston. My sister, her husband and my niece and nephew were on their way from Norwich. We had arranged a meeting point: 6pm at The Bay Tree on Great Western Road. As the clock ticked down, I could feel them all getting closer and an immense amount of relief started sweep over me: a year's worth of tension was gradually beginning to dissipate. We were all safe.

For me, the most unethical part of the project had always been the way in which it embroiled my poor family. I had been walking on a tightrope all year: if anything had happened to any of them, that tragic event would have become part of this whole nasty thing. I had been consciously highlighting and challenging what was my single greatest fear by writing it into the rules. I would remain in Glasgow, 'except in the event of the ill-health/ death of close relative or friend'. In the 'Risk Assessment' section of my original funding application, I wrote:

> The fact that this is the only caveat to the project's strict rules, highlights what is perhaps the greatest human anxiety – explored in [my talk] *Ethics: Extremism & Compromise* (2015) as the only ongoing tension between my life in Scotland and family roots down south.

If necessary, I would have to address this event as part of the project.[1]

At around 5.30pm, I packed up all my stuff and jumped on my bike. It was pissing with rain, great swathes coming horizontally from all directions. I had all my waterproofs on. I pedalled as quickly as I could; taking one of my usual routes along the side of that roaring motorway through Sighthill Park – which the council had just started to bulldoze, transforming it into a mountain of mud.[2] I just had time to pop up to the flat to hang up my wet clothes, change my shoes, compose myself and walk back down the stairs again and out across the road. I walked through the door of the restaurant, and there was my mum, sat smiling in the middle with a bottle of Prosecco. She had been looking forward to her first Christmas in Glasgow and to the end of this bloody project. Now finally we could all relax. It was a brilliant Christmas. My mum, dad and uncle stayed at a hotel down the road which was more accessible to the elderly and disabled than an attic flat (my dad could no longer climb the stairs). My sister's family crammed in with me. We had great fun making a Christmas tree out of an old green sleeping bag and a clothes rack, as I hadn't had the time or the inclination to bother with any decorations until they arrived. We went to watch *The Muppet Christmas Carol* at the Glasgow Film Theatre (GFT) and laughed about my former incarnation as Scrooge... not much had changed it seems! All the able-bodied ones walked with me and the others took a taxi.

Just before Christmas, Diane's daughter Martina had arrived from New York to look after Diane over the holidays. She had been rushed back into hospital. The doctors were concerned that the radiotherapy had not stopped the tumour growing and so had put her on more medication. She was out again by Christmas Day and so I invited Diane, Martina and her cousin round to have Christmas dinner with us all. We borrowed the ground floor flat in my block from my neighbour who was away, so that everyone could get in.

My family all left the day after Boxing Day. Then I went round to see Roanne for what turned out to be the last time before she died on 31 January 2017 at the age of 51. She was very weak by then, hooked up to a portable morphine syringe driver. But she was still so keen to hear about everything I'd been working on. She wanted to know how my ideas were developing for the talk I'd arranged to do the following Sunday at the GFT, as one of key public outcomes of the project. We sat at either end of her big sofa, facing inwards towards each other; our legs meeting in the middle beneath a big duvet. I felt incapable of properly expressing my gratitude to her. If she hadn't put her head above the parapet to support me in January (when so many others took the easy option of joining the angry mob) and organised *The Glasgow Effect: A Discussion* event with Darren and Katie, it would have been a very different year. This 'premature mortality' that we'd been discussing was now all too painfully present in all our lives.

It was nearly dark by the time I left Roanne's; there were only seven hours of daylight at that time of year. I cycled back to my studio. I had to get back to work. By arranging to do this public talk, I had set myself a near impossible task. On top of all the projects and campaigns I had been developing, I had to try to summarise my year's activity and research – drawing it all together succinctly and articulating exactly what it all means. All in front of a potentially hostile audience. I had just 11 days left to prepare. So much had happened, I had learnt so much – how could I take account of everything and bring it all together? I started to reread all my notes – I had six and a half notebooks full of research and ideas.[3] How on earth could I fit it all in and do justice to all the issues the project had raised? I looked back at all the books and reports that I'd been reading across the year, which had informed my thinking and action. On New Year's Eve, I uploaded a 'Reading List' album to Facebook to document them all and offer a taster for the talk.

I had a quiet one that evening round at my friends Deniz and Tom's. At 11.50pm, my mum texted me: 'Here's to finale of Glasgow Effect and your release. Happy New Year!' A small smile began to creep across my face – it was nearly over. But I didn't feel I could celebrate just yet, if at all. I certainly wouldn't be leaving Glasgow on New Year's Day. I had too much work to do. The next day I stayed at home, as I had done exactly one year

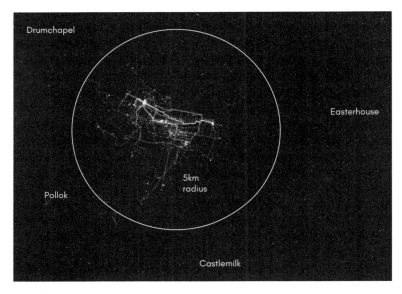

Annual Heatmap
1 January – 31 December 2016
Time moving: 185:07:00
Distance travelled: 3,753.20km
Average speed: 20.3km/h
Energy output: 29,157kJ

before (but with less of a hangover). The annual subscription on my Trackimo device was set to expire at 4.10pm. I downloaded all the data from their server – 33,584 individual GPS points. I wanted to create 'heatmaps' of the city to visualise everywhere I'd been each month throughout the year. I found a little online app to help me and made 12 different images, plus one showing the entire year. I calculated that I'd travelled a total of 3,753.20km under my own steam (the same as cycling to London and back three times), burning 29,157kJ in the process. I'd slashed my carbon footprint for transport from 3.48 tonnes in 2015 to zero.[4] Given the carbon reduction targets the Scottish Government has signed us up to (now net-zero emissions by 2045),[5] surely we need to start valuing this sort of outcome in its own right? Just before 6pm, I posted my 'Annual Heatmap' (above) up on Facebook.

Back on *The Glasgow Effect* event page, things were kicking off again – people were coming back in expectation of the 'great reveal'. 'Not just that she's "up her own arse" if she wanted to have some sort of poverty safari then she should do it in her own town/city. Don't do it in my City' wrote Richie Smyth at 11.01pm on New Year's Eve. 'Was it peng yeah?' asked Adam Lamont in Maryhill at 11.53pm. 'Fucking cowwwwwwwwwwwwww' added Robbie Sith at 3am. And at 10.09am on New Year's Day, Iain MacDhòmhnaill wrote: 'One year later and still nothing to show for it, I see. Fucking swindler'. The invitation was open for any of these friendly folk to come to my talk on Sunday. I had 'pinned' the details to the top of the event page on 5 November 2016 when they were first released. I wondered if they would dare be that aggressive face-to-face. I guess I'd soon find out.

That first week of January was perhaps even more stressful than the one the year before. I was testing the limits of the 'one woman band' mentality that self-employment and the 'creative industries' have encouraged; the privatisation of work. I was organising all the practicalities of the talk – the venue, the documentation: filming, photos and a live webcast (to make it as accessible to as many people as possible), as well as co-ordinating the discussion panel, volunteer ushers and finding someone to chair the Q&A. I was managing media enquiries – not quite as many as in 2016, but I had to pick quite carefully what I was going to respond to and plan what I wanted to say. I was also organising more national demonstrations against train fare increases and helping with the launch of a new campaign for the Public Ownership of ScotRail, speaking at a rally at Renfield St Stephen's on 5 January 2017. Oh, and I was also meant to be trying to condense a year's worth of material, research and ideas into a coherent lecture of about an hour. I had to put a note on my studio door saying I was 'on leave' as I didn't think I would be able to cope with any unexpected interruptions. I had one good night's sleep that Wednesday. The morning afterwards was productive and I thought I was getting on top of everything. I agreed to do a live interview with BBC Radio Scotland the following day. That night was not so good, riddled with anxiety that I would not be able to articulate the complexities of what I'd been doing, coupled, of course, with its simplicity as a 'symbolic act of resistance'.

I was exhausted by the morning. I cycled back down to the BBC studios at Pacific Quay, where we were going live at 10.20am. I got taken upstairs to the studio and this time was sitting opposite the presenter Stephen Jardine (who was standing in for Kaye Adams). It was nervewracking, but I'd made notes about the key points I wanted to get across. I was inspired by Cuban artist Tania Bruguera to try to 'turn my cultural capital into political capital' to draw attention to the causes of inequality and climate change in Glasgow and beyond – namely our car-centric culture and the privatisation of our buses. Stephen Jardine still took quite an accusatory tone: 'But that's not why Creative Scotland gave you the grant, Ellie'. To which I replied 'Have you read my funding application?' to which he bashfully admitted 'No'. When I left the studio, I felt quite relieved. I got a nice text from my friend Katherine Trebeck, who worked in Oxfam Scotland's research team rethinking our economy to deliver 'social justice, good lives, vibrant communities and protect the planet'.[6] She said: 'We listened to your BBC chat – you got sooooo many important points over!! Good luck on Sunday. And Monday. Katherine xx'. Phew, I guess it went quite well? I could still hear the programme broadcasting in the waiting area outside. My ear tuned in:

> Wow, what a response to Ellie Harrison who has just been on the programme with me. The controversial artist who spent a year living in Glasgow funded by Creative Scotland to the tune of £15,000. Massive response from you. All of it negative I have to say. Here's a statement from Creative Scotland: 'We have supported Ellie Harrison to research and develop her work over a 12-month period. As with all our funded projects we look forward to hearing the final outcome of Ellie's work so that we can best understand the impact and benefits our funding has had on her practice. As such, Ellie's talk this Sunday will begin that process'. If you want to have a look at the full funding application that Ellie made to get the cash, it's at: glasgoweffect.tumblr.com. More of your comments coming up in the next hour of the programme...[7]

'All of it negative' – really? It appeared I was still as 'toxic' now as I had been at the start of the year. This did not bode well for the talk on Sunday, shit! I cycled up to the GFT where I had to check the PowerPoint was working on the big projector in screen one, and to meet the lovely guy Stuart Platt who had agreed to film the talk for me. Then I headed back to my studio – I still had some serious work to do. I made some important decisions. I knew I could not talk about any of the specific projects and campaigns I had been working on directly, as it was just too much of a risk to tarnish them with the 'toxic Ellie Harrison brand'. I would have to take another big hit. Having worked flat-out, seven days a week for the whole year (with only ten days off when friends and family visited to drag me away from my desk), I would have to sacrifice myself to the painful accusation that I was 'lazy' and that I'd 'done nothing' for the sake of preserving the integrity of those projects and campaigns. I would have to find a way of talking around the issues they addressed without naming, and therefore shaming, them directly.

By 9pm on Saturday evening, I was still in my studio working on the talk. I spoke to Peter McCaughey on the phone. He is a really interesting Irish artist who I'd met when studying at Glasgow School of Art. He had led a discussion group on the book, *Why Are Artists Poor? The Exceptional Economy of the Arts*, by Dutch artist and economist Hans Abbing, which shows how the artworld is structured as a 'winner takes all' economy, where you have millions of artists and arts workers supporting the whole system with their free or low-paid labour whilst only a tiny number of superstar artists such as millionaire Anish Kapoor become rich. It's a bit like Steve Bell's 'Tower of Babel' cartoon (described in Chapter 1) – a microcosm of the wider global economy.[8]

Peter had agreed to chair the Q&A following the talk, alongside my friend, curator Michelle Emery-Barker and the President of the Scottish Artists Union, Janie Nicoll. I remember trying to explain to Peter all the dilemmas I was facing. I think he told me to go home and get some rest. That was probably a good idea. I was too tired to cook, so stopped at the takeaway and bought a massive box of greasy vegetable pakoras on my way back to the flat. They didn't make me feel much better. I thought my best bet was to get an 'early

night' and set my alarm for 6am. I'd get up and finish it all off then.
I must have slept for less than two hours. I was so fearful of what
to expect. I was awake before the alarm went off, so I just got up,
put my old dressing gown on, pulled my thick red woolly jumper
over the top for an extra layer of warmth and walked the two steps
to the desk opposite my bed. I'd agreed to meet everyone at the GFT
at 10.30am, but when it got close to that time, I was still working
on the talk. I just needed another hour. I sent them a quick text:

> Morning Peter and Michelle! I'm still working at home
> and need a bit more time. Can you still meet at 10.30am
> at GFT and meet Malcolm (the projectionist) and Kevin
> (the webcast guy) and Stuart (the guy filming) and start
> getting everything set up? I'm bringing a list of attendees
> for the welcome desk in the foyer. I'll be there at
> 11.30am at the absolute latest!! Sorry! Ellie xx

I had to print out the bloody thing for starters. By then it ran to
27 pages. No time to edit, now I really had to go. I jumped on
my bike. By the time I rounded the corner from Sauchiehall Street
onto Rose Street, there was a queue starting to form outside.
The 440 free places had booked up in less than two weeks
after I released the details on the Facebook event page back in
November. Two hundred and twenty two people had registered
on the waiting list. Before I left my studio the night before,
I had contacted everyone on that list to say that they should just
come along if they still wanted to, as there would likely be free
seats. People never bothered to show up for things that were free,
especially that early on a Sunday morning (I wonder what that
says about our value systems). Michelle was there in the foyer
waiting patiently when I arrived, and as soon as I handed over the
list she began slowly ticking people off.

   As I'd been researching and writing, I was thinking about
ways to pacify the audience at the start. I didn't want to launch
into any of the antagonistic stuff straight away or appear aloof
or superior. So I decided that I would join the other ushers on
the door. I had been working as a volunteer usher at the GFT for
six and a half years by then, taking tickets and picking up litter

one evening a week. If I was ushering, I would be able to look every single person in the eye before they came in. I wondered how many of those trolls I would recognise from Facebook, and how many of them would actually realise it was me. I looked very different from the images which were splashed all around the internet during the shitstorm in January 2016, especially as I'd grown my hair. In honour of the Taiwanese artist Tehching Hsieh who allowed his hair to grow to mark the passing of time during all his famous 'one year performances', I'd decided not to get my hair cut at all in 2016. Plus how could I justify spending any public money on something that frivolous?

Once the auditorium had filled up, you could sense the anticipation of 'Ellie Harrison' coming onto stage. Those in the know were glancing over at me, puzzled at my surreal attempt to be in two places at once. My heart was racing. I reached for the handheld mic which I'd stashed near the door, turned it on and started speaking as I made my way down the steps to the front. My voice was quivering slightly but, once I got behind the lectern, the fear and exhaustion sent me into autopilot. I found myself, once again, standing on a stage with everyone looking at me expectantly and not being able to perform quite in the way I wanted. But I just had to get on with it. About halfway through the talk the projector went blank. It was a massive technical hitch that took more than 15 minutes to fix, but I barely noticed. There was too much adrenaline. I just tried my best to entertain the crowd with a little costume change – taking off my black uniform to reveal the 'F**k First Buses: Public Ownership Now!' t-shirt which I had designed in May as a way of helping me channel all my anger, 'punching up' at a more deserving target – the privatised bus system, the profiteering bus companies who operate within it and the people in power who are doing nothing to address the huge social and environmental injustices that result.

By the time the talk was finished, I had been going for nearly two hours – it ended up being an endurance marathon to end an endurance marathon. The technical problems meant there was not as much time for discussion as I would have liked. But the GFT let us stay in a little bit late and people were invited to come and join us for a drink at the CCA afterwards. I tried my best to chat, but I was ready to drop to the floor. I left about 5pm, I needed to get home to get everything ready to go back to work tomorrow – we had a staff 'away day' starting in Dundee at 9.30am. I was psyching myself up to get back on a train.

When I got back to the flat, I checked my emails and some of the tweets. I shouldn't have done really. There had been an arts journalist in the audience 'live tweeting' throughout the event. This is what arts criticism has come to now: instantaneous soundbites rather than any pause for serious reflection. One read: 'what, so the point is "privatisation is bad?"' That really pissed me off. Yes, the point is that privatisation is bad – it is a key driving force for inequality but, since I realised that eight years ago, I have dedicated my life to fighting back! Relentlessly making the case for the public ownership of public transport, energy and more. If everyone else wasn't wasting so much time writing pointless tweets, they could be doing the same. Aaaarrrrgggghh! I shouldn't have been looking at Twitter anyway, I really did need to get an early night. But then I saw an email from the *Daily Mail*.

There had been so much scrutiny of the £15,000 grant I'd received, I thought I better explain exactly how I had spent the money as part of my talk, making good on my public service commitments to 'accountability' and 'openness'. One of my slides featured a pie chart showing how the money had been divided up: £8,400 for living expenses (what I'd calculated to be bare minimum I would need to survive), £6,600 invested in the various projects and campaigns. Still the *Daily Mail* journalist persisted:

> I think some people remain concerned with the fact that
> so much money was spent purely on living expenses
> – particularly as you are probably substantially better
> off than most people in the city by virtue of owning
> property. I understand that residencies are common
> in the artworld, but what is it that you would say the
> public/public audience got in return for paying for your
> living expenses for a year?

'So much money', I thought. Suddenly it all made sense – now I could see why the right-wing press spent so much time and effort reporting on and stigmatising individuals receiving small amounts of state benefits: *the smaller the income, the greater the scrutiny*. It was 'divide and rule' again: a massive deflection tactic from those getting away with millions of pounds of public subsidies at the top of the pile. I used Oxfam Scotland's online Income Inequality Calculator and discovered that, after my housing costs were taking into account, £8,400 meant that I was in the sixth-lowest percentile; eleven percentage points below the so-called 'poverty line'. Even on my normal salary of £18,884, I am only in the 33rd percentile, hardly a member of the '1 per cent'. Meanwhile, take someone like Gordon Maclennan, the Chief Executive of SPT, the public body which is meant to be overseeing our region's transport network and 'delivering better public transport for all'.[9] He gets at a £143,265 salary from our taxes, which does put him in that '1 per cent'. And I'll tell you, out of the six SPT Board Meetings that I attended as an observer in 2016, he didn't say one word. I did and am doing far more to try to improve public transport for Glaswegians than that man has ever done, and for less than 6 per cent of the cost.

Nobody should be getting paid that much money. We need much tougher rules on pay ratios. My union UCU says one to ten at the very most, though EF Schumacher argues in *Small is Beautiful* for one to seven. I would go even lower – why not? I heard Natalie Bennett, former leader of the Green Party of England & Wales, advocate this policy when we both spoke at We Own It's conference on the public ownership of all public services in November 2018. When asked how it would work with Premiership football clubs, she said, 'You'll have some very well-paid cleaners'. That is what we need to reduce inequality! They're the ones doing the toughest work anyway – we must meet in the middle. The more money you get paid, the less ethical you become, there's no two ways about it. You just end up buying more houses than you need and renting some of them back to those less fortunate for a profit, or you invest it in all manner of unethical and corrupt industries: fossil fuels, weapons, tobacco – these people only care about making a financial return. We need to realise that the most sustainable way to reduce relative poverty is by reducing the standard of living of those at the top. It is their lifestyles which are completely unsustainable: the richest 10 per cent of the world's population cause half of all carbon emissions.[10] They are the ones we need to stop. As Schumacher writes:

> Excessive wealth, like power, tends to corrupt. Even if the rich are not 'idle rich', even when they work harder than anyone else, they work differently, apply different standards, and are set apart from common humanity. They corrupt themselves by practising greed, and they corrupt the rest of society by provoking envy.[11]

Aaaarrrgggghh! I really need to stop ranting and get into bed. The next morning my alarm went at 6.30am. I leapt up out of bed. I have never been so happy to get up so early in all my life. I felt like I'd got away with it. Almost escaped unscathed. I packed my bag, my Thermos of instant coffee, my flask of water and my tupperware of nuts and walked down to Queen Street station to catch the 7.42am. I crept nervously onto the train. It had been 375 days since I last travelled on that one heading back up to Glasgow on 31 December 2015. I just didn't know how my body was going to react to the movement (even going in lifts had started to freak me out). I tucked myself away in one of the double seats by the window. As the train chugged out of the platform, into the Queen Street tunnel and that familiar stench of diesel fumes hit the back of my throat, it felt as though no time had passed at all. As we emerged into that early morning light, I couldn't stop smiling. I had survived.

Impact agenda

As we pulled into Dundee station, I had the weirdest sensation that I'd been in a time machine. I had left work in December and had returned after the Christmas break in January. Except that a whole year had completely disappeared in between. I had lost a year of my life to *The Glasgow Effect*. The only thing that gave away the passing of time was the building work on the waterfront. They'd still not finished the new station, but they'd made a lot of progress on the new V&A. I walked into the 'away day'. It felt like the punchline of the longest practical joke in history. Some colleagues were pleased to see me, others seemed less impressed. I sat quietly listening to the presentations. I gazed out of the window at the beautiful sunshine reflecting on the ripples of the Tay. For the first time in over a year, I had the feeling of 'dossing about' – now I actually was being paid to sit on my arse and do nothing. But I felt like I deserved a break. On the train on the way home, I made a note-to-self to attempt to capture my mood: 'I'm on the train now and I'm so fucking happy!!! Listening to some music and getting organised. I missed this time just to sit and think and chill out. I've 'clocked off' and can do whatever I want!! NOT PUBLIC PROPERTY ANYMORE'.[12] The sense of release was immense.

Two days later on 11 January 2017, we had a 'research day' and I had to make a three-minute presentation about my 'research' to all the research staff and vice-principals at the university. I took in the press cuttings from the first week of *The Glasgow Effect*, which Creative Scotland had sent me. I'd had them spiral bound into a hefty A4 document, which was nearly an inch thick. I introduced the project which I said had been 'successful' in receiving a research grant of £15,000 from Creative Scotland, but that the university had decided not to publicly support it, despite the 'impact' it had had. I used the word 'impact' as a cue to slam down the document on the desk in front of the Dean of Research. I showed a slide paraphrasing economist Tim Jackson's well-known criticism of consumer culture from *Prosperity without Growth*: 'We spend money we don't have, on things we don't need, to impress people we don't like'.[13] And concluded my short presentation with a warning that 'unless we apply the same level of critique to the systems within which we operate in higher education, we run the risk of: applying for funding we don't need, to do things we don't want to do, to impress people who hate us'.

My frustration was never with the art school itself – I loved that place and so many of the things it stood for. I was passionate about the critical and creative education we were able to provide. My frustration was with the wider university with which it was merged in 1994. In that era many of our once independent art colleges were subsumed by their nearby universities, losing their freedom and autonomy in the process. In *Small is Beautiful*, Schumacher shows that 350 employees should be the maximum size for any company or organisation. If they are allowed to grow larger than that, the bureaucracies running them become too removed from the day-to-day realities of the staff and even more so from the students who we are there to serve. To retain its integrity, our art school should have remained independent from the university beast, which now employs more than 3,000 staff. It's at that point where you get inappropriate and generic rules applied across many varied disciplines, which are completely irrelevant and often counterproductive.

On 14 January 2017, I went in a car for the first time in 380 days, since I got a cab to London Euston with all my luggage on 31 December 2015. I was with Diane and two of her friends, Nina and Julie. It was Julie's car and we were heading to a dance club in Scotstoun. All Diane wanted to do in the last few months of her life was dance. It was so warm and cosy being sat in the back seat on that cold, dark winter night. It felt nice to be so close to other people; to travel together after all that time going it alone. It felt luxurious. There was a familiar smell inside that car, which reminded me of my childhood – being sat in the back of my parents' car in the '80s and '90s and driven between England, Scotland and Wales to visit relatives or go on holiday. It was a smell of the past. This was a technology of the past. It was lifestyle of the past. It was more private luxury than we can be afforded, if we want our species to survive on planet earth. The privileged people who can afford these polluting metal boxes were going to have to learn to give them up.

I was on a high for those first few weeks of the New Year. On Wednesday 18 January 2017, I was cycling back to the SPT offices for Transport Scotland's Rail Infrastructure consultation event and a fellow cyclist pulled up alongside me at the traffic lights. I felt his stare on my cheek, so I turned to offer a smile.

'Are you that person that did that study?' he said.

Hmmm, I'd never thought of referring to it as a 'study' before, but I guess in many ways it was.

'Yes', I said tentatively.

'I was watching the live stream of the talk and it was great...
I was sceptical at first, but I think it's amazing what you've
done', he said.

'Thank you', I said, quite bowled over. 'I really appreciate it!'

The lights turned green and we both went our separate ways.
Maybe it had been meaningful to some people after all.

The report

But there was still something big looming over me. I knew I had to
write an 'End of project monitoring report' for Creative Scotland
in order to satisfy the terms of the grant. It was due within 12
weeks of the end of the project, which would be 26 March 2017.
I wanted to get it done before then, while everything was still
fresh in my mind, so I could finally put it all behind me. I'd
initially set aside four days in my calendar in which to blitz the
report. But the more I thought about it and the more I thought
about everything I'd felt silenced from saying during my talk,
the more I realised that the key task of the report should be to
put on the record all the hidden activity that I had undertaken
throughout the year. When I said that I had worked 'flat-out,
seven days a week for the whole year, with only ten days off',
how were people meant to believe me unless somewhere there
was evidence of *a hell of a lot of work*? I came to realise that
this report had to attempt to capture it all. I wanted it to be the
longest report that Creative Scotland had ever received. I wanted
my Creative Scotland Officer to suffer almost as much reading
it as I had suffered doing it. I was going to need more than four
days!

In the end I spent 18 full days and wrote more than 30,000
words. It was therapeutic. It all just kept flowing out. It was so
satisfying to see all these disparate critical and creative activities
'on a spectrum between art and activism; self-interest and altruism'
being drawn together and documented in a coherent form. It almost
felt like closure. I knew I couldn't publish it immediately, as there
were too many people's names mentioned – the 134 individuals
I had had meetings with to plan projects and campaigns, plus the
many others who had influenced me or I'd collaborated with in

some way. Years would need to pass before this could go into the public realm. At present, only five people have read it: my mum, my dad, my tireless Officer at Creative Scotland and two friends. My mum and dad read it the weekend before I submitted it on Monday 27 February 2017. My dad said he found it 'exciting'. I'm not so sure about that, and my mum said that when she finished she went down to the living room where my dad was sat in the armchair by the window and said, 'I can't believe she's been doing all that work while we've just been watching telly!' I reread the report several times in 2017, as it helped me to reflect on everything I'd done and why, and to refocus on what still needed action next. Nearly all the projects and campaigns I'd helped to set up in 2016 were ongoing.

There's a bragging quality to the report, not a million miles from those loathsome academics entranced by 'extrinsic' values who list all the sums of money they've been awarded for their 'research', as though they are badges of 'success'. But I was reporting back to my funders after all, and so meticulously listed all the facts and figures: the amount of partnership funding I raised for various projects and campaigns and the number of 'views', 'likes' and 'shares' various outputs received. But the report has two unwritten qualities, which I feel are just as important. The lifestyle I'd led in 2016 was simply a localised and intensified version of the lifestyle I had been living ever since I arrived in Glasgow in 2008. It was a lonely and isolated life, in which I learned to fill the vast voids with work. It was a vicious circle, where I worked because I was lonely and I was lonely because I was working (having to turn down social invitations because I had too much to do). The greatest irony of *The Glasgow Effect* for me was that it was a supposed experiment in 'sustainable living', but the way in which I had chosen to use my time – with the added pressure of the public money and the public scrutiny it brought with it – was totally unsustainable. I no longer had the social networks that are 'the very immune system of society'.[14] Instead it was self-exploitation to the max.

On 2 February 2016, on one of my ushering shifts at the GFT, I got to see the film *Spotlight*, which I found fascinating. The film focuses on reporters at *The Boston Globe* newspaper who, through a persistent investigation, expose a vast sexual abuse scandal within the Catholic Church. At the time, I thought that Scotland's media could have learnt a thing or two from those reporters who resisted the instantaneous news splash in favour of the more patient task of investigating the systemic failings, which meant the same problems kept recurring. I was particularly interested in the lawyer character, who does most of the investigative work. He is an immigrant to the city. He therefore has 'no life' of his own; no 'core economy' of family, neighbourhood and community and so dedicates himself to work. With the 'outsider's view', he is able to spot the abnormalities and injustices which someone bought up within that culture may never notice; the things which Carol Craig would describe as 'hiding in plain sight'.[15] The irony is that it's the 'dislocation' and discomfort that is felt by the isolated immigrant which becomes the powerful motivational force. Provided they have 'education, organisation, and discipline' to use this in positive directions, then they can become an unstoppable force for social change (even if they sacrifice their own happiness in the process). As Edward Said writes in *Reflections on Exile*:

> Much of the exile's life is taken up with compensating for disorienting loss by creating a new world to rule. It is not surprising that so many exiles seem to be novelists, chess players, political activists, and intellectuals. Each of these occupations requires a minimal investment in objects and places a great premium on mobility and skill.[16]

The second unwritten element of the report is its anger. When I wrote it, I still felt so hurt at the abuse that I endured and so used the report to lash out at some of the local people and organisations who I felt most abandoned by. Rejection is so much harder to bear when it's 'close to home'. After rereading the report, I began to feel slightly embarrassed about my tone (as I'm sure I will with this book eventually too), wondering whether after a year stuck in this city, I had lost my sense of perspective. Evidence shows that in the parts of the city where people cannot escape for socio-economic reasons (their bus routes are cut or the fares are too expensive), they have the lowest well-being and are more violent and (self-)destructive as a result. I wondered, after a year of tunnel vision where all I saw was Glasgow, come rain or occasional shine, whether I had become 'small-minded'? For all my talk of 'big picture' thinking (fighting for systemic change and a long-term vision), perhaps I'd started to lose sight of it myself. Perhaps Patrick Geddes only had the foresight to proclaim that we should 'Think Global, Act Local', because he was an extremely privileged man who'd had the opportunity to travel and live all over the world first.

Homecoming

On Tuesday 28 February 2017, the day after I emailed the finished report to Creative Scotland, I caught the last train heading south from Dundee station after work. It was to be my first visit home since 2015. My mum was so excited. So was I. We texted each other from the train:

'Are u homeward bound?', said my mum at 4.44pm.

'Yes! I'm in Gordon Brown's patch, heading south! xx', I replied.

'Wonderful, The Beautiful South', she said.

'Down towards Brexit land! Brace myself xx', I replied.

'You'll be safe in Ealing, bastion of remain', said my mum.

'Phew, see you soon! xx by the way – I did a word count and that report was 31,891 words! Like a PhD! xx', I replied, then I plugged my laptop in and started to catch up on some work.

'Still on time? Can pick u up at S or N Ealing', texted my mum at 8.13pm.

'I'm on home soil!!' I replied when I finally made it to Euston station at 10.36pm.

We had the most beautiful week together. I had to work in the daytimes. There was so much I'd had to put on hold until that report was finally submitted, but we managed to make time to do all our favourite things. We got up for our early morning swim. Driving down to the pool in my dad's old diesel-guzzling Volkswagen, my mum joked: 'This must be quite a change for you being chauffeur-driven round Ealing?' We laughed off the whole traumatic experience, genuinely starting to wonder what all the fuss had been about. As my mum had maintained from the beginning, 'I don't see why everyone's so angry, you're only doing your job!' And indeed I was. Part of my motivation was to illustrate the often unethical consequences of blindly doing what you're told without questioning why. Or as Schumacher puts it, being 'ruled by rules, that is to say by people whose answer to every complaint is: "I did not make the rules: I am merely applying them"'.[17]

My mum's New Year's resolution had been to sort out the house, which, having been her home since 1977, had accumulated far too much stuff. In those first two months of the year, she had undertaken a massive project of sorting through all her books – giving all those she no longer wanted to charity shops and starting to arrange the others in alphabetical order. As a former English, French and 'Communication Studies' teacher, she had a massive library. She sorted through all her old records, which she'd bought as a student in North Wales and in France in the '60s and had bought a new record player so that she could start to listen to them again. We spent a couple of evenings listening together. I remember the intensity of a recording of French poetry by Rimbaud and Verlaine, the crackling of the needle and the pained and dramatic delivery of the words (my mum could understand them, I could not).

One night when she was upstairs replying to emails on the computer, we started having a heart-to-heart. In my anxiety about mortality, I blurted out: 'How long do you think you think you've got left?' She answered without much hesitation: 'Nine years'. That would have made her 82, the same age her mum was when she died of pancreatic cancer. We had both been at my granny's bedside at Heath Hospital in Cardiff when she died on 2 October 2000. A sudden panic hit me. Why the hell was I wasting this precious time stuck the other end of the county? Over breakfast the next morning we flicked through the *Ealing & Acton Gazette* together looking at flats in the local area – they were all ridiculously expensive, but the sentiment was there: we wanted to be closer together.

Saturday 4 March 2017 was my mum's 73rd birthday. I'd helped arrange a meal for family and friends at a Persian restaurant in

West Ealing. My sister and my niece travelled down from Norwich. It was a beautiful sunny day and we all walked through the park together on the way back from the underground station. When we got to the restaurant, I sat next to an old friend of my mum and aunty's who they'd known since they were kids. I remember her asking me what I was up to. I answered: 'I don't know really, I'm a bit of lost soul'. I felt more in limbo then than I ever had before I started *The Glasgow Effect*. Was that ever going to change?

On Monday 6 March 2017, we went swimming again early at Gurnell Leisure Centre: the same pool where I'd first learnt to swim back in the early '80s. We didn't have time to have breakfast together as my mum had to get down to the Weight Watchers meeting where she was volunteering at the 'weighing in' station. She loved that job. We said our goodbyes and I packed up my stuff, before heading off to Euston station on the tube. I needed to be back in Dundee to meet my students first thing on Tuesday. The following Saturday was my birthday and I went on a trip to Greenock in Inverclyde with my friend Neil – we did the Parkrun along the Clyde and then hung out in the sauna at the Leisure Centre before lunch at the Beacon Arts Centre. I had a little get together at my flat that evening and a very sore head the following day. I didn't leave the house until the evening, when I met my friend Sarah for dinner at the Hug & Pint. We were talking about our mums: about how much we relied and depended on them. How would we ever cope without them? I remember saying: 'I'd probably have to kill myself'. It made me so anxious, I needed to get home to call my mum for a chat.

She'd just finished watching *Call the Midwife* and was relaxing on the sofa. I was lying on my bed with my hands-free set on. We had the most brilliant conversation which went on for nearly an hour. I recounted my birthday in minute detail. She was desperate to hear about every moment of the day including when Neil and I sneaked into Labour's 'The New Economics' event at the Concert Hall on our way back through Glasgow and got to see Jeremy Corbyn and John McDonnell. The whole thing had been inspired by the New Economics Foundation's chief economist James Meadway who was now seconded to Labour HQ. That was definitely an exciting development. Maybe now, after 16 years, I could finally vote Labour again (at the next General Election at least).

My mum had been learning German for several years and for Christmas 2015, I'd got her a selection of German films installed on a little media player, which she could plug into her old telly. We'd watched *Run Lola Run* together the previous weekend and, after another lesson from me about how to operate the media player (it was a bit fiddly for a not very tech-savvy septuagenarian), she'd managed to watch *The Bitter Tears of Petra von Kant* with her friend Lisa on Thursday night. She described the entire plot to me in such vivid detail, I don't think I ever need to watch it myself. It sounded totally intense. I looked at my watch and it read 10.17pm. We decided we both better get to bed. I said goodnight and thanked her again for my birthday presents: two spatulas and two new pairs of shoes (one black and one silver).

'I love them so much', I said.

'That's what happens when you choose them yourself', she quipped back, giggling.

I turned my little Nokia mobile phone off and drifted off to sleep.

Worst nightmare

My alarm went off at 7.30am. I was planning to head to the swimming pool. As I ran down the stairs, I reached into my pocket and turned my phone back on. A message came through straight away. That's weird, I thought. I stared at the screen. It's was from my dad: 'Hi Ellie. Please ring'. That's all it said. Oh god, I thought, what's happened? My heart started to pound. I rang and I heard my dad's voice. He's okay, phew. It must be my uncle.

'Hello', he said. 'Here's your sister', and then passed over the phone.

What the hell is she doing there? What's going on?

I heard my sister's voice, breaking slightly: 'It's mumma, she's had a stroke'.

I froze on the staircase. I felt myself beginning to walk in reverse back up to the flat as if on autopilot. My whole world was spinning.

'She's in Charing Cross Hospital, you need to come'.

'OK, I'm coming', I said.

I was already shoving a few clothes into a bag. I had no idea how long I would be away. An hour later I was on the train. It was the longest train journey in history. I'd had nightmares about this train journey. Tears streaming down my face, I just did not know how I was going to get through it on my own. That's when I started writing, it was the only thing I could do. I was just talking to her yesterday. I just saw her last week. One week ago exactly we were at the pool together swimming! This just didn't make any sense. If I wrote it all down perhaps it would. At least then I would be able to hold onto it. To process it at a later date and to keep those happy memories forever.

I wrote and I wrote, pages and pages recounting everything that we had done. The New Year's resolutions, the record player and the records and the silver shoes. Still the tears kept coming but, with the help of my pen and notebook,[18] I'd somehow made it down to London. I went straight to the hospital and made my way to intensive care. My dad and my sister were in the waiting room. They were exhausted. They had both been up all night. My sister had got a taxi all the way from Norwich arriving at 2am.

At around 11pm on Sunday night, just after we'd been speaking, a tiny aneurysm on a blood vessel in my mum's brain had popped, causing a massive haemorrhage. She had made it up to her bedroom and called out to my dad next door: 'I feel funny', to which he shouted back something sarcastic. 'No, I feel really funny' she said. By the time he had made it in she was beginning to be overcome and fell backwards onto her bed. He called the ambulance at 11.07pm.

It was nearly 2pm by the time I arrived at the hospital. I wasn't allowed to see her straight away, as they were doing a second procedure to try to stop the bleeding. I had an ominous feeling. When I finally got into the ward, it was a shock. There she lay, my beautiful mum linked up to all manner of machines. They had shaved a patch on the side of her head, where a tube was inserted to keep draining the fluid. She was unconscious, but I sat next to her and held her hand, speaking over and over the only words that seemed important: 'It's Ellie here, mumma. We're all here and we love you so much. We don't want you to worry'. She had a dedicated nurse and regular visits from consultants. Every few hours they would boot us out for this procedure or that. We went back to the waiting room, where we got to know some of the other families in similarly traumatic situations.

That night my aunty and I slept on the cold benches in that room just so we could be close to her, popping in and out of the ward to see her, hold her hand, lean on her side and repeat the same words: 'It's Ellie here, mumma. I love you so much. We don't want you to worry'. It went on for five days. I remember pacing slowly and calmly through those sterile corridors staring down at my feet. I had the strangest sensation. I wasn't me anymore. I was her. Those were her legs and her feet. I wasn't controlling them anymore: instead it was me that was fading away. On the Friday afternoon, they wheeled her downstairs for another brain scan to see if her condition had stabilised. Even that short journey itself was a risky procedure, as if any one of the machines had become detached, no one knew how long she would survive.

At about 10pm that evening, we had a visit from the consultant, a German neurologist who was most angry about Brexit. My mum would have loved her. She took us into the 'family room' with another doctor we'd not seen before who was deliberately straining her face to look sad. I knew what she was going to say. It was too much like a soap opera.

'I've got some bad news', said the new doctor.

The brain haemorrhage had been so big and created so much pressure under her skull that it had caused more and more parts of her brain to become 'infarcted' – that meant slowly dying off bit by bit. They showed us the two scans: one from Monday and the one from that afternoon. You could see where more and more areas of the brain had just been greyed out. Although she was lying there, still breathing, still warm to touch and still comforting to hold onto and to smell, her brain and all the brilliance, kindness, wit and knowledge inside it that she had accumulated over 73 years was gone. I knew. I had felt it – my worst nightmare slowly becoming a reality. But to my dad the news from the doctor was a total shock. He had never for one moment thought that he would outlive my mum. He was the invalid, this was not how it was meant to be.

Our family now faced the most hideous of moral dilemmas, which only modern medicine could have created. We knew we had to switch off the machines. It took us until Sunday to pluck up the courage. She kept on breathing for about 20 minutes, every one of them I clasped onto her hand so tightly and repeated the words, 'I love you so much'. I stared at her beautiful face, as her breathing slowed and slowed until it eventually stopped: that final moment of release when a wave swept over her whole body, changing its colour from head to toe, transforming her into a lifeless form. I knew then that I had lost something irreplaceable: that was unconditional love. A love so powerful and so great that it could never be replaced. This woman knew me better than anyone in the world. She had watched me my entire life, through every trial and tribulation she was there for support. How could all of that, every conversation, joke, argument, dilemma, happy memory or trauma we had shared, just be gone, like that? Suddenly everything had changed. Everything was put into perspective.

I cancelled all the stuff I had planned and was given two more weeks off work, which led up to the Easter holiday. The Deputy Dean of the art school wrote me a really kind message saying, 'I know that you are very close to your mum so this must be extremely hard to deal with and we are all thinking of you'. I was really touched. How did she know I was so close to my mum? I guess I must have talked about her a lot. My friend Sarah went into my flat and dug out more clothes to post me down a bundle of essentials. By the time we had organised the funeral, started dealing with the paperwork and began sorting out some support for my dad, I had been away for nearly six weeks. I dreaded the train journey back up north, but my friends rallied around to meet me and Diane's daughter Martina moved into the spare bedroom in my flat. I went back into work to help with the end of semester assessments. I don't know how I got through it. It was spring fortunately, the days were getting longer and Glasgow was a slightly more bearable place to be.

We spent time looking after Diane. I took her to the polling station to vote in the local elections on 4 May 2017. That Saturday, we had a beautiful day in Pollok Park and went for dinner and more dancing in The 78. Diane just did not want to stop dancing. The next morning she took a turn for the worse, the ambulance came and they took her to the Hospice in Yoker. I began to visit her several times a week, normally cycling along the canal or on the old railway line near the Clyde which had been closed by Beeching in the '60s and was now a cycle track. I was well aware of the moment when I crossed the line from Glasgow to West Dunbartonshire. If this had been 2016, what would I have done?

We listened to music and watched old films of Diane's work as part of the feminist art band *Disband* in the '80s. We had grown very fond of each other. I found it comforting to spend time with someone so close to death and to have Martina there to share this life-changing experience with. On my second to last visit I was wearing my 'Lesbians & Gays Support the Miners' t-shirt which my cousin had given me. It was a replica of those sold at the 'Pits & Perverts' benefit concert in 1984 depicted in the film *Pride* – a great story of solidarity between marginalised groups fighting back against those in power (the opposite of 'divide and rule').[19] Diane was drifting in and out of consciousness. We listened to *The Very Best of Dusty Springfield*, the soundtrack she had used in her last performance about her much-loved brother Donald who had died of AIDS in 1992. Dusty was also a favourite of my mum's. Diane woke up briefly, smiled at me and said, 'Cool t-shirt'. That was the last thing I heard her say. She died the following evening, aged 68.

I had never experienced grief like I was forced to confront in 2017. It sent me spinning out of the orbit of my normal world; all my priorities were transformed overnight. 'Work' was no longer 'the meaning of life' – the phrase which had been my Welsh grandpa's mantra. Work was now something that you had to do or could just use as a distraction from the pain. My identity was re-shaped around family and friends (as my ex-love Isla had so wisely advised), and most importantly around my mum. I went on a pilgrimage to visit her student home in Bethesda in North Wales where I got to hear the Penrhyn Choir sing 'Calon Lân'. There was barely a day when I wasn't in tears. I read all her travel diaries from her trips to Uganda, where she ran a charity helping some of the poorest children in the world get access to education. It was so comforting to hear her voice in my head as I read. I finally came to appreciate how her values and life experiences had shaped the person I'd become. I owed her everything, I just hoped she knew.

And yet, amongst all the disorientation, there was something curiously familiar about the grieving process. I realised that the constant feeling of absence, of loss, that something profound was missing from my life, had haunted me ever since I arrived in Glasgow in 2008. It was a feeling that was only ever ameliorated when I travelled back to Ealing to the family home. In August 2017, I found myself there again. My sister had taken my dad to stay with her in Norwich and I was left totally alone. I rattled around the house. The silence was intolerable. The comfort that place had provided my whole life was now no longer there. The emptiness would be with me always from now on, no matter where I went. There was now no point in even dreaming about leaving Glasgow, as there was no home to return to anymore. I had followed the artworld 'rat race' to Glasgow and I lost my soul in the process. Was there really any hope for me now?

I was forced to take a break and adapt to a different pace of life. I didn't think I would ever be able to make any art again – not after all of this. Writing seemed a more therapeutic process and a simpler means of communication. I realised that optimism and hope – the things which Harry Burns, Scotland's former Chief Medical Officer, says are lacking from so many Glaswegians' lives[20] – can only come from *the possibility of change*. It's the possibility of moving towards something better which gives us all a reason to carry on. That was the project I had begun in Glasgow in 2016 – fighting for a sustainable city of the future in which we can all live happy, healthy and creative lives – and it was to that project which I would eventually have to return. My 'core economy' was in crisis, but maybe through the act of fighting for a fairer city, I would be able to rebuild it, find 'local love', and finally make this brutal place feel like home.

Part 3

The Sustainable City of the Future

CHAPTER 9

Think Global

Climate emergency

SO MUCH HAS changed in the short time that I've been writing this book, since summer 2018. As I've been holed up in my flat in Glasgow relentlessly tapping away on the keyboard of my old laptop,[1] it's been inspiring to watch the rest of the world finally waking up to the state of 'climate emergency' our species is in. 16-year-old Swedish activist Greta Thunberg has done much to mobilise young people around the world since staging her first protest outside the Swedish Parliament in August 2018. She initiated the global 'school strike for climate' movement that November, encouraging young people everywhere to walk out of their classrooms by saying:

> For way too long the politicians and the people in power have gotten away with not doing anything to fight the climate crisis, but we will make sure they will not get away with it any longer… we are striking because we have done our homework and they have not.[2]

In October 2018, the United Nations Intergovernmental Panel on Climate Change (IPCC) met again in Incheon in Korea to plan how to implement the Paris Agreement of 2015 – the international pact which aims to limit global temperature increases to 1.5°C above 1990 levels. The IPCC concluded that 'rapid, far-reaching and unprecedented changes in all aspects of society' are required over the next 12 years (before 2030) in order to achieve this goal and maintain a climate on our planet which can continue to support human life. As well as governments taking urgent action at global, national and local levels to facilitate this transformation, they say we must all make changes to our own lifestyles – eat less meat, use less energy and use public transport (as opposed to planes and private cars). Debra Roberts, the co-chair of the IPCC said 'the decisions we make today are critical in ensuring a safe and sustainable world for everyone, both now and in the future'.[3]

When the UK's Extinction Rebellion activist group launched in October 2018, they used the IPCC's evidence as their rallying cry: '12 years left to save the earth!' The message from both was clear: we only have until 2030 to radically change the way we live. If we fail to act now, we 'will significantly worsen the risks of drought, floods, extreme heat and poverty for hundreds of millions of people',[4] running the risk of causing our own species' extinction. We must start treating the 'climate emergency' with the same sense of urgency as we have done other global crises, such as the dramatic reorganisation of our economy and our use of resources during the Second World War. Factories currently producing unnecessary carbon-intensive stuff, must be transformed to help build our new sustainable infrastructure instead (world-class public transport, renewable energy, insulated homes and decarbonised heating). As Greta Thunberg said at the European Parliament on 21 February 2019, 'we need to focus every inch of our being on climate change, because if we fail to do so, then all our achievements and progress have been for nothing'.[5]

Closer to home, in November 2018, Climate Ready Clyde, a consortium of public and private sector organisations published a report said to be 'the most comprehensive for a city region in the UK', which shows how temperature increases will directly affect the Glasgow area – with many major roads, railways and hospitals at risk from flooding and coastal erosion in the next 30 years.[6] The report talks about financial cost of climate change to our city region, 'estimated to be £400 million each year by the 2050s' with increased risks of 'large economic shocks from major weather events', which could cost many millions of pounds more per event. 'In many cases', they say, 'these impacts will fall on disadvantaged and vulnerable groups'.[7] It talks about 'climate justice' and the need to support 'those who are disproportionately affected by climate change risks due to socio-economic status, race, gender, or disability'. In December 2018, Glasgow Live also published maps on its website showing large areas of our city region along the Clyde which will be underwater if we hit 2°C temperature rise, worse still if and when we hit 4°C.[8]

After Extinction Rebellion's global 'week of rebellion' in April 2019, which saw activists shut down many major roads in Edinburgh and London, demanding that governments 'tell the truth' about the climate and ecological emergency by 'communicating the urgency for change', Scotland's First Minister Nicola Sturgeon became the first leader to declare a 'climate emergency'.[9] Later that week, the Committee on Climate Change (CCC), the independent body advising all the UK's governments on 'building a low-carbon economy and preparing for climate change', laid out a plan for the whole of the UK to 'reduce its greenhouse gas emissions to net-zero by 2050' (Scotland's target is now net-zero emissions by 2045).[10] The UK Parliament promptly followed by declaring a 'climate emergency' on 1 May 2019.[11] Glasgow City Council also did so on 16 May 2019. Although Extinction Rebellion and the IPCC maintain that the CCC's new targets are not ambitious enough to avoid climate breakdown, they are at least a step in the right direction. But at a local/regional level 2030 must remain our goal.

I finally feel vindicated in the action I took in 2016, and the dramatic reduction in carbon emissions I was able to achieve by localising my existence for that one year (evidenced in my Carbon Graph on pages 138–9). In order to address the climate emergency we must now make those 'rapid, far-reaching and unprecedented changes in all aspects of society', to enable this same dramatic reduction for everyone currently living a carbon-intensive lifestyle, but also, more importantly, to ensure it becomes permanent. Although middle-class guilt can be a powerful motivational force, we cannot rely on it alone to solve the climate crisis. We need governments to act. And because of all the wealthy vested interests at play (oil companies have spent more than $1 billion lobbying against action on climate change since 2015)[12] we must continue to force them to do so. We must take the action which is 'scientifically necessary', and not just what the people in power deem to be 'politically possible', and we must transform our broken political systems in the process.

As we saw in Chapter 1, it has been the neoliberal policies

of privatisation, deregulation and trade liberalisation implement-ed across the world since the '70s and '80s, which have been the key accelerating factors towards the triple crises of increas-ing social inequalities, catastrophic environmental destruction and frequent financial instability and uncertainty that we now face. And so in order to deliver 'climate justice' with the holis-tic solutions which can address all three of these crises at once, it is those neoliberal policies which we must urgently challenge:

- Privatisation must become *democratically accountable public ownership* of all our essential goods, services and infrastructure, managed at local/regional or (occasionally) national level.[13]
- Deregulation must become *reregulation* of all aspects of the economy that are causing unnecessary social and environmental harms.
- And in order to reduce carbon emissions and create local jobs, trade liberalisation must become *localism and protectionism*. But instead of just saying 'No', we need positive alternatives and incentives to encourage everyone to 'buy local'.

The aim of the third and final part of this book, is to return to some of the many issues and ideas that have been raised so far, to sketch a variety of ways we can 'unfuck' the system (as Robin McAlpine from Common Weal says),[14] and to start to build 'the sustainable city of the future' which will enable us all to live happy, healthy and creative lives. As well as transforming our broken political systems, challenging neoliberal policies and creating positive alternatives, we must also urgently address our materialistic value systems and the way our societies measure 'success', and indeed, 'failure'.

Downward mobility

In a world where the richest 1 per cent own two-thirds of global wealth and the richest 10 per cent cause half of global carbon emissions,[15] it is wealthy people who need to change their lives the most. That was the key message of *The Glasgow Effect* – highlighting the one thing which privileged people take for granted the most: their freedom of movement. The freedom to travel around the world for work or for leisure – to capitalise on the opportunities of the globalised 'knowledge economy', thereby exacerbating inequalities as those who are less 'mobile' for socio-economic reasons are left further behind. International travel is the one thing which isn't accounted for in most intergovernmental agreements on climate change, including the Paris Agreement of 2015.[16] As you can see from my Carbon Graph on pages 138–9, it is aviation that is by far the most damaging form of transport, because of its high carbon conversion factor and the comparatively longer distances it enables us to travel.

As writer George Monbiot explains: 'there is no technofix to the disastrous impact of air travel on the environment... the only answer is to ground most of the aeroplanes flying today'.[17] If we want to stand any chance of meeting the 1.5°C target, this is a loophole that must urgently be closed. Aviation will 'produce 22 per cent of the world's carbon emissions by 2050, unless there is a sharp change in policy'.[18] Research published in 2018 showed that 'Tourism is responsible for nearly one tenth of the world's carbon emissions', and, of course, it is the world's wealthiest that are taking all the holidays.[19] We must urgently localise our industries and economies so we can all travel less – as I said in my talk about *The Glasgow Effect* on 7 January 2017, 'it is privileged people who are gonna have to take a hit'.[20]

The UK is the fifth largest economy in the world, and according to the UN Special Rapporteur on Extreme Poverty & Human Rights, Philip Alston, who visited us in November 2018 to investigate the impact of government austerity, it is a place of great contrasts:

> Britain is one of the leading financial capitals in the world, a thriving industrial and financial centre – contrasted with the fact that a fifth of the population (14 million people) are living in poverty, four million of those are more than 50 per cent below the poverty line, 1.5 million are destitute.[21]

It's not that the UK needs more money or more economic 'growth' to tackle poverty. We already have more than enough to provide good lives for everyone. What we must urgently do is to decentralise and redistribute that wealth. One of the answers is simple. It was beautifully articulated by the Dutch historian Rutger Bregman when he confronted the billionaires at the World Economic Forum in Davos on 25 January 2019. He said we need to 'stop talking about philanthropy and start talking about taxes'.[22] It shouldn't be down to the wealthiest people to decide if/when they 'donate' a bit of money to whatever 'charity' they choose. We all know they have dubious ethical codes or they'd never have acquired so much money to begin with. As Rutger Bregman says, the only immediate solution to the problems caused by neoliberalism is 'Taxes, taxes, taxes. All the rest is bullshit in my opinion'.[23] We need a massive reinvestment of funds into the poorest parts of the country to start equalising our living standards across the board.

Equality means downward mobility as well as upward mobility, until we meet in the middle. And 'climate justice' means that we must also be equalising wealth and resources on a global scale. There is already a 'shift in global economic power' away from Europe and America to other parts of the world.[24] We are the ones writhing around at 'the end of an empire' (the value of the pound has been falling steadily for nearly a century)[25] and through a few final altruistic acts of self-destruction, particularly Brexit, our transformation back to the small insignificant island that we are will be complete. The most important thing to note about the decade that followed the economic crisis of 2008 is that it was only in the west where living standards fell; elsewhere around the world they improved. As John Lanchester writes:

At the time of the crash, 19 per cent of the world's population were living in what the UN defines as absolute poverty, meaning on less than $1.90 a day. Today, that number is below 9 per cent. In other words, the number of people living in absolute poverty has more than halved, while rich-world living standards have flatlined or declined. A defender of capitalism could point to that statistic and say it provides a full answer to the question of whether capitalism can still make moral claims. The last decade has seen hundreds of millions of people raised out of absolute poverty, continuing the global improvement for the very poor which, both as a proportion and as an absolute number, is an unprecedented economic achievement.[26]

You could argue, therefore, that the hardship that has been experienced in the west since 2008 is simply part of the necessary global redistribution and equalisation of wealth. The problem is that whilst this has been happening for the majority of the UK's population, the super-rich continue running away with more and more. This is something that we can only address through global cooperation, cutting pay packets for the highest earners and through 'Taxes, taxes, taxes'. Taxes on consumption, taxes on pollution, taxes on long-distance, high-carbon travel, taxes on wealth, property and land, taxes on all the things that are causing unnecessary social and environmental harms. We must urgently tackle the 'Pareto Principle' and the 'unfair system that gives the greatest freedom and mobility to the biggest polluters and those with the most resources'.[27] Then, once inequality has been effectively reduced, we must ensure wealth is 'predistributed', so that we never reach this crisis point again.[28]

Back to the future

It is our moral duty here in Britain – the place which has benefited most from exploiting fossil fuels, other resources and human labour expropriated from other parts of the world – to allow those places to continue to increase their living standards while we reduce our consumption. If you look again at *The Other Forecast* graphs on page 98, you can see clearly how global carbon emissions and energy use fell immediately after the financial crash of 2008 because we in the west were producing and consuming less. It's nearly impossible for 'growth' and carbon emissions to be 'decoupled'.[29] But what does that decrease actually mean? Does it mean that we can no longer buy loads of unnecessary crap to fill our homes? That we can no longer jet off on holiday? That we can no longer own and run our own cars? Is that really going to be that bad? Not if we build a city, society and economic system, where these things are no longer necessary.

When I'm interviewed about the Bring Back British Rail campaign, critics often say that we don't want to go 'back to the '70s' and the days of the 'soggy British Rail sandwich' (conveniently ignoring fact that many of our current private Train Operating Companies have now removed catering altogether because it's a service they claim is 'no longer sustainable').[30] Often, ironically, these are also the same conservative critics who are taking us 'back to the '70s' anyway – to before Britain voted to join the European Economic Community in 1973 – by pushing through Brexit.[31] But going 'back to the '70s' is actually what Naomi Klein (author of the book, *This Changes Everything: Capitalism vs the Climate*) says we need to do in order reduce carbon emissions to the extent that is 'scientifically necessary':

> The truth is that if we want to live within ecological limits, we would need to return to a lifestyle similar to the one we had in the 1970s, before consumption levels went crazy in the 1980s. Not exactly the various forms of hardship and deprivation evoked at Heartland conferences [an American neoliberal think tank]. As Kevin Anderson [climate scientist] explains: 'We need to give newly industrialising countries in the world the space to develop and improve the welfare and well-being of their people. This means more cuts in energy use by the developed world. It also means lifestyle changes which will have most impact on the wealthy...'[32]

I admit I was only alive for eight and a half months of the '70s, but from what I have garnered from my parents and learnt about since, there are many things we could now learn. 'Social mobility' was less of an issue because class divisions were less pronounced. Income inequality was at its lowest levels for the whole 20th century between 1975–8.[33] This was because the top rate of income tax was 83 per cent or as much as 98 per cent on certain categories of 'unearned income'.[34] Now the top rate is 45 per cent in England and Wales (on incomes over £150,000) and just 1 per cent more in Scotland. And we wonder why there's inequality? An income tax rate of 98 per cent on salaries over £150,000 is essentially a 'maximum wage' and that is what we need to reduce inequality and redistribute wealth to those who need it most to the extent that is necessary. I only know one person who'd have to pay it anyway – the principal at my university whose salary is now £300,000.[35] Perhaps then the university would finally realise that pay packets like that really are obscene.

The '70s were a time when the 'costs of living' were affordable – there was far more social housing available and if you bought a house the average price was just three times the average salary, whereas now it is more than eight times.[36] My parents were able to buy their house in Ealing, which would now be near impossible for anyone with a regular teaching job. Because living costs were lower, people had more time to invest in volunteering and creative and political activity, which did not offer any financial return. It was this 'labour of liberation'[37] (coupled, of course, with that more oppressive culture my parents lived through) that gave rise to an emancipatory social climate and some of the most inspiring political movements of recent times – Women's Liberation, the grassroots network of local women's 'consciousness raising' groups all around the UK and across the globe, and the Gay Liberation Front. Without both these movements we would not have many of the rights and privileges we take for granted today. The Gay Liberation Front's manifesto was one of solidarity, ending with the demand 'All power to oppressed people!'[38]

The emancipatory social climate of the '70s also enabled artists to be more radical and community-focused,[39] working in tandem with these political campaigns.[40] In Glasgow, the Third Eye Centre established in 1974 by Tom McGrath was far more radical than its refurbished and relaunched shiny successor the CCA. Incidentally, following my talk about *The Glasgow Effect* on 7 January 2017, somebody tweeted: 'It reminds me of conversations in the café at the Third Eye Centre in the '80s

#Bless'.[41] *The Cheviot, the Stag & the Black, Black Oil*, written and performed to the principles of radical 'popular theatre' by John McGrath (no relation to Tom) in 1973, toured schools and community centres around Scotland, with the direct political aim of alerting people to the coming capitalist exploitation of North Sea oil. Had it not been for the incoming neoliberal policies of privatisation, deregulation and trade liberalisation and had we kept the proceeds of all the oil extracted from the North Sea over the last 40 years for the common good, we could have used it now to fund the urgent transition to a low-carbon economy where the burning of fossil fuels to fulfil our energy needs is no longer necessary. This is what Norway did after it discovered its own oil in the North Sea in 1969 and their country's people now have more than $1 trillion sitting in their sovereign wealth fund.[42]

The first two months of 1974 also saw the implementation of the 'three-day week' in order to save energy. Although this caused turmoil as wages fell and unemployment rose, it did save energy and cut carbon emissions. In their report, *21 Hours: Why a shorter working week can help us all to flourish in the 21st century*, the New Economics Foundation also show that:

> when the [three-day week] crisis ended, analysts found
> that industrial production had dropped by only 6 per
> cent. Improved productivity, combined with a drop
> in absenteeism, had made up the difference in lost
> production from the shorter hours.[43]

People work better if they work less. Had this 'three-day week' been handled in a more collective fashion, making allowances for the poorest who would be hit the hardest through lack of income, rather than being imposed in a top-down way by a failing Tory Prime Minister, Edward Heath, NEF show that this is one of the key ways to fight climate change by creating a society where we can all 'work less, consume less, live more',[44] and find the 'time to live sustainably, to care for each other, and simply enjoy life'.[45]

And from a public transport perspective, the '70s were also a time of positive change. Following the destruction caused in the previous decade by Ernest Marples, Richard Beeching and the short-sighted and self-serving leaders of our local authorities who ripped up all our trams, Barbara Castle (the first and only woman to be Transport Secretary, until Ruth Kelly in 2007) aimed to salvage what she could by enshrining public transport as a social good. In her epic radical Transport Act 1968, she created:

> Passenger Transport Executives for the major
> conurbations [Glasgow, Manchester, Newcastle,
> Liverpool and Birmingham] whose job it would be
> to produce master plans for transport in their areas,
> run local bus services and turn around the urban rail
> networks that had survived the Beeching era.[46]

As a result, in Glasgow in the '70s under the Greater Glasgow Passenger Transport Executive (an ancestor of SPT set up in 1972 as a result of the Act, and then replaced by the Strathclyde Passenger Transport Executive when the Strathclyde Regional Council was set up from 1975–96), we had the closest we've ever come to the 'holy grail' of fully-integrated public transport – with buses, trains and Subway working together in harmony. A 1979 telly advert proudly presented the outcome of the policy:

> Trans-Clyde: Strathclyde's new interlocking, interlinking,
> integrated transport system. Trans-Clyde – brand new
> trains at brand new stations on the brand new under-
> ground, interchange with trains and buses, ride the brand
> new cars around… link-up, link-up, link-up! Trans-Clyde
> links you Clyde-wide![47]

It was a brilliant vision to serve and reconnect all our region's citizens who had been displaced and disempowered during that short-sighted post-war policy of deliberate urban sprawl. The Passenger Transport Executives (PTEs) were constituted to contin-ually expand and improve their network for the benefit of pas-sengers, not to allow millions of pounds to seep out of the system in shareholder profits (FirstGroup made £50.2 million in profits from their UK bus businesses in 2018 alone).[48] In Manchester, under the SELNEC PTE, they introduced their first pollution-free electric bus the 'Silent Rider' in 1973.[49] As we saw in Chapter 7, the extensive powers these Passenger Transport Executives had were stripped away when Thatcher's government deregulated the buses in 1986 – a brutal act of vandalism which the Scottish Government, in its 20 years of power over our transport network, has done absolutely nothing to address.[50] Bring back Trans-Clyde I say!

It's the myth of progress that says we're always heading towards something better – that technological innovation is always going to be good for us. It's often the opposite. In *Small is Beautiful*, EF Schumacher shows that 'the amount of real leisure a society enjoys tends to be in inverse proportion to the amount of labour-saving machinery it employs' – something Gertie and Gus the bears learnt the hard way.[51] The truth is that the neoliberal policies of privatisation, deregulation and trade liberalisation have destroyed many of the great social achievements of the post-war period and have created a deeply divided society in Britain, with much lower levels of well-being. For my entire lifetime, in many respects, we have actually been going backwards, despite what our mainstream measures of economic 'success' may say.

Prosperity without growth

As we saw in Chapter 6, it was feminist economist Marilyn Waring who first began to highlight the root cause of many of our problems. She aimed to show how it is the Gross Domestic Product (GDP) measure of economic 'growth' (as dictated by the United Nations System of National Accounts) that is pushing our species towards (self-)destruction. Infinite growth on a finite planet is inherently unsustainable.[52] As Marilyn Waring says, this system

> recognises no value other than money, regardless of
> how that money is made. This means that there is
> no value to peace. This means that there is no value
> to the preservation of natural resources... for future

generations. This means there is no value to unpaid work, including the unpaid work of reproducing human life itself, including the unpaid work of women who feed and nurture their own families. This system cannot respond to values it refuses to recognise. This system leaves out the work of half the population of the planet... and the planet itself. It is the cause of massive poverty, illness and death of millions of women and children and it is encouraging environmental disaster. This is an economic system that can eventually kill us all.[53]

In their 2019 book, *The Economics of Arrival: Ideas for a Grown Up Economy*, Katherine Trebeck and Jeremy Williams take a more nuanced approach. Rather than reinforcing what they call the 'false binary of painting all growth as either good or bad',[54] they celebrate the achievements of economic growth globally in bringing more and more people out of poverty (the same rising living standards in other parts of the world, which John Lanchester highlights above).

Trebeck and Williams show that in some parts of the world, 'growth' is still necessary to raise material standards of living for the world's poorest, however, in 'developed' countries like the UK, 'beyond a certain point, economic increase ceases to be meaningful'[55] and often 'growth is bringing a diminishing suite of benefits and often even increasing harm'.[56] In the UK, one pound in every five of current government spending now goes to mitigate the damage being caused by our broken economic model.[57] This is known as 'defensive expenditure' – completely unnecessary costs which we must now strip out (interestingly 'commuting' is considered to be one of them). GDP also creates 'perverse incentives', such as the incentive to rip up world-class sustainable public transport networks to 'create a market' for oil and cars. Trebeck and Williams also highlight all the unnecessary expenditure spent on 'consolation goods' – people buying crap to cheer themselves up because our lives are so miserable. As the fifth largest economy in the world, the UK is already 'GDP-rich'. That means that, as Trebeck and Williams would say, we have 'Arrived' and therefore no longer need further 'growth':

The idea of Arrival does not imply that all problems are solved. It does not suggest that everything is resolved and everyone has what they need. Rather, it is the idea that a society collectively has *the means* for this.[58]

The task then becomes what they call 'making ourselves at home' in our economy: about sharing out that wealth – back to Rutger Bregman's 'Taxes, taxes, taxes' – and also, about redefining how we measure 'success' in order to halt further harm. The 'growth agenda should give way to new priorities that are qualitative rather than quantitative'.[59] We need alternative measures of 'prosperity', which take into account the things which actually bring meaning and happiness to our lives – the 'core economy' of family, neighbourhood and community, of care and of love, volunteering or creative and political activity, and not just those things with a price tag attached. Or at the very least, we need alternative measures of 'prosperity' which actually take into account the real damage that continual growth and 'overdevelopment' is doing.

The Genuine Progress Indicator (GPI) is one of these alternatives, which aims to measure the true cost of economic 'growth' to people and the planet (by measuring the 'externalities', as economists would say) – particularly the impact of 'growth' on people's well-being. The historical GPI calculations tell a very different story from that told to us by those under the spell of neoliberalism. As measured across 17 different developed countries, they show 1978 to be the peak of 'genuine prosperity', the year before I was born. 'Since then there has been a downward trajectory due to growing inequality of incomes and environmental degradation'.[60] Some places in America and Canada are now actually starting to use GPI as a planning tool and we must do the same. But as well as these generalised global measures of 'success', we also need ones that are specific to our localities, as many of the world's problems arise from our tendency to try to shove everyone into the same ill-fitting box.

In 2012, Oxfam Scotland began the task of creating a new measure of 'prosperity' called the Humankind Index. Introducing the project, they state that: 'the model of the economy that has dominated the UK for most of the last century is outdated and has failed to address poverty', and that 'in order to achieve sustainable livelihoods for all, the range of assets that are important to people must be recognised and taken into account at all levels of government decision making'.[61] Their new Humankind Index is based on discussions with almost 3,000 people across Scotland in some of our most marginalised communities. Economic measures of 'success' take a backseat. People are clear that sufficiency and security of income is much more important than a high income. Instead, the three key strands that people value the most are identified as:

- An affordable, decent and safe home and good physical and mental health.
- Followed by living in a neighbourhood where you can enjoy going outside and having a clean and healthy environment.
- Followed by having satisfying work to do (whether paid or unpaid); having good relationships with family and friends; feeling that you and those you care about are safe; access to green and wild spaces; and community spaces and play areas.[62]

Is that really too much to ask? While our national governments are obsessed with competing with one another for economic prowess on a global stage, the people who contributed to the Humankind Index clearly show that it's the day-to-day stuff which unfolds in our specific localities and our 'immediate neighbourhoods', which is what really actually affects our well-being. And yet, as Darren McGarvey clearly articulates, in Glasgow, these are the things which have consistently been most overlooked:

In working class communities symbols of culture and identity are ripped out, renamed, sold off, mysteriously burned down, gentrified and/or demolished routinely in the name of progress. This progress usually comes in the form of a road which connects affluent towns and suburbs to shopping destinations in cities.[63]

Deconsumerisation

In order to build 'the sustainable city of the future', we must return to the municipal vision of our early city 'Corporations' (described in Chapter 7), which prioritised the direct provision of local goods, services and infrastructure to serve local needs, instead of pandering to the artificial 'wants' that our global capitalist system creates. With new measures of 'success' that are 'qualitative rather than quantitative', we can create a society which is no longer driven by profit and growth. That's how we resolve John Maynard Keynes' 'paradox of thrift', so that 'every human being [actually can be] an artist' – learn to love living a thrifty lifestyle – without the whole economic system falling apart.

The insidious pressure to spend, which corrupts our own value systems, can then be removed. This is what it means to 'deconsumerise' our economy so we can refocus on what really matters instead. In 2007, in São Paulo in Brazil, they introduced the 'Clean City' law that banned billboards from the urban area altogether, protecting its 22 million population from this wide scale disinformation.[64] Glasgow, the city which is 'famous' for its advert-smothered buses, should do the same.[65] 'Happy people don't shop!' says Robin McAlpine from Common Weal, and in their report on 'deconsumerisation', they show:

> In a post-consumerist Scotland there is a need to provide greater access to life-long learning… through more part-time college places and evening classes.[66]

We must urgently move beyond the absurd situation where we have to 'create a market' for something which is actually killing us – oil, cars, sugary foods and so much more. When qualitative growth replaces quantitative growth, the aspiration to learn and help change our city for the better will replace the aspiration to buy and escape to somewhere else. Community participation will replace monetary wealth.[67]

Act Local

City as a site for social change

THE MESSAGE OF Patrick Geddes' command to 'Think Global, Act Local' is that we must all strive to understand the global context (described in the previous chapters). And, that we can all affect change by applying this knowledge to our local contexts: the places – towns and cities – where we live. Writing about Patrick Geddes for the exhibition *Politics of Small Spaces* at the Cooper Gallery in Dundee in autumn 2018, Lorens Holms says:

> Our actions are always place-based, we build this city or that one, not cities generally or globally, but every local action should be the result of global reflection and it will have global consequences.[1]

The intractable problems that humanity faces become much easier to tackle at a local level, as long as all the powers necessary to address them lie with the local or regional authority, and not in the hands of a distant administration (in Edinburgh, London or Brussels). As writer George Monbiot says, 'the smaller the scale, the more meaningful democracy is'.[2] Then the city really can become the ideal site for social, environmental and economic change. American activist Debbie Bookchin argues that municipalism (the system where cities and regions have freedom and autonomy from the nation-states in which they're located) 'isn't just one of many ways to bring about social change, it is really *the only way* that we will successfully transform society'.[3]

As Jimmy Reid warned in his famous 'alienation' speech, it's not just that people in Glasgow feel like 'victims of blind economic forces beyond their control', they also feel 'excluded from the processes of decision making', because our political systems are broken as well. In Glasgow we have two key problems. Firstly, that our current political structures are completely dysfunctional. And secondly, that toxic tribal politics often get in the way of solving practical problems. Both present barriers to addressing the urgent task at hand – the 'rapid, far-reaching and unprecedented changes in all aspects of society', necessary before 2030 to keep global temperature rise under 1.5°C.

As Gerry Hassan said at our *The Glasgow Effect: A Discussion* event at the Glad Café on 3 February 2016, 'only a lunatic would think they would [affect change in this city] by going through Glasgow City Council or becoming a councillor'.[4] Our local democracy is completely ineffective. Glasgow City Council in its current form is an unwieldy size – far too big to have a meaningful relationship with its citizens and far too small to be able to coordinate strategic regional infrastructure and services, such as our transport network. Transforming these broken political systems is therefore an urgent part of the challenge we face. The only way to achieve this may be for citizens to 'win back the city' and to change the system from within.[5]

Regional power

As we saw in Chapter 7, Glasgow was brutally reshaped in the post-war period as a result of those short-sighted policies of deliberate urban sprawl. It was transformed from a compact, manageable, walkable and sustainable city, which the Glasgow Corporation could easily govern, into a 'conurbation' – a term also defined by Patrick Geddes in 1915 to refer to a region in which many smaller towns and cities merge together to become one 'travel to work area'. Four hundred thousand commuter journeys (loads more 'defensive expenditure') were created overnight. When Transport Secretary Barbara Castle was charged with sorting out the mess our public transport was in as a result of the terrible decisions made by all those men in the decades before, she realised, in her 1966 White Paper, that in the new era of the 'conurbation',

> the fragmentation of public transport and highways,
> traffic control, parking and land use planning... could

only be put right by a major reorganisation of local government.[6]

Political structures had to evolve in line with our new physical structures. And so on 15 May 1975, the Strathclyde Regional Council was born. With powers over education, social work, police, fire, water and sewage, strategic planning and roads and transport, its job was to deliver goods, services and infrastructure to all the new places where people now lived, ensuring wealth was decentralised and redistributed around the region. Strathclyde Regional Council had responsibility for more than two and a half million people (nearly half of Scotland's population). Our new region was bigger and more unwieldy than EF Schumacher would have advised – he said the ideal size for any city 'is probably something of the order of half a million inhabitants',[7] but it was a *necessary compromise*, given the damage done by deliberate urban sprawl. Beneath the Strathclyde Regional Council sat 19 District Council areas which had responsibility for housing, libraries, leisure and other local amenities (of which Glasgow District Council was one).

Strathclyde Regional Council had the powers necessary to fend off the privatisation of Scotland's water, holding a referendum in March 1994 which showed 97 per cent of people were against the policy being imposed by Tories in Westminster (1,194,667 people voted 'No').[8] That's why we still have publicly-owned water here today. Unlike 'Scottish Power' and 'ScotRail', Scottish Water is actually still Scottish! And, as we saw in the previous chapter, Strathclyde Regional Council had the powers necessary (until bus deregulation in 1986) to deliver the 'holy grail' of fully-integrated public transport to reach all corners of the region and reopen many new train lines and stations which had been axed by Beeching.[9] This meant everyone in the 'travel to work area', could actually travel to work in an affordable and sustainable way.

The problem was that Strathclyde Regional Council was seen as 'too powerful' by some. It appeared to have too much autonomy and control. So in 1996, John Major decided to abolish it. Shortly after he privatised our railways with the Railways Act 1993, his Local Government (Scotland) Act 1994 legislated to remove our regional layer of democracy. Strathclyde's 19 District Councils (there were 53 in total in Scotland)[10] were replaced with just 12 unitary authorities including the new Glasgow City Council (there are now only 32 in total in Scotland). In 1996, all the lines on the map around Glasgow were redrawn and people across Scotland were left further disempowered. Scotland became 'the least democratic country in the European Union', possibly even 'the least democratic country in the developed world'.[11] With no regional authority to oversee the decentralisation and redistribution of wealth, or to undertake the strategic planning of our regional infrastructure and services, we have been left with

> the absurdity of the Glasgow City Region's wealthiest suburbs carved up into self-contained enclaves, where the residents enjoy relatively low rates of council tax, while the urban core of the city itself is home to the poorest communities in the region, yet those residents must carry the burden for maintaining and operating all the core services and amenities that are enjoyed by its wealthier suburban 'free riders'. Not only has Glasgow been stripped of its residential tax base through historic depopulation and relatively recent gerrymandering of its suburbs, but the advent of the Scottish Parliament has seen a continuing war of attrition against the power of local government.[12]

These are some of the reasons why, when I moved to Glasgow in 2008, I found a city riddled with potholes which I had to learn to navigate on my bike. And these are some of the reasons why we have some of the worst inequalities in wealth and health in Western Europe; why one of Glasgow's neighbouring councils East Dunbartonshire was voted the 'best place to live in Britain if you are a woman' in 2017,[13] in stark contrast to nearby parts of Glasgow, such as Calton, where your life expectancy could be nearly 30 years less.

We are suffering in Scotland as a result of austerity politics from Westminster – that is what's caused the sharp rise in homelessness forcing the Scottish, Welsh and Northern Irish governments to spend some of their national budgets mitigating the worst effects of malicious Tory welfare reforms.[14] But we're also suffering in Glasgow as a result of decisions made at Holyrood. The Scottish Government's cuts to local government budgets have been more severe in many cases than the Tories' 'down south'. In fact, Glasgow has been the ninth worst hit out of all 60 British cities in the decade of austerity since 2009, losing 23.3 per cent of our budget.[15] Glasgow's lack of local/regional freedom and autonomy has come at the expense of the Scottish Government's obsessive centralisation of power in Edinburgh.

Following devolution in 1999, Scotland has become 'the most centralised system of government of any country in Europe'.[16] Rather than urgently addressing the damage done by John Major in the '90s (neither his 'gerrymandering' of local government, nor his privatisation of our railways),[17] the Scottish Government, in its 20 years in existence, has continued to strip away Glasgow's local/regional power. Despite the fact that 'there is a link between the absence of strong local democracy… and the prevalence of inequalities'.[18] Just as British national identity was strengthened in the post-war period through the nationalisation and naming of all key services and industries as 'British', the Scottish Government know all too well that with centralisation comes national power. They have been busy building brand identity by labelling everything as 'Scottish', at the expense of our local democracy.

In 2013, both the Strathclyde Police and the Strathclyde Fire & Rescue Service were abolished in favour of centralised services: Police Scotland and the Scottish Fire & Rescue Service. But in terms of fighting climate change and investing in sustainable transport infrastructure, it's the centralisation of transport planning which has been most devastating. Transport Scotland – the centralised body responsible for strategic transport planning – came into being on 1 January 2006, as the result of the same Transport Act which turned SPT into a 'partnership', stripping away yet more of our regional powers. Since then, Transport Scotland, without any local/regional accountability, has invested billions of pounds in road building, locking in car-reliant, carbon-intensive lifestyles for decades to come.[19]

We're still living in the 'conurbation' that those arrogant post-war planners created. We can't just uproot everyone again and move them back into the city in order to make it more sustainable, as we now know the 'psychosocial' damage that would do. It is therefore clear that we still need a regional authority to effectively decentralise and redistribute wealth across the region and to over-see regional services and strategic planning in the whole 'travel to work area', particularly, of course, our transport. What to name this regional authority is another matter. And the confusion that arose during *The Glasgow Effect* project over the area in which I had to remain – was it 'Strathclyde', 'Greater Glasgow', 'Glasgow City Region' or 'Glasgow City'? – was perhaps symbolic of the state of limbo our region's governance is in.

Strathclyde Regional Council was perhaps too big, and must still have appeared a 'distant administration' to those living in its extremities in Campbeltown or the Isles of Islay or Mull. The 'Strathclyde' brand is also too associated with Labour and the 'wee hard men'[20] who betrayed Glasgow's people in the '80s,[21] which is another reason why the SNP in power at Holyrood want to erase that name from the map, at the expense of the people who happen to live here. In recent years, there have been attempts to reclaim some autonomy for our region, with the smaller area covered by the 'Glasgow City Region' brand, but this is a wholly undemocratic venture (with only eight council leaders steering it) which seems most concerned with pandering to business interests.[22] Maybe it's only 'Greater Glasgow' as a name which isn't tarnished with toxic tribal politics in some way. So from now on I will proceed with using that term to define everyone who now lives in Glasgow's 'travel to work area' and benefits from the goods, services and infrastructure in the city centre, and the Greater Glasgow Authority for the democratically accountable regional authority, which we desperately need to govern it.

Community control

Below the regional level, if we are going to proceed with the local Council areas that John Major created, then they also need many more powers devolved from Westminster and Holyrood.

The only real purpose of national governments should be to redistribute wealth to the places that need it the most and to devolve power over how to spend it down to the lowest level possible. Research by the Centre for Cities shows there are four ways to end local government austerity, and to 'help cities to strengthen their position as creators of prosperity'. They are:

- Freedom to raise new funds and decide how to spend them
- Fairer funding to ease the pressure on services
- Social care funding reform to reduce the burden on cities
- Long-term budgets to provide certainty[23]

It is clear that freedom and autonomy, and financial security and stability, are just as important to a Council's well-being as they are to an individual's. But it could be argued that with a new Greater Glasgow Authority (GGA), Glasgow City Council (John Major's creation that it is) could be abolished. As Lesley Riddoch shows in her book *Blossom: What Scotland Needs to Flourish*, echoing the findings of Oxfam Scotland's Humankind Index, it is at the community level where we most need power to be devolved. Before the local government reform of 1930, there were 871 civil parishes in Scotland who provided all local amenities for citizens. Compare that to the 32 Councils that we have now and you'll see it's no wonder people in Scotland feel so disempowered. As Lesley Riddoch says, 'Scotland is run by units that are far too big'.[24]

The civic infrastructure of Scotland's Community Councils has been in place since the Strathclyde Regional Council was set up in 1975, and they cover similar areas to the old civil parishes. But these have never had any real power and each only has an annual budget of £400.[25] If instead of Glasgow City Council, we re-empowered all our region's Community Councils to provide for our local amenities: housing, libraries, leisure facilities etc (in cooperation with the GGA which would oversee transport, health, education etc), then we would finally return to 'the average unit of local government in Europe' in terms of population size.[26] In France, the ratio of councillors to citizens is one to 125; in Scotland, it's one to 4,270 (compared to one to 2,860 in England).[27] This matters because there is a direct correlation between these ratios and voter turnout. If people know their councillors personally, they're more likely to vote (or even run for election themselves) and can more easily identify where the levers of social change lie. As we saw in Chapters 6 and 7, political engagement improves well-being. In a sustainable 'deconsumerised' society, community participation is what must replace monetary wealth.[28] This is how we return politics to a more human and accessible scale.

Fearless Cities

It's not just about getting the structures right. We need to transform how politics is done. And the latter can, of course, help us achieve the former. 'National political dogma and ideology are not relevant at a local level'.[29] That was the message of the group of local residents in Frome in Somerset who were so frustrated with the toxic party politics in their town that they launched a campaign to take back control of their council in the 2011 elections. 'Independents for Frome' had no manifesto, just a description of how they aimed to operate in a more open and participatory way, to facilitate the views, ideas, skills, knowledge and expertise of local people. They won a majority of seats and have been running the council in the interests of Frome's people ever since, with a focus on delivering sustainability, well-being and prosperity.

Glaswegian activist and anti-poverty campaigner Cathy McCormack could also see the urgent need to transform how Scottish politics is done. She coordinated a People's Parliament in 1999 (the same year the Scottish Parliament was opened). It aimed to bring people together who felt like 'silent witnesses, excluded from the policy-making structures that are meant to tackle poverty', and to give them a voice.[30] She said:

Apart from the gap between rich and poor there is a very big gap between the Government's policy makers and the reality of life people are living. I don't think that will ever change unless they really listen to what people are saying.[31]

We need a political system which gives everyone a chance to be heard. The idea of a Citizen's Assembly has been gaining ground, promoted by the Extinction Rebellion activist group as the only way we'll make the urgent decisions which are 'scientifically necessary' to fight climate change, and also by the progressive think tank Compass to resolve UK Parliamentary gridlock over Brexit. It is based on the idea of 'sortition' – that people are selected at random from within a certain area to serve on our assemblies, councils or governments, as they are our juries (instead of through elections). They are then presented with impartial evidence and asked to deliberate and make decisions collectively. This is a system that ensures we have autonomous individuals doing their stint of public service and then going back to 'real life', not the 'career politicians' desperate to cling on to power that Glasgow has been dogged with. This is a system which ensures we have people involved in decision-making who are 'At the Sharp End of the Knife'[32] of neoliberal policies day-in day-out (as Cathy McCormack would call it), not the small number who are reaping the benefits at the expense of everyone else. If we can't have 'sortition', then limits on the number of terms Councillors can serve is a must.

The 2019 book *Fearless Cities: A Guide to the Global Municipalist Movement* documents inspiring examples of groups of citizens all over the world, such as Independents for Frome, who have taken back control of their towns and cities to implement policies that actually serve their needs. In 2015, a group of people in Barcelona in Spain (a city with a population similar to Greater Glasgow) with no money and very little experience 'wrest the city from the entrenched political caste that had been running it for the past 40 years'.[33] Barcelona en Comú (Barcelona in Common), as they're called, ran a people-led campaign, which was deliberately not a political party, and got enough candidates elected in the May elections to take control of the city's administration. They remunicipalised many of the city's public services and infrastructure which had been privatised (bringing them back into local ownership and control), as well as transforming the governing structures of the city to make them far more 'transparent and participatory',[34] so as to improve how politics is done in the long-term (there's more on this in Chapter 12).

In Vancouver in Canada, when a group of local people took control of the city's government in 2008, they made the pledge to make Vancouver the 'greenest city in the world' by 2020. Because they had the powers necessary at a local level to make the changes they needed, by April 2015 they had already achieved their target of 50 per cent of journeys being made on foot, bicycle and public transport. In just 12 years, they have already gone from not even being in the top 500 to being the world's fourth greenest city, with a per capita carbon footprint of just 3.9 tonnes (compared to the Canadian average of 15 tonnes). This has all been achieved despite a Conservative national government withdrawing Canada from the Kyoto Protocol and pursuing carbon-based development elsewhere. Andrea Reimer, Vancouver Councillor for Environmental Action, says 'That's the difference a city can make'.[35]

With local/regional autonomy and power, we can fight climate change. Even though Donald Trump has withdrawn America from the Paris Agreement, better devolution of power at a federal level and specifically the autonomy of America's cities under their mayoral system, mean 'the failure of national government' may not have that bad an effect on plans to reduce America's carbon emissions.[36] In April 2019, America's second largest city Los Angeles launched its own Green New Deal, a plan to transition to a carbon neutral economy by 2050.[37] There are many elements to the plan aimed at 'protecting the environment, strengthening the economy and building a more equitable future', including building a zero-carbon electricity grid with the goal of reaching an 80 per cent renewable energy supply by 2036.[38]

In Hamburg in Germany, when local people became frustrated at the slow progress in decarbonising their energy sector and the lack of cooperation from the privately-owned energy companies and privatised energy grid, they ran a successful campaign to take back control of them all.[39] They forced a local referendum in 2013 and people voted 'Yes' to the remunicipalisation of their energy network. They then brought their gas companies back into local public ownership and control a few years later. Claire Roumet, director of Energy Cities, says that these remunicipalisations enabled Hamburg to bring together 'the different elements of the municipal energy sector and [they] can now carry out in a much more straightforward fashion a policy of energy transition to help fight against climate change'.[40]

Now there's a referendum I could finally enthusiastically get behind! One that does actually affect our 'immediate neighbourhoods' and our everyday lives. Just as Strathclyde Regional Council was able to save our water in 1994, with local/regional autonomy and power, we could ask the people in Greater Glasgow whether they would like public ownership and control over our public transport network and our energy production and supply. I am sure, just as in 1994, people would vote against privatisation.

Looking south on a clear day in Glasgow, you can see hundreds of wind turbines. None of them are publicly- or community-owned.[41] Two hundred and fifteen sit in Whitelee windfarm owned and run by so-called Scottish Power (aka Iberdrola of Spain), generating and selling electricity to our city's residents at hiked-up prices to turn a nice profit for overseas investors. This is an absurd and unjust situation, which could be easily remedied if we adopted the 'residency criteria' principle which has been so crucial to the development of Denmark's world-leading renewable energy sector since the early '80s. Denmark's regulations aimed to bring 'decentred and localised forms of collective ownership to the fore' and their 'residency criteria' (or distance regulation) laws state that no energy generation can be owned by an organisation or individual who does not live within a certain radius (within the municipality) of the development.[42]

If we want to re-empower people in our city to improve their well-being, so that they are no longer 'victims of blind economic forces beyond their control' or 'excluded from the processes of decision making' as Jimmy Reid warned,[43] then the answer is simple. We must ensure that it's the people who actually live in a local area who own and control its local assets and that they are the ones that deliberate and decide on the policies that will affect them. Like the famous disability campaigners' slogan goes, it must be: 'Nothing About Us Without Us!'

Non-material pathways out of poverty

The previous chapter looked at how our society must rethink its measures of 'success' by moving beyond profit and growth to measures that are 'qualitative rather than quantitative'; where community participation replaces monetary wealth.[44] The final part of this chapter aims to rethink how we measure 'failure', in order to build 'the sustainable city of the future' in which we can all live happy, healthy and creative lives. The people with the smallest incomes are the ones living in the most sustainable way. They are far more conscious of energy use because of the high cost of bills and they are far more likely to use public transport for lack of other options. This is the exactly the behaviour we must celebrate and reward, not deride and marginalise.

As we saw in Chapter 7, Glasgow Centre for Population Health's 2016 report acknowledges that 'inadequate measurement of poverty and deprivation' is one possible reason why Glasgow appears to have such poor health.[45] One of these measures is the Carstairs Index of Deprivation developed in 1991 specifically for Scotland by two academics in Aberdeen. It uses four variables from our ten-yearly Census data to determine the 'material deprivation' in a local area: 'no car ownership', 'overcrowding', 'unemployment' and 'low social class'.[46]

This measure is inadequate because it fails to acknowledge that it is the city, society and economic system that we have created which means that being or having one of these variables ensures that you are 'deprived'. And that it's the very fact that this measure exists which causes stigmatisation and discrimination. It causes stress and desperate attempts to buy a car, get a 'better job' or move house to get away from an area, and yet more stress and anxiety if you're not able to do so. It either forces 'social mobility' through shame or it inscribes 'failure' to those who chose to stay. This is the opposite of what we want. Instead our task is to rebuild our city, society and economic system so that these variables no longer mean that you are 'deprived', and that, no matter where you live, you have equal access to the social, economic and cultural life of the city. This is what Cathy McCormack calls 'non-material pathways' out of poverty.[47] If we fail in this task, we will only create more 'diseases of affluence' (heart disease, obesity, cancer and mental health disorders),[48] and ultimately cause our own species' extinction.

World-class public transport

Cars are killing us. They promote the inactive and unhealthy lifestyles which are fuelling our obesity and diabetes epidemic.[49] They are responsible for illegal levels of air pollution in our cities causing 40,000 premature deaths a year across the UK.[50] They produce the most carbon emissions out of all sectors of Scotland's economy.[51] They promote intolerance towards others and stress, anxiety and animosity in the form of 'road rage'.[52] And as well as all this slow-burn damage, they can also kill you instantly, with traffic accidents being one of the leading causes of death and injury globally.[53] Electric cars are not the answer. They still 'run on money and make you fat' and continue to cause nearly all of these problems, as well as increasing society's energy demand when we need to be doing the opposite. Surely out all these Census variables used in the Carstairs Index of Deprivation, it is 'no car ownership' that should be celebrated and rewarded the most?

In building 'the sustainable city of the future', our most urgent task is to rebuild the world-class public transport network that was robbed from us in the post-war period. The ultimate aim must be that everyone in our region, no matter where they live in the 'travel to work area', can travel for work or leisure, quickly and at all times of day and night without need or aspiration to own a car; so that not owning a car no longer holds you back.[54] We must rekindle the ambition of the Glasgow Corporation in first half of the 20th century, which successfully oversaw our city's extensive publicly-owned tram, bus and Subway networks, to reach all corners of the city. But because we now live in a 'conurbation' it is essential, as Trans-Clyde understood back in the '70s, that we go region-wide.

With proper regional governance we can prioritise investment in public transport (not roads as Transport Scotland has done) so that we can finally extend our Subway (the only one in the world that has never been extended) to Castlemilk,[55] the East End and many other places and build the Strathclyde Tram joining Easterhouse and Drumchapel, which have all been proposed over the years.[56] It shouldn't just be privileged people in the West End who get to travel into town smoothly, underground in less than 15 minutes, while others have to make do with long, tedious and expensive journeys on the bus.[57] This just perpetuates the 'Pareto Principle'. As you can see from the carbon Conversion Factors listed on page 141, Subway travel is the most sustainable form of urban transport after walking and cycling. Although extending the Subway would create carbon during construction, it would help facilitate low-carbon lifestyles for centuries to come, and it is this long-term thinking that we urgently need.

Transport investment does not increase economic development, it simply 'moves the economy around on the map'.[58] It therefore becomes a vital tool in our aim to redistribute and decentralise wealth to the places in our region which need it the most. Building an 'outer circle' to link up all four of our biggest post-war housing estates – Easterhouse, Pollok, Castlemilk and Drumchapel – would be a good place to start. This would encourage communication, exchange, opportunities and solidarity to blossom between these places – helping to release all the 'latent, untapped potential' on our periphery.[59] And it would stop the centralisation of economic development in the city centre (something that's been going on far too long), which then forces everyone to travel in. Our ultimate aim must be to decentralise jobs and opportunities so people don't have to travel so much in their daily lives.

Transport for Greater Manchester (TfGM) has just released plans for 'Our Network', its new fully-integrated public transport network.[60] It maps all the fixed transport infrastructure in their region – trams and trains – and then plans the bus routes, not to compete with these (as they currently do in Glasgow), but to fill in the gaps and integrate seamlessly at the train stations and tram stops. Nearly all the proposed bus routes are 'radial' – linking up communities – rather than just going to the city centre. We need a new Transport for Greater Glasgow body (TfGG, to replace the failed SPT) which is accountable to the new Greater Glasgow Authority. TfGG must have the powers necessary to reregulate the region's public transport network and do this vital planning work, as well as investing in a fleet of electric buses to cut our illegal levels of air pollution.[61] Like the transport authority in charge in Munich City Region in Germany, the task of planning our network must be approached on the principle: 'one network, one timetable, one ticket',[62] so that our buses, trains and Subway can once again work together in harmony to serve everyone in our region. It's no coincidence Munich was voted the 'world's most liveable city' in 2018.[63]

Sharing is more sustainable

Returning to the Census variables used in the Carstairs Index of Deprivation, the next, 'overcrowding', is perhaps the most contentious given Glasgow's history of squalid housing conditions and the impact this is known to have had on our population's health. However, by the time of the 2011 Census, Glasgow had caught up with Manchester and Liverpool with comparable household sizes[64] and we are now swinging to the opposite end of the spectrum, which will also have a severe impact on mental and physical health:

> In Glasgow, the percentage of single households is expected to reach 50 per cent by 2039. In comparison, households with children in the city are predicted to shrink overall and account for 17 per cent of all households by 2039.[65]

A 'great demographic change' is underway and the health impacts of social isolation could end up being much worse than what came before. In January 2018, the Scottish Government launched its own social isolation strategy, *A Connected Scotland: Tackling Social Isolation & Loneliness & Building Stronger Social Connections*. It showed the extent of this growing problem: 48 per cent of people 'exhibit a degree of social mistrust, which is connected to level of social contact and feelings of belonging to the local community'.[66] Loneliness increases mortality by 26 per cent by increasing our risk of developing heart disease and stroke, high blood pressure, cognitive decline and dementia, depression and suicide.[67]

Our world-class public transport network will of course enable people to get out and see friends and family more often, going some way to addressing this crisis. But our housing system and the way we are living is even more of a problem. The BBC Radio programme, *The Anatomy of Loneliness*, estimated 'the cost of loneliness to employers might be as much as £2.5 billion a year' as lonely people are more likely to be mentally or physically unfit to work. Speaking on the programme, historian David Vincent says that our problem with loneliness

> runs down the 20th century. It begins with the
> increasing emphasis on individualism, and seeing
> loneliness as the point at which individualism fails to
> deliver. It is associated with the great demographic
> change after the Second World War. Before 1945, most
> people spent all their lives in company, in other people's
> households. We've seen a very large growth in single
> person households, since then, now about 30 per cent.
> And it also, I think, is associated with the growth of the
> welfare state. In 1945, the state took responsibility for
> the well-being of society: psychological and medical.
> And at exactly that moment, the numbers of people
> living by themselves, and the numbers of people growing
> old started to climb remorselessly.[68]

In 1929 writer Virginia Woolf described the necessity of a woman having 'a room of one's own' – like the 'studies' which were common for privileged men – in order to concentrate without distraction and develop and pursue creative and political activities. Carol Craig also notes that lack of 'space to read or concentrate on homework' in overcrowded homes was one reason working class Glaswegians fell behind at school.[69] But in our attempts to solve the problem of 'overcrowding', we're now heading to the opposite extreme (total alienation).

'Single person households' are absurd. They are 'the point at which individualism fails to deliver' – a completely inefficient use of energy and resources, with everything – boilers, fridges, kettles, toasters and more – needing to be duplicated again and again (an issue shared district heating systems aim to address).[70] Sharing is not only more social, it's more sustainable. I'll return to our broken housing system in the next chapter. I don't have all the answers (like I do for public transport!), but I know balance is urgently needed.

Motivational structures and meaningful work

The worst possible job for your health is no work at all; being 'unemployed'.[71] This is because of the lack of income and the poverty that results, but also because of the lack of a 'sense of purpose and meaning in life'; the lack of a motivational structure, and again the stigmatisation. This explains why 'unemployment' is the next Census variable used in the Carstairs Index of Deprivation. However, having been trained as an artist, I was surprised when I found this out. It is most artists' dream not to have a job, so that we can fill all our 'free time' with creative and political activity instead. It was my creative education which helped me to develop the skill of self-motivation.

If the previous deindustrialisation in the '70s and '80s wreaked havoc on Glasgow's economy and population health, just wait for the next one (addressed in Pat Kane's article in Chapter 7) caused by 'the coming automation'. Research by the Centre for Cities estimates that 'Scotland will lose 230,000 jobs in the next 12 years – 112,000 are likely to be lost in Glasgow alone'.[72] That's a fifth of our city's population. It may sound like a looming disaster, but it doesn't need to be. Not having a job doesn't need to be bad for your health, as long as you have access to 'those things we all need to live happily and well' (described in the next chapter) and can create your own 'motivational structure'; with endless ideas about productive and fulfilling ways to use your time.[73] Not only can a creative education provide everyone with the creative skills that are least at risk of automation, it can also enable everyone to thrive when there's less paid work to go round.

EF Schumacher takes a different perspective – demanding that we reject the whole notion of 'the coming automation' altogether. At the start of *Small is Beautiful*, he mocks economists and thinkers like Pat Kane who believe 'the problem of production' has, or will soon be, 'solved',[74] saying it's an 'absurd and suicidal error' to believe that we'll soon no longer have to work ourselves as machines will produce everything for us. He believes that it is only through work that we can find meaning and purpose, a sense of community and contentment in our lives. He argues against mass production in factories by machines and instead for 'production by the masses',[75] where we all pitch in to help produce what our society needs using 'intermediate' or older forms of technology which require more labour, rather than the cutting edge. Schumacher writes:

> If our intellectual leaders treat work as nothing but a
> necessary evil soon to be abolished as far as the majority
> is concerned, the urge to minimise it right away is hardly
> a surprising reaction, and the problem of motivation
> becomes insoluble.[76]

Instead we must create meaningful work which will allow people to contribute to the transformation of our society: providing essential goods and services or building our new sustainable infrastructure – world-class public transport, renewable energy, insulated homes and decarbonised heating – which will enable sustainable lifestyles for us all. As Schumacher says:

> Think of the therapeutic value of real work: think
> of its educational value... There would be little
> need for mindless entertainment or other drugs, and
> unquestionably much less illness.[77]

'Having satisfying work to do (whether paid or unpaid)' is one of the things the people who contributed to Oxfam Scotland's Humankind Index said they valued the most.[78] This is also something I'll return to in the next chapter.

Variety is the spice of life

The final Census variable used in the Carstairs Index of Deprivation is 'low social class' (Classes IV and V). In the five-class version of the National Statistics Socio-economic Classification (NS-SEC) scale introduced in Chapter 1,[79] Class IV and Class V are the two above 'unemployment' and are defined as 'Lower supervisory and technical occupations' and 'Semi-routine and routine occupations' respectively. They are the 'unskilled' or manual jobs, which have largely been created by the 'division of labour', which paved the way for mass production in factories. Work which, as Schumacher describes, is:

> That soul-destroying, meaningless, mechanical, monotonous, moronic work [which] is an insult to human nature which must necessarily and inevitably produce either escapism or aggression, and that no amount of 'bread and circuses' can compensate for the damage done...[80]

This is not because the work is 'manual' – quite the opposite. But rather because in a global capitalist system the individual has no meaningful connection to the fruits of their labour: 'When he feels as nothing more than a small cog in a vast machine and when the human relationships of his daily working life become increasingly dehumanised...'[81]

As we saw in Chapter 1, the reason nearly all work in the creative and cultural industries are classified as I and II in the NS-SEC – as 'middle class jobs' – despite the fact the work is often 'low-paid and precarious',[82] is because of the high levels of creativity and autonomy the work allows. Creativity and autonomy are things which our present system sees as a 'privilege', but which should be central to every human's life. These are two of the six core things which Harry Burns, Scotland's former Chief Medical Officer, says create wellness.[83]

The real issue here is our work falling into 'classes' in the first place and the lack of variety we are all missing as a result. It is variety that enables us to develop both practical and theoretical skills: to become 'generalists' like Patrick Geddes was and be fully-rounded human beings. Writing about the medieval Italian philosopher Thomas Aquinas (1225–74), Schumacher says that he defined the human as 'a being with brains and hands'; one which 'enjoys nothing more than to be creatively, usefully,

productively engaged with both his hands and his brains'.
There is so much 'modern neurosis' Schumacher says, because
'the type of work which modern technology is most successful
in reducing or even eliminating is skilful, productive work of
human hands'.[84] The point is that we need to share out the work
more evenly, so that everyone gets to do a variety of tasks. As
Common Weal say, we need a

> democratic distribution of work and leisure. This would
> serve to rebalance the current situation whereby neither
> those with a job, but no time; nor those with time,
> but no job can experience the pleasure derived from
> alternative modes of living.[85]

I found working as an usher and picking up other people's litter,
as well as the many other voluntary roles I've been doing, is what
has helped me to 'keep it real'. Nobody wants or needs full-time
or 'professional' artists – it's the other jobs that most artists do
to subsidise their work which help them maintain a connection
to the 'real world' and ensure they have something meaningful to
say about it. If 'the coming automation' does hit us, then we must
share out the remaining work more evenly so that none of us is
doing more than 21 hours of paid work a week. New Economics
Foundation have shown that this is one key way to create a society
where we can all 'work less, consume less, live more'.[86]

In *The German Ideology*, one of the early texts by Marx and
Engels written two years before they published *The Communist
Manifesto* in 1848, they write about how the 'division of labour'
has forced us all to specialise in ways that are detrimental to
our health and well-being and which create misunderstanding,
mistrust and disrespect across disciplines and occupations, and,
of course, across classes:

> For as soon as the distribution of labour comes into
> being, each man has a particular, exclusive sphere of
> activity, which is forced upon him and from which he
> cannot escape. He is a hunter, a fisherman, a herdsman,
> or a critical critic, and must remain so if he does not

want to lose his means of livelihood; while in communist society, where nobody has one exclusive sphere of activity but each can become accomplished in any branch he wishes, society regulates the general production and thus makes it possible for me to do one thing today and another tomorrow, to hunt in the morning, fish in the afternoon, rear cattle in the evening, criticise after dinner, just as I have a mind, without ever becoming hunter, fisherman, herdsman or critic.[87]

In 'the sustainable city of the future', we will no longer be defined by our jobs. Instead, working in multiple roles will help increase understanding and create a better integrated society. It's the variety of labour, and a balance of work and leisure, which can make us all complete human beings.

CHAPTER 11

Universal Luxurious Services

Those things we all need to live happily and well

NONE OF US 'needs' more money. Money is the ultimate 'means to an end'. Take away the economic infrastructure (the shops to spend it in) and it has no value at all. It's worthless, especially in its far more common dematerialised digital form (which now makes up 92 per cent of all global currency): then it is, quite literally, nothing. If 'the love of money is the root of all evil' as the Bible suggests,[1] then to think that giving people more money (in the form of increased welfare payments or as a Citizen's Basic Income) will solve all our problems is a mistake. It's more likely to create them elsewhere. One reason to be sceptical of the Citizen's Basic Income idea is that Milton Friedman, one of the architects of neoliberalism, came up with a similar proposal back in the '60s, called Negative Income Tax.[2]

Milton Friedman specifically set his Negative Income Tax rate at $1,500 per year in 1968 (around £8,500 in today's money – coincidentally almost exactly the 'living expenses' I used from my £15,000 grant for *The Glasgow Effect*). His rate was deliberately half the minimum you would receive if working full-time, so as not to 'destroy the incentive of people earning'.[3] However, his idea was really premised on the neoliberal goal of making everything in our lives a 'commodity' – with price tags attached that must be bought and sold with money, so that all further state support could be removed. Suddenly £8,500 does not go very far when you have to pay for all your basic needs: private transport, private housing, private healthcare, private education and more. With a Citizen's Basic Income, we run the risk of doing both of these things: giving government the excuse to cut our public services and disincentivising work – not for the reason Friedman suggests but because, as soon as we attach a price tag to something, we think about it in a different way. Paid workers are often less motivated than volunteers who are doing it off their 'own back', working with the 'labour of liberation' because they believe in the cause.[4]

When we're thinking about how to rebuild a more equal, sustainable and connected city, we must instead be thinking about how we enable people to get what they actually do *need* to live happily and well. In 2017, in response to the increased interest in the idea of a Citizen's Basic Income as the 'silver bullet' solution to all our problems, the Social Prosperity Network published an alternative proposal called Universal Basic Services.[5] It proposed that rather than getting 'free money', all our basic services, 'the fundamental building blocks for life',[6] should be provided free at the point of use. In addition to the three services that are (just about) provided free already – 'healthcare', 'education' and 'democracy and legal services' (in spite of recent Tory privatisations and cuts to Legal Aid)[7] – we would add free access for everyone to 'shelter' (housing and energy), 'food', 'transport' and 'information' (mobile phone and internet). The universal aspect is important as it removes the stigmatisation of receiving benefits, ensuring everyone is treated in the same way, yet it still disproportionately benefits the poorest and most marginalised. The outcome would be a society where anyone could live a basic standard of life, without needing any money.

Universal Basic Services sounds like a brilliant idea. But it isn't exactly a new one. It's a bit like socialism rebranded for the 21st century, rekindling the 19th century's famous radical artist-socialist William Morris' concept of 'a moneyless society',[8] or the idea of replacing 'exchange values' directly with 'use values' as David Harvey (author of *A Brief History of Neoliberalism*) suggests. David Harvey considers money as the 'great corrupter of our desires and our needs' and says the sooner we can remove it from our lives the better.[9] His view chimes with Tim Kasser's theory of universal human values, which shows that the more we prioritise the 'extrinsic' goal of 'financial success', the 'less happiness and life satisfaction' and 'fewer pleasant emotions (like joy and contentment)' we have, and the more likely we are to behave in 'manipulative and competitive' and 'unethical and antisocial' ways.[10] David Harvey quotes Thomas More's 1516 book, *Utopia*, which says 'in a moneyless economy, you would sleep very well at night'.[11]

The Universal Basic Services report shows that it would be well within our grasp as the world's fifth largest economy to provide all these things for free. In fact, it would only require a 5 per cent increase on our current public spending to deliver them (from the 41 per cent of GDP that we currently spend to 43.1 per cent).[12] When you look at the detail of the report, however, you can see how this calculation has been made, most unimaginatively, by basing costs on our current privatised model, which makes services far more expensive than they need to be.

Information

For example, for 'information' services, the report assumes that to provide internet free to every citizen you would have to pay an average of £20 per month to one of the current private providers (BT, TalkTalk, PlusNet etc).[13] Every single house would have its own plastic wireless router, beaming out its own exclusive signal – when that's clearly a totally inefficient way of providing a universal service. It would be much cheaper and more sustainable if a big conurbation like Glasgow just had wi-fi signals broadcast centrally, which we could all then connect to for free. Job done.

Transport

Similarly, with free public transport (and it is important to add that this is free *local* transport only), the calculations are based on our completely inefficient present privatised system, in which we still subsidise private bus companies with nearly £300 million of public money every year in Scotland.[14] In their brilliant report, *Building a World-class Bus System for Britain*, Lynn Sloman and Ian Taylor show that if all Britain's buses (outside London) were 'remunicipalised' and run in the interests of the people who they served (like Edinburgh's Lothian Buses), we would save £506 million per year. That would be more than enough to reinstate the thousands of routes which have been cut across Britain in the period of austerity since 2010.[15] Across Europe, in France and elsewhere, many towns and cities (including our 'twin town' Marseille) are now 'remunicipalising' their buses to save money *and* provide a better service.[16] I'm sure it would be possible to provide free public transport to everyone in Scotland for the £300 million that we're currently wasting on the private bus companies every year.[17] And indeed, because they have public ownership and control of their networks, more than 100 towns and cities around the world (in America, France, Poland, Sweden, Italy, Slovenia, Estonia, Australia, Luxembourg and elsewhere) are already providing free public transport for their citizens. They know it's the *only way* to shift people onto sustainable forms of transport and cut carbon emissions and air pollution at the speed that we need.[18]

Capitalism is good for some things – for producing the non-essentials of life, those 'ornaments of dress, trinkets, fine furniture or jewellery', which economist Adam Smith (1723–90) described.[19] Things people might 'want', but don't actually 'need'. Capitalism is not good for providing our fundamental basic services – the things we all need to survive. With a logic based on maximising profit, it's always going to be 'more efficient' to cut back the services and charge more for the remaining ones (that's certainly the story of our bus services since deregulation in 1986). Public services must always only be publicly owned and controlled – the clue is in the name. Public. To build 'the sustainable city of the future', we need to take back control of all our key public services, infrastructure and assets and once we have them, we must ensure that they are all held in trust to the citizens they serve at a local/regional level and so they can never be privatised again.[20]

Food

On food, the Universal Basic Services calculations are based on rolling out the current 'food banks' provision, rather than dealing with the huge inefficiencies of our present profit-driven system of food production and supply – responsible for the dual crises of the obesity epidemic and the scandal of food waste which sees more than 15 million tonnes end up in landfill every year, whilst so many are still going hungry. The free-market will always result in an uneven distribution of food outlets, and quality and price. This is why Glasgow has a severe problem with 'food deserts',[21] and the absurd situation where the West End can have 'ten Tescos in two miles', but if you live in Castlemilk it can take 75 minutes to walk to the nearest supermarket.[22] And it's why we have so many retailers like Blue Lagoon, playing to our weaknesses by promoting high fat or high sugar foods so as to turn a greater profit. If there's one industry where we need more strategic planning and better use of resources, it's in our food production and supply. We need more democratically-run food co-operatives and we need to be growing more ourselves. 'Subsistence farming' – growing and eating our own food – is the ultimate 'freedom from the market',[23] which is why feminist economist Marilyn Waring says it has been completely ignored by GDP.

Healthcare

In 1948, the World Health Organisation defined health as 'a complete state of physical, mental and social well-being and not merely the absence of disease or infirmity'.[24] The biggest criticism of our National Health Service is that it's actually a 'National Illness Service' – that people only access it when their lives have reached crisis point and they've fallen ill.[25] And because so many people now no longer have the social networks that are 'the very immune system of society',[26] they're more likely to turn to the NHS for help.

If you're going to get cancer (and one in two people now do),[27] then the west of Scotland is probably the best place in the world to get it. We have amazing facilities for sufferers: the beautiful, 'starchitect' Rem Koolhaas-designed Maggie's Centre, the Beatson Well-being Centre with hairdressing and relaxation facilities and Cancer Support Scotland, all on the Gartnavel Hospital campus alone, which my friends Roanne and Diane both benefited from. But as we know from 'the Glasgow effect' itself, if you live in the west of Scotland then you're more likely to get cancer in the first place. Instead, we need to develop a system which is *preventative*: addressing the causes of cancer rather than the symptoms (diet, inactivity, air and other pollution, drinking, smoking and stress, and, of course, the connections between these).

People often give the example of the 'community doctors' in Cuba, who, working within a specific community over long periods of time, get to know their patients intimately and check up on them regularly to ensure they're keeping well. Or the Chinese system, which 'aligns' the aim of wellness with the way community doctors are paid – they continue to be paid until a member of the community falls sick, then they cease to be paid until they're better again![28] It might sound a bit extreme, but it certainly focuses doctors' minds on developing and engaging the community in preventative measures (eating healthily, exercising, socialising; engaging our 'hands and brains').

As much as I dislike the Weight Watchers corporation's business model which, since 1963, has sought to capitalise on people's body anxieties, ostensibly making them worse, I really enjoyed my time with the group at Maryhill Central Hall.[29] Participating in a weekly wellness support group is a great way to motivate one another to make cheap and healthy food and exercise choices (especially if they're not centred on buying lots of Weight Watchers products!). Ironically, I only stopped going to the meetings because they clashed with the Saturday morning Parkrun (which my sister got me doing in 2014). Parkrun is a free initiative of weekly 5km community 'fun runs' that's spread across the globe since it was founded in 2004, which writer Aditya Chakrabortty says is a symbol of a new society not powered by profit but instead by 'love and fun'.[30]

And so it is clear that we must also place 'leisure' firmly within 'healthcare' in our bundle of Universal Basic Services. Our leisure centres – like Victoria Baths in Nottingham, like Govanhill Baths in Glasgow (shut in 2001 despite a massive community campaign and the longest occupation of a public building in British history), and Whitehill Baths in the East End and Gurnell Leisure Centre in Ealing – both of which are currently threatened with closure – must be held in trust to the people who use them, so they're no longer seen as the 'private property' of the council to do with as they please.[31] If we had much stronger local democracy and power lay with our Community Councils, then facilities could no longer be 'ripped out, renamed, sold off, mysteriously burned down, gentrified and/or demolished', as Darren McGarvey describes, without the collective agreement of the users and residents of the local area. That would soon stop these mindless 'efficiency savings', which have massive knock-on impacts on healthcare spending – all that 'defensive expenditure' we must now avoid.[32] We need to develop a holistic approach to health and well-being, like the famous Peckham Experiment in London, which ran from 1926–50 and was a health, leisure and community centre rolled into one. Ironically, its closure was put down to its innovative model not fitting into the standardisation of the new NHS.

Housing

The people who contributed to Oxfam Scotland's Humankind Index said 'an affordable, decent and safe home' was the thing they valued the most.[33] Certainly, not having a roof over your head is one of the most detrimental things to human health.[34] The Universal Basic Services proposal works on the premise of building more social housing (which we certainly need to do), but does not tackle the fundamentally broken housing system that we have in Britain (both private and publicly-owned), in which the sky-high rents being extracted by wealthy private landlords mean that our housing benefit bill is more than £20 billion a year.[35] At the end of *The Tears that Made the Clyde*, Carol Craig gives her policy recommendations for a healthier and happier Glasgow, one being 'Housing developments encompassing a greater range of social classes',[36] as a way of breaking up the 'estate-ness' of Glasgow's housing estates, which Lynsey Hanley has shown to be a cause of deep social divides. You could argue that the Thatcher government's 'right to buy' scheme did achieve this to a certain extent, in mixing a new breed of 'home-owners' into council estates. But we are now left with the most absurd two-tier system running in parallel, where some homes are 'affordable' (and therefore exceptionally scarce), and all the rest are 'unaffordable' – both are, therefore, increasingly difficult to access.

This two-tier system entrenches inequalities and division for life, as if you get on the 'housing ladder' you're more likely to leave inheritance to your children, who can then get on the 'housing ladder' more easily themselves. It's the opposite of integration. It's the 'Pareto Principle' in action once again. In places like Glasgow (which in the post-war period had the single largest amount of social housing outside the Soviet Bloc) the people who remained social housing tenants were left further behind as Margaret Thatcher's materialist ideology of creating 'a nation of home-owners' slowly became normalised, and people like me from England took advantage of the 'comparative cheapness of Scottish property',[37] snapping up private flats and creating more division.

There's clearly a good argument for saying there should be no privately-owned housing at all, although even I (the idealist) would say that's a hard goal to achieve. EF Schumacher is not against private ownership per se. In *Small is Beautiful*, he instead makes a very useful distinction between:

a) property that is an aid to creative work and
b) property that is an alternative to it. There is something natural and healthy about the former – the private property of the working proprietor; and there is something unnatural and unhealthy about the latter – the private property of the passive owner who lives parasitically on the work of others.[38]

He quotes the economist RH Tawney (1880–1962), who says 'For it is not private ownership, but private ownership divorced from work, which is corrupting to the principle of industry'.[39] If people were only permitted to own the one home in which they live, there would be no more foreign speculators, no more 'buy-to-let landlords', no more holiday homes. It would free up all the overpriced and underutilised housing in the UK, which is the *true cause* of our 'housing crisis'. It would send house prices plummeting so that they become better realigned with their actual 'use value', as they were back in the '70s (as opposed to the artificially inflated 'exchange value' houses have now). Using high house prices as a measure of 'success' is another collective delusion of a world under the spell of neoliberalism. Rising house prices have been one of the core drivers of inequality over the last 40 years.[40]

Failing this, then rent controls, as fought for and won by the famous Glaswegian activist Mary Barbour (1875–1958) after she coordinated a peaceful rent strike of more than 20,000 of the city's tenants in 1915, are essential. Mary Barbour's rent controls were in place across Britain for more than 70 years and were only abolished in the UK by Margaret Thatcher's Housing Act 1988 – it's no wonder both sale and rental prices have skyrocketed since. Scottish tenants union Living Rent advocate 'a points-based system of rent controls', which links the property to its actual 'quality and amenities' and not simply to 'market rates'.[41] If your flat is not being looked after by the landlord or it's in a food, cultural or transport 'desert', then you pay less. Although, of course, our aim is to eradicate these 'deserts' altogether and ensure everyone is reconnected to social, economic and cultural life of the city.

Everywhere we can see the inefficiencies of the neoliberal marketplace that must be addressed in order to enable everyone to get 'those things we all need to live happily and well' in the fairest, most affordable and sustainable way.

Co-production

But when redesigning our new system for delivering social protection to all, it's essential we don't fall into the same traps of the 20th century, which created a dependency culture, where we lost many of the skills we previously had to look after ourselves and became further socially isolated and atomised as individuals. As New Economics Foundation show, the 20th century welfare state undermined 'people's sense of personal responsibility for tackling common problems'.[42] Our new system of Universal Basic Services must be designed and delivered in a way that helps to build and to grow the things that bring meaning and happiness to our lives – the 'core economy' of family, neighbourhood and community. Universal Basic Services need to be delivered in a way that will nurture and support the six core things which Harry Burns, Scotland's former Chief Medical Officer, says create wellness:

- Sense of control and internal locus of control
- Sense of purpose and meaning in life
- Confidence in ability to deal with problems
- Nurturing family
- Supportive network of people
- Optimistic outlook[43]

In *Co-Production: A Manifesto for Growing the Core Economy*, New Economics Foundation show that 'neither markets nor centralised bureaucracies are effective models for delivering public services based on *relationships*', and it's real human relationships that are the 'crucial dimension that allows doctors to heal, teachers to teach and carers to care'.[44] Instead we need a system of 'co-production' of our services which:

- Provides opportunities for personal growth and development to people, so that they are treated as assets, not burdens on an overstretched system.
- Invests in strategies that develop the emotional intelligence and capacity of local communities.
- Uses peer support networks instead of just professionals as the best means of transferring knowledge and capabilities.
- Reduces or blurs the distinction between producers and consumers of services, by reconfiguring the ways in which services are developed and delivered: services can be most effective when people get to act in both roles – as providers as well as recipients.
- Allows public service agencies to become catalysts and facilitators rather than simply providers.
- Devolves real responsibility, leadership and authority to 'users', and encourages self-organisation rather than direction from above.
- Offers participants a range of incentives which help to embed the key elements of reciprocity and mutuality.[45]

Co-production plans for a 'low-growth or no-growth econo-my',[46] 'both because it is likely and because the planet cannot sustain continuing growth, least of all in rich countries'.[47] It requires that we strip out all the inefficiencies and leakage caused by the so-called free-market and the 'defensive expenditure' that it creates. The more locally services are delivered (within Community Council areas), the easier it will be to build relationships and the less people need to travel. Just as in our own individual lifestyles, we must also strive to find the 'win-win-win' sweetspots for our public services, where public funds are only used to deliver the best social, environmental and economic outcomes for all.

Foundational economy

We need people to literally feel ownership and pride over their core services and to pitch in and get involved in running them. This is what's sometimes referred to as the 'foundational economy'. As Dave Watson from Unison Scotland explains, because the 'oil and finance sectors' are in decline (a fact we should be celebrating given the unsustainable and unethical nature of that work):

> There is a strong case for a new approach to local economies in Scotland, grouped around the concept of the foundational economy... The foundational economy is built from the activities which provide the essential goods and services for everyday life, regardless of the social status of consumers. These include, for example, infrastructures; utilities; food processing; retailing and distribution; and health, education and welfare.[48]

We should be mobilising everyone in the city to contribute their skills and experience to the 'foundational economy' – especially those living transient and disconnected existences within a global 'knowledge economy', that's what will increase everyone's sense of 'belonging'. We must create as many meaningful jobs in the 'activities which provide the essential goods and services for everyday life' as possible, which will help people to build relationships with others who live nearby, and make our city's economy less vulnerable to the whim of international markets.

As Robin McAlpine from Common Weal says, 'we created

the poverty by lengthening the supply chains'.[49] People became unemployed locally, because we exported so many of the jobs providing for our basic needs to other parts of the world. In *The Tears that Made the Clyde*, Carol Craig writes:

> Many commentators on Glasgow remark that it was a hard culture made by hard men but the economy they created became increasingly fragile: the city's economy was too interconnected and overly specialised… It was also heavily dependent on exporting to international markets which were notoriously volatile.[50]

She contrasts the shipbuilding industry on the Clyde with Edinburgh's economy in the 19th and 20th centuries built around 'small-scale consumer industries', which were catering for 'local demand'. This meant there was 'much more stability and security for the Edinburgh working class',[51] which was much better for well-being. Whilst the skill of shipbuilding must have given people an immense sense of pride and mastery in their work, the jobs 'were insecure and subject to the vicissitudes of international markets which meant that whole yards could shut as a result of orders drying up'.[52] Very few people actually need a massive boat, especially if our ultimate aim is to start cutting imports and exports on 'carbon-spewing container ships' to reduce emissions. Instead, we must start manufacturing the sustainable infrastructure that we do actually urgently need. When London was building its Crossrail underground line, they set up the Tunnelling & Underground Construction Academy (TUCA) in 2011 to train local people to help with construction. To start rebuilding our world-class public transport network we must do the same. And set up many other academies to train people to work in other vital elements of the 'foundational economy'.

It is working the 'foundational economy' which will help improve well-being by allowing us to all see the fruits of our own labour in front of our very eyes and enabling us all to realise what EF Schumacher says are the three key purposes of work to give a man a chance to utilise and develop his faculties; to enable him to overcome his egocentredness by joining with other people in a common task; and to bring forth the goods and services needed for a becoming existence.[53]

Public luxury

If we work together to 'co-produce' the 'essential goods and services for everyday life', they actually don't just need to be 'basic'. As writer George Monbiot argues, public services can and should always be 'luxurious'. His principle is 'private sufficiency' and 'public luxury'. That it is the elements of our lives which are private where we must be thrifty – living an austere life at home with the 'waste not, want not' mentality should be encouraged and celebrated. But our public services, on the other hand, they can and should be luxurious. In an interview with We Own It campaign founder Cat Hobbs, George Monbiot elaborates:

> If we try to all enjoy private luxury and all have our own swimming pool and tennis court, play barn and art collection and the rest of it, very quickly we literally fill the whole world up. And you suddenly discover that only a few people can have that because there's simply not enough physical space or ecological space for everyone to do it. If you all try to live like the ultra-rich today, we burn the planet up and we use all its resources and we turn it into a dystopia. But there is enough space for everyone to enjoy public luxury – fantastic quality public amenities, where everyone gets to play tennis if they want to play tennis, and everyone gets to swim if they want to swim. And there are amazing playgrounds and wonderful parks and beautiful nature reserves and everything that you would look for in your own life, to enhance your luxury is found in the public or the common sphere... [Then] you genuinely make this a land of luxury for all, rather than luxury for some.

The other thing about public luxury is it's far more efficient than private luxury. So for example you can say: 'oh, I'm sick of being stuck in traffic in my old banger, I'm going to get a really fancy car. I'm going to spend 20 thousand quid on a fancy car'. You're still stuck in traffic and you're 20 thousand quid down. Now imagine if £1,000 of your taxes, and everybody's taxes were invested in a really great mass transit electrified system, which is super-efficient, you're going to get there at ten times the speed and it's cost you one 20th of what that fancy car cost you to get stuck in traffic. So you can actually use resources far more efficiently when you're seeking public luxury, than when you're seeking private luxury.[54]

Echoing the idea of 'public luxury', in the '80s novel *A Very British Coup*, when the main character, newly-elected Prime Minister Harry Perkins, is interviewed sitting in second class on a train heading to Westminster, he is asked, 'Do you intend to abolish first class rail travel?' He answers, 'No, no, we will abolish second class rail travel. I think all people are first class. Don't you?' That's the attitude we need for Glasgow – we need to stop wealth and investment being centralised in the city centre and we need to get the best, most luxurious public services and amenities out to all parts of the city region. It's through Universal Luxurious Services that we can start to create a city that is just so pleasant to be in, that people actually don't want to leave.

Localism and protectionism

The vision for a thriving local economy which EF Schumacher presents in *Small is Beautiful* also allows private enterprise to play its part. As you may guess from the title of his book, he's only against private enterprise when businesses grow to an unwieldy size and their interests are no longer aligned with the needs of the local community. He writes:

I have no doubt that it is possible to give a new direction to technological development, a direction that shall lead

it back to the real needs of man, and that also means:
to the actual size of man. Man is small, and, therefore,
small is beautiful.[55]

Schumacher has two simple propositions for enabling
companies to retain their social and environmental values and
ensure they serve the community in which they are based. The
first is that companies are democratically owned and controlled
as co-operatives and (as mentioned in Chapter 8 in relation
to the university) that they are not permitted to grow beyond
350 employees or pay any employee more than seven times the
lowest paid.[56] If there is more demand, then another separate
autonomous organisation should be set up to help supply this.

His second proposal is for businesses which are already of
a larger scale and which represent foreign interests – like the
majority of the businesses operating in Glasgow today – such
as our food suppliers: Tesco, Lidl, Sainsbury's, Morrisons etc,
plus all the other clothing and other retailers (selling the things
people might 'want', but don't actually 'need'); our transport:
ScotRail (aka Abellio), FirstGroup, Stagecoach etc; and our
information services: BT, TalkTalk, PlusNet etc. Schumacher is
clear that because this 'private enterprise in a so-called advanced
society derives very large benefits from the infrastructure –
both visible and invisible – which such a society has built up
through public expenditure',[57] they must therefore give back. He
demands that exactly one half of the shares in any large private
company operating in a local area should be given over to the
people of that area and held on their behalf by the 'public hand'.
In return, Corporation Tax would be abolished. The public
would then benefit from sharing exactly one half of company
profits (to fund our Universal Luxurious Services) but, more
importantly, our new role as major shareholder would create
total transparency in the way these organisations are run; the
social and environmental values they embody, ensuring local
interests are prioritised. If Glasgow City Council had made this
proposal to Barclays bank before it began moving its premises to
Glasgow, then we, the people of Glasgow, could have had a say
in what the bank chose to invest in – locally, where employees

could actually see the positive impact of the money (like Birmingham's Municipal Bank), rather than investing in deeply destructive fossil fuel extraction projects in other remote parts of the world.[58] Or if First Bus was half-owned by the people of Glasgow, then I'm sure we would all forgo dividends in order to ensure that our most isolated communities were provided with a decent bus service.

There has never been a time when action like this is more important for Scotland, as ownership of the majority of our big so-called 'Scottish' companies has now fallen into the hands of foreign investors. As John Foster from the University of the West of Scotland has shown, almost all major companies controlled from within Scotland 'now have predominant shareholdings from outside Scotland'.[59] In the age of globalisation, Schumacher's ideas about the local ownership and control of industry become all the more important. As the book *Fearless Cities: A Guide to the Global Municipalist Movement*, shows:

> Cities with more local development have less inequality, healthier citizens, more social capital, more diversity; the companies based there have more of an interest in the development of the wider community.[60]

Positive alternatives

In Glasgow there are so many more innovative ideas we could be introducing to localise our economy and supply chains to a more human and accessible scale. In 2016, building on my learning from the Say No To Tesco campaign, I was involved in a working group exploring possibilities for a local currency for Glasgow to enable wealth to circulate within our local area and not leak out the 'leaky bucket' of the local economy. If we can't have a completely 'moneyless society', we should at least be thinking about how to redesign the present money system, so that it better serves the needs of local people.

The New Economics Foundation's book, *People Powered Money: Designing, Developing & Delivering Community Currencies*, gives many inspiring examples of places where alternative currencies have helped to build stronger local economies. It tells the story of the 'miracle of Wörgl' – a town in Austria where in 1932, in the midst of the Great Depression, the mayor issued 'free money' in a local currency which could only be spent in their town and which had a negative interest rate (known as demurrage), meaning the longer you kept it the less it was worth (a currency designed to stop the Scrooges hoarding it all up!). In Wörgl, 'unemployment was vastly reduced; streets were repaved; new houses were constructed; major infrastructure projects were accomplished' – all financed by the town's 'free money' miracle, before the Austrian government stepped in and made community currencies illegal in their country.[61] In the UK they are still legal and therefore present a powerful tool for addressing the social, economic and environmental challenges that we face.

NEF also give the example of Switzerland's WIR Bank, again founded during the Great Depression in 1934. Still running today, the WIR Bank has over 60,000 members – local businesses and service providers trading in a currency which can only be spent within their geographically defined network. This alternative currency has a

> countercyclical effect – that is, expanding during recessions and contracting during booms. This allows [businesses] to survive and even prosper during times of uncertainty in the mainstream cash economy.[62]

The WIR Bank counteracts the frequent financial instability and uncertainty which has been caused by globalisation, by retaining local control of the money system. There are also successful examples much closer to home. In 2016, we worked with Gloria Murray who runs the Castlemilk Timebank, an initiative founded in 2001 which aims 'to promote community involvement and to rebuild a sense of community spirit in Castlemilk'. They do this by 'helping people to exchange skills, services and support. One hour of your time will give you an hour of someone else's. Essentially, our project turns spare time into shared time'.[63] Castlemilk Timebank is a fine example of 'co-production' in action – a framework which enables people to participate without needing any money, and which helps build the 'core economy' of the local area based on meaningful relationships. There is a micro 'moneyless society' already blossoming right here in Castlemilk! And more recently, the Glasgow Chamber of Commerce have also launched a 'Circular Economy' initiative. If this intends to do anything meaningful it should help us to implement some of Schumacher's ideas described above and develop a radical project like Switzerland's WIR Bank. Actual geographical restrictions on wealth and 'residency criteria' over ownership is what we need to make the local economy work in the interests of local people.

The story in Preston in north-west England is also inspiring. In the face of the global financial crisis and the Tory austerity that followed, in 2011, the city council adopted a policy of 'guerrilla protectionism'.[64] With the help of the progressive 'think and do tank', the Centre for Local Economic Strategies (CLES), they realised that only one pound in every 20 spent by the city's public services went to local businesses – with the rest leaking out the 'leaky bucket' of the local economy to firms based in London or overseas. So they got six of the city's major public bodies to adopt a policy of local procurement instead. In just eight years, the city's fortunes have been dramatically transformed.[65] Preston 'bought local' and the city mitigated the damage caused by austerity and brought its local economy back to life.

In the face of global capitalism, many of our big cities – especially Glasgow – are suffering in the way Harry Burns says our citizens are, from a lack of an 'internal locus of control' and from a lack of 'confidence in [their] ability to deal with problems'.[66] Yes, we need the powers necessary to deliver the essential goods, services and infrastructure for our citizens devolved to a local level, but we must also think creatively and fully utilise the powers that we already have (but perhaps don't even realise), as Preston has done so well.

Car-free future

Finally we need to think about the environment we are living in and how we can make this far healthier and more people-friendly. Those who contributed to Oxfam Scotland's Humankind Index also said 'living in a neighbourhood where you can enjoy going outside and having a clean and healthy environment' is one of the things they valued the most.[67] We are never going to achieve that in Glasgow until we start to get rid of the cars and roads. As we saw in the previous chapter, it's cars that are killing us, for so many reasons. They are also a key driver of inequality:

> People who travel by car are more likely to be male and white, and to have higher than average incomes, meaning that car-based models of urban and territorial development reinforce gender, ethnic and economic inequalities.[68]

But once we have rebuilt a world-class public transport network for Greater Glasgow in which we can all make luxurious, seamless and accessible free journeys anywhere in the region we need to go, cars will no longer be necessary. All that road and city space currently given over to these polluting metal boxes will be freed up, and we can finally redesign our city as Patrick Geddes suggests, to help bring people together – to build and nurture social groups – not to rip them apart.

In Oslo in Norway, they have banned parking in their city centre altogether, so people can now only travel in via public transport, on foot or by bike. The result is a far more pleasant and liveable city in which 'Parking spots are now bike lanes, transit is fast and easy, and the streets (and local businesses) are full of people'.[69] A 'Car-free Future' is not some utopian idea, it's a necessary one: to tackle toxic air pollution, cut the carbon emissions caused by transport and address all the other health problems caused by car use (described in Chapter 10). Many other cities around the world have acknowledged this fact and are now taking urgent action. In 2018, France's capital Paris announced it would close its city centre streets to cars on the first Sunday of every month, to allow people to 'enjoy the sights… without the noise, smell and obstruction of traffic'. On Car-free Days noise levels drop by three decibels in comparison to other Sundays.[70] And in Edinburgh, in March 2019, they announced an 18-month 'Open Streets' trial closing city streets around the Royal Mile to give pedestrians priority and reduce air pollution.[71]

These sorts of events have been happening on a weekly basis for decades in many cities in South America (including, of course, forward-thinking Bogotá)[72] and in south-east Asia.

'Glaswegians who don't own a car currently contribute least to the air pollution but suffer the most from it', says Glasgow City Council's 2018 Connectivity Commission report.[73] Regular Car-free Days across our city would begin to address this injustice and create inclusive spaces to bring people together. But in Glasgow, it's essential we also address 'the elephant in the room': the motorway, which cuts right through our city's heart (yet which is controlled by the centralised agency Transport Scotland). In 2016, Dutch architects MVRDV were invited by Glasgow City Council to develop the 'district regeneration framework' for the Broomielaw area of the city centre which the M8 cuts through on the Kingston Bridge. The draft version of their report described the motorway as a 'highway scar' in need of healing. They suggest that by diverting through traffic on the new M74 extension,[74] you could now feasibly close the city centre section of the M8.[75]

MVRDV also recommended temporary 'highway take-overs' where sections of the M8 are closed for street parties and other events.[76] Just like the Car-free Days described above, the aim of these 'highway take-overs' would be to bring people together, giving everyone a shared vision of how great Glasgow's 'Car-free Future' could be. In this city, which has been dominated by cars and roads for far too long to the detriment of all our health, this is the scale of the Car-free Days we should be going for. In 2010, in the Ruhr region of Germany (an area with a population similar to the whole of Scotland), they closed a 60km section of motorway for one day for an event known as 'Still-Leben' (Still Life). Three million people (three-fifths of the region's population) came out to join what became the 'world's longest street party' (pictured opposite).[77] Let's do this on the M8 between Glasgow and Edinburgh so that everyone, regardless of whether they own a car or not, gets the chance to reclaim that space and begin to imagine that a happier, healthier and more sustainable Scotland is possible for us all. In a time of accelerating global social, environmental and economic crises, we really must 'Be realistic, demand the impossible!'

60km of the A40 motorway in the Ruhr region of Germany was closed to traffic on 18 July 2010 for Still-Leben – the 'world's longest street party' – with three million guests. (Bettina Strenske/Alamy Stock Photo)

If there's ever a place that Glasgow should learn from, it's Seoul in South Korea. At the heart of their city lay an 11km stretch of motorway carrying vast amounts of traffic, polluting the air and creating a hostile environment for all of their citizens. In 2005, the mayor made a bold decision and began a massive regeneration project to take away the motorway and restore the Cheonggyecheon River which had once run below it. Twenty nine months later, the 'outmoded' motorway was gone and in its place they had created a huge urban riverside park, which is now used by more than 60,000 people per day – creating a green lung for the city, enabling it to breathe again.[78] Also in Seoul in 2017, MVRDV completed the transformation of another 1km section of urban motorway which they decided to convert into another city park now known as the Sky Garden.[79]

Glasgow must 'Think Global, Act Local' and look and learn. When our M8 was built in the late '60s, the section from Easterhouse to Townhead was built right over the Monkton canal (you can see the canal on the 1940s Transport Map in Chapter 7). As a bold signal of 'the sustainable city of the future' we want to be, Glasgow should now repair its 'highway scar' and restore the Monkton canal. These are the 'rapid, far-reaching and unprecedented changes' to our transport network required before 2030 to avert climate catastrophe. In the coming years, we're going to be faced with some big decisions – the concrete flyovers that run through our city are not going to last forever. They were built in the same era as the Morandi Bridge in Genoa in Italy which collapsed in 2018, killing 43 people. Although I'm sure Transport Scotland has a more diligent maintenance programme than the privatised Autostrade per L'Italia company whose failings caused Genoa's tragic collapse, our concrete bridges aren't going to last forever. Are we really going to spend billions of pounds repairing or rebuilding them? No, we need to invest that money in sustainable public transport infrastructure instead, remove or reappropriate the M8, reconnect the communities in the north of the city which have been severed off for too long. Let's let our city breathe again. It's only then that we will truly create a city that is just so pleasant to be in, that people don't want to leave.

CHAPTER 12

Travelling Without Moving

Equalising mobility

RUMOUR HAS IT that the German philosopher Immanuel Kant (1724–1804), who in the 18th century wrote several vast tomes, never left his hometown of Königsberg. He lived a life that personified the 'education, organisation and discipline', which EF Schumacher says is the key to personal and societal development. Kant took his daily walk around the town at such regularity that locals came to set their watches by him. He didn't need to travel the world physically because he was reading and writing so intensively that he was travelling all the time in his imagination. This book has given me the chance to 'travel without moving', to relive my whole life from the safety of my desk. To help me understand how all this knowledge and experience has shaped the person I am now. It has been an essential 'creative outlet' – something every human being must have to help process everything happening in the world around us and create meaning and purpose.

To address the climate emergency we must urgently rebuild our city, society and economic system so that we can all enjoy physically travelling less. In Glasgow this means addressing the key things which currently make people so desperate to escape:

- the hostile physical environment created by all those misguided town planners who chose to put cars before people.
- the lack of local services and amenities in many places, exacerbated by an appalling public transport network which is *causing* poverty[1] and 'literally holding people back'.[2]
- the non-existent or poor quality work which, as Schumacher says, means people are 'working solely for their pay packet and hoping, usually forlornly, for

enjoyment solely during their leisure time'.[3]
- and the lack of 'education, organisation, and discipline', which prevents people channelling their energy (alienation, anger and frustration) in creative or productive ways.

I want to begin this final chapter with a reminder of what one friendly Glaswegian wrote on *The Glasgow Effect* event page back on 5 January 2016:

> The only reason anyone is interested in 'the glasgow effect' is that the patter is on fire on the comments! We as a nation, will rip the absolute shite outta ye hen, untill you piss back aff to whatever entitled shire you hail from. If ah want to go on holiday, i have to save up ma minimum wage (a cheeky wee 3 star all inclusive to magaluf does the rest of us... We think you should do the same... [sic]

This comment is relevant to this chapter for two key reasons. Firstly, because of its 'piss back aff' to where you came from attitude, which was echoed in so many of the comments I received (something I'll come back to shortly) and, secondly, because it was also one of many comments equating what I was doing to a 'holiday'. Local journalist Brian Beadie (who ironically used to blog under the moniker 'not enough sun' – another key reason people are so desperate to escape!) was the first to acknowledge the absurdity of comparing *The Glasgow Effect* to a 'holiday'. In his article for the Kiltr website to mark the end of the project on 7 February 2017, he wrote:

> Certainly, one aspect that critics of the project ignored was that Harrison was actually taking a pay cut in order to fulfil the project, and not being paid to stay in the city, but paid not to be able to leave. This is one under recognised and crucial aspect of the project, in that most people, when they gain access to free time and spare cash, use it to get as far away from their communities

as possible, or go on holiday. And, of course, one of the perks of being an artist is getting to work and exhibit abroad, which compensates for the low incomes that most artists actually have to live on. Indeed, the piece was originally entitled 'Think Global, Act Local' and Harrison used her freed-up time (and some of the public money) to engage more fully with local projects and activism.[4]

Admittedly, I did really feel like I needed a holiday by the end of *The Glasgow Effect*, because the public scrutiny and pressure made it such an intensely stressful and exhausting experience. However, had it not been for these less than ideal circumstances, the 'low-carbon lifestyle of the future' I was pioneering should be one that is so rewarding and empowering, with so much creativity and autonomy, that it breaks down the barriers between work and leisure, whilst also, simultaneously, effecting positive change in the city which will help improve everyone's quality of life.

I don't normally take holidays (not more than a day or two anyway), because I normally love my work so much. That is the urgent task at hand – to create the social conditions so that this sort of work/life becomes the norm, so there is no longer need for the 'escapism or aggression'[5] Schumacher warns against, nor the feeling of 'hardship' for those who don't get to go away. Instead of having an economic strategy based on 'increasing tourism from two million to three million tourists a year', we must have one based on including everyone who's already here – releasing all the 'latent, untapped potential'[6] on the periphery – by reconnecting everyone to the city's social, economic and cultural life (physically, with world-class public-transport links, and psychologically, with inclusive arts and culture). This will also stop large chunks of our city's disposable income leaking out the 'leaky bucket' of our local economy to places like Magaluf.

To address the climate emergency, the cost of travel must urgently be made to reflect the actual cost of each journey to the environment (to include the 'externalities', as economists would say). The cost of travel must be made to 'align' with the sustainable 'transport hierarchy', which has been designed to help us plan fairer, more sustainable and people-friendly cities.[7] This puts pedestrians and cyclists at the top, then public transport users, followed by taxis, then shared and private cars. The 'transport hierarchy' reflects the carbon Conversion Factors on page 141, where walking and cycling don't even feature as they have zero impact and are naturally free (once you have a bike that is). This is how we address the 'Pareto Principle' and celebrate and reward those who are already doing the least damage to our environment. But in Greater Glasgow we must make special allowances, because we are part of that 'conurbation' created in the post-war period, which demands that so many more people must now travel in and out for work. We must follow the lead of those other forward-thinking towns and cities around the world and make all travel on public transport within the Greater Glasgow 'travel to work area' free,[8] so removing the cost of 'mobility' for those stuck on the periphery. This is the *only way* we'll shift people onto sustainable forms of transport and cut carbon emissions and air pollution at the speed that we need.

Just as a new regional currency for Greater Glasgow would be a positive way to encourage people to 'buy local', free regional public transport (provided by our new Transport for Greater Glasgow body) would also encourage people to 'travel local' for work and for leisure. Both would disproportionately benefit those who can't normally travel beyond the region for socio-economic reasons. And we would all benefit twice from what New Economics Foundation call the 'local multiplier effect'.[9] That is the more the same pound is spent and respent within a local area, the more its benefit to local people is multiplied. This is why we must plug the holes in the 'leaky bucket' of our local economy, as the city of Preston has so successfully done.

Short, quick, efficient and sustainable regional journeys (on foot, bike or public transport) could be seen like these regional transactions in our new currency – the more that are made within the area, the more people's social and cultural life flourishes, the more the local economy flourishes – the 'core economy', the 'foundational economy' and the mainstream one. People would no longer be 'stuck in the hoose' because they 'cannae get a bus', which is often the case at present. 'Loneliness, hopelessness and lack of connectedness', the three things Chris Harkins from

Glasgow Centre for Population Health says causes ill-health,[10] would be addressed. That's how we build a strong local economy and encourage people to enjoy themselves in the region – taking advantage of all the beautiful countryside that we're blessed to have, rather than jetting off overseas.

And we may well find that with these two things – a new regional currency and free regional public transport – our regional economy also 'grows' as measured in the old, crude monetary sense. But this new model, which encourages people to 'buy local' and 'travel local' to cut carbon emissions from imports and exports and from excessive global travel, may actually prove to be the only way economic 'growth' and carbon emissions can be 'decoupled'.[11]

It's not that we don't want people to travel the world and experience other cultures at all, but this travel must be less frequent and more meaningful when it does happen. A 'frequent flyer levy' as proposed by the campaign group A Free Ride would exponentially increase taxes on air travel based on the number of journeys that you make each year. Those who travel most often would be penalised and those who make the occasional trip would not.[12] Revenue raised could be used to build and run our free regional public transport network. This would help *equalise mobility* – to stop the middle classes flying about the world for academic conferences and art biennales and give others the opportunity.

Just as Patrick Geddes was able to make his 'Think Global, Act Local' pronouncement because he had first travelled so extensively and lived and worked in many countries (Scotland, England, France, Mexico, Cyprus, America, India and Israel),[13] Cathy McCormack developed her global perspective through the trips she was able to make to South America (where she discovered the 'popular education' of Paulo Freire) and to South Africa. She made connections with other oppressed people across the world and began to see her activism in Easterhouse as part of the international struggle against a global capitalist system which is centralising wealth in the hands of the few. And in *The Tears that Made the Clyde*, Carol Craig describes an exchange programme between Ruchazie in Glasgow and Malawi run by the local church, which helped build solidarity between the two places and make everyone more appreciative of what they do have already. For the Glaswegians, 'their exposure to Malawian culture and way of life helped them to see that the poverty they experience in Scotland is not so much material as spiritual and cultural'.[14]

It's this sort of experience which is truly educational and enables the acceptance of difference that is essential for overcoming the 'piss back aff' to where you came from attitude towards migrants. It is that *very fine line* between localism and protectionism, and bigotry and xenophobia, which was exposed in *The Glasgow Effect* debate. The key question is: how do we balance the need to cut carbon emissions from imports and exports and from excessive global travel with the need to remain outward-looking and welcoming to 'outsiders'?

Minimising migration

There's no getting away from the fact that to address the climate emergency we need less movement of goods and people and, ultimately, less migration. As my Carbon Graph on pages 138–9 clearly shows, the further you move from the people you love, the more you end up travelling back and forth to visit them. This is evidenced in the increasing number of flights to and from Eastern Europe since 2004, when eight new countries (Czech Republic, Estonia, Hungary, Latvia, Lithuania, Poland, Slovakia and Slovenia) joined the European Union and many people migrated to the UK.[15] EF Schumacher echoes the concerns of Edward Said (quoted in Chapter 4), blaming the advent of 'mass transport and mass communications' systems for making this migration far too easy, saying this is what has made people 'footloose'.[16] Rather than giving us 'freedom', these technological advancements actually 'tend to destroy freedom, by making everything extremely vulnerable and extremely insecure'.[17] Those who migrate, no matter how hard we try, may never be able to find or make the 'home' we're searching for, even if we do eventually 'piss back aff' to where we came from. As Glasgow's history as a city of migrants shows, we will always suffer from the psychologically and socially damaging consequences of 'dislocation'. In a globalised world, we are now faced with what Schumacher calls the rapidly increasing and ever more intractable problem of 'drop-outs', of people, who, having become footloose, cannot find a place anywhere in society.[18]

Ultimately, we must strive to equalise wealth, resources and opportunity around the world so that migration is not so necessary and people are not forced to move away from their vital family support networks – their natural 'core economy'. This is a task which is becoming ever more challenging as climate change begins to make large parts of the world completely uninhabitable. The United Nations Refugee Agency (UNHCR) says climate change has already caused at least 22 million migrations,[19] and that there will be more than 200 million 'climate refugees' by 2050 (that's three times the population of the UK on the move).[20]

It is the UK which, more than any other country in the world, has cumulatively contributed the most to global warming per capita since the start of the Industrial Revolution.[21] It is our carbon-intensive lifestyles that are responsible for most of the environmental destruction (80 per cent of the world's population has still never even set foot on a plane).[22] And so it is our moral duty to welcome the people who are suffering in other parts of our world as a result. The only way to ensure that this influx of migrants does not cause more tension – as evidenced in the 'piss back aff' attitude – is by actually improving the quality of life of the people who are 'from here' as well (in the numerous ways described in the previous chapters). We must therefore strive to strike the perfect balance between being inward- and outward-looking.

It is only the Fearless Cities: Global Municipalist Movement that has been able to achieve this balance so well. Ada Colau, the housing activist elected to serve as mayor of Barcelona when the people's platform Barcelona en Comú (Barcelona in Common) successfully 'won back the city' at their local elections in 2015, says that the 'fear of the other' that is fuelling the rise of nationalism and fascism is caused by an increasingly uncertain and unstable world. It is the 'brutal neoliberalism that has managed to take hold of the global economic system' that means 'more and more people are living in fear: of losing their job, their home, their pension or even their life at the hands of an abusive partner or in a terrorist attack'.[23] Instead Ada Colau advocates Municipalism because it

> seeks to do away with these divisions [by] starting from the place where we all recognise one another as equals: the community. Our neighbourhoods, our towns, our cities. Municipalism is an emerging force that seeks to transform fear into hope from the bottom up, and to build this hope together, in common.[24]

Barcelona en Comú have been able to do this by 'working at the local level through concrete actions and policies that not only improve people's lives, but also show that there is an alternative, and that politics should work for the many and put people at the centre'.[25] A fairer and more sustainable city is an inclusive one, which makes everyone feel as though they are welcome and that they 'belong', regardless of where they might be from. The 2019 book, *Fearless Cities: A Guide to the Global Municipalist Movement*, highlights many successful initiatives run by cities striving for this inclusiveness. In New York, since 2015, they have been issuing municipal ID cards to anyone who lives in the region regardless of their 'immigration status' as a tool for 'equity and inclusion'. These enable everyone to easily access public services and to participate fully in civic life. The IDNYC programme has shown to increase citizens' 'sense of belonging in the city'.[26]

The vision of the Fearless Cities: Global Municipalist Movement is one where the sense of belonging to your specific locality, your local community, becomes far more important than belonging to any nation-state.[27] With an international network of towns and cities which have the freedom and autonomy necessary to provide for their citizens, the nation-state becomes less relevant. It is through this international network of towns and cities that we can 'defend human rights... fight together against climate change and misogyny and against all the policies that only benefit the few and condemn the rest to uncertainty and fear'.[28] It is through this international network that we can work together to effectively tax the global tech giants who are scooping off the world's wealth, and abolish the tax havens where they are stashing it (currently roughly $8 trillion).[29] It is with this money that we can provide Universal Luxurious Services for everyone on earth. That is how we stop the uncertainty, instability and injustice which fuels fear and hatred and forces so many to move.

Paradox of repopulation

Glasgow is at a key turning point in its history and *The Glasgow Effect* exposed the tensions this is causing. Having become known as one of only two 'former million cities' in the world for shedding nearly half its population in just 50 years (along with Detroit in America which suffered an even greater decline as a result of deindustrialisation), Glasgow's population is now finally on the rise again.[30] Since 2007, the birth rate has exceeded the death rate and inward migration from the rest of the UK and elsewhere has now finally exceeded the numbers leaving in search of a better life elsewhere. The populations of our neighbouring councils Inverclyde and West Dunbartonshire have fallen in recent decades and we are finally starting to repopulate the city centre and recreate the more compact, walkable city which will help us all to live more sustainable lives. This repopulation is happening slowly and 'naturally' this time, rather than by force.

The number of households in Glasgow is predicted to rise 16 per cent by 2041 (half of these are predicted to be 'single person households' unless we do something to intervene). The ethnic diversity of the city is changing (from 5 per cent of people from non-white backgrounds in 2001 to 12 per cent in 2011), as is the class demographic:

> Since 1981 the proportion of Glasgow's population in the top two social classes has more than doubled. This means that four out of ten of Glasgow's citizens are now classified as Social Class I or II.[31]

Glasgow is becoming more middle class.[32] As we have seen, this recent transformation has in part been encouraged by using the narrative of the 'creative city' as a marketing technique. In his book, *Against Creativity*, Oli Mould explains how the concept of 'creativity' has been hijacked by neoliberal policy makers in pursuit of profit and 'growth'. He describes the 'creative city' as an

> intense (re)branding of a city's cultural offerings along 'cultural' and 'creative' lines. Nightlife, art scenes, café cultures and so on are aggressively used in marketing material for cities and neighbourhoods as a means to attract 'creative class' workers...[33]

'The Glasgow miracle' has been harnessed 'to woo these footloose, agile and energetic creative types',[34] like me when I first arrived in 2008, to take up opportunities in the 'knowledge economy' and the 'creative industries'. This has meant that in Glasgow, over the last 30 years:

> The use of a non-politicised, contemporary art 'scene', coupled with a cosmopolitan night-time economy [has] completely glossed over the social strata that characterises neighbourhoods.[35]

The neoliberal 'creative city' is a policy proven to exacerbate inequality and breed the 'resentment and scepticism'[36] that fuels the 'piss back aff' attitude which was so evident in *The Glasgow Effect* debate. As both Darren McGarvey and Alasdair Gray have articulated, artists with little understanding of Glasgow's history or present are often 'parachuted in' to fulfil somebody else's gentrifying agenda. Forced into this position they become inherently compromised, as Oli Mould describes, 'caught in the middle, struggling to balance the need to earn a living and the social ethics of their practice'.[37] The abuse they may receive just for 'doing their job' only makes them less likely to commit and contribute back to their community and makes them more likely to pack up and leave.

If we want to continue to grow our city's population and our local economy/ies in a sustainable way,[38] that is inclusive, encouraging integration and easing tensions, then we need to drop the meaningless marketing campaigns altogether. No more

'Glasgow's Miles Better' (launched in 1983), no more 'Scotland with Style' (launched in 2004) and no more 'People Make Glasgow' (launched for the Commonwealth Games in 2014), no more promoting 'the Glasgow miracle'. Instead there is a simple answer. We 'advertise ourselves' to the outside world by providing the Universal Luxurious Services and building the Car-free Future that really will improve everybody's quality of life. In Tallinn in Estonia (population 440,000), where they made all their public transport free in 2013 and are now rolling this out across the whole country, the policy paid for itself.[39] As Friends of the Earth show:

> The €12 million loss of fares income to its municipal public transport operator was more than offset by a €14 million increase in revenues, more people moved to the city, increasing its tax-base.[40]

People moved to Tallinn because they could see it really was a place that provided for all its citizens. I know I'd rather have an East End Subway extension (as was promised by SPT in 2007)[41] than a pink 'People Make Glasgow' logo, as the 'legacy' of the Commonwealth Games. If we prioritise *actually* improving everybody's quality of life, rather than just making it look like we are through vacuous marketing campaigns, then the accolades (if we care about them) will follow. People will want to visit, and people will want to stay and make the place their home and, hopefully, contribute back to the community, which has shown them such kindness and generosity.

Rather than making headlines around the world for being the 'Sick man of Europe',[42] let's make headlines around the world like Estonia and Luxembourg did when they made all their public transport free; like Oslo did when they banned car parking in their city centre; like São Paulo did when they banned advertising; like Manchester did when they announced plans to reregulate their entire public transport network to 'put passengers first';[43] like the Ruhr region of Germany did when they closed 60km of motorway for a day just to hold the 'world's longest street party', or like Seoul did when they permanently replaced the hideous big motorway which tore right through their city centre with a beautiful riverside park instead. These are the policies which will make global headlines for the right kind of reasons, and these are the policies which will address the climate emergency and improve our population's health.

Education for life, not for work

I want to return one final time to the wise words of Glaswegian trade unionist Jimmy Reid. Speaking in 1972, on the cusp of the wave of automation that caused mass unemployment in Glasgow's industrial economy, he said in his famous 'alienation' speech:

> If automation and technology is accompanied, as it must be, with full employment, then the leisure time available to man will be enormously increased. This being so, then our whole concept of education must change. The object must be to equip and educate people for life, not for work.[44]

Jimmy Reid's vision has since been echoed by the New Economics Foundation[45] and by Common Weal in their call for a 'democratic distribution of work and leisure'.[46] If there's going to be less work to go around (and as Schumacher would say, that's a choice for us to make), then we must share the remaining work out evenly so that none of us are doing more than 21 hours of paid work a week. The question then does become: how do we fill our 'free time'? This is where education comes in.

The students I work with at the art college might be studying 'Fine Art', but this is really just a framework through which they acquire that eclectic mix of essential life skills – critical thinking, practical skills, confidence and self-motivation. They are not really there to learn how to be 'artists', but rather to learn how to learn, and to learn how to live. They are there to 'clarify their central convictions',[47] as Schumacher advised, and to develop their ethical codes so that they can produce 'wisdom' as opposed to 'cleverness',[48] and work out how they can best contribute back to society and become fulfilled human beings. They are there to learn that education, action and reflection is a life-long process, and that this is just the start.

We need everyone in Glasgow to have this opportunity or our city's 'metaphysical disease'[49] – the so-called 'diseases of despair': alcohol, drugs, suicide and violence – will only get worse.[50] So we cannot allow this education to remain the preserve of elite establishments – especially a higher education sector under the spell of neoliberalism obsessed the 'EEE Agenda' (Employability, Enterprise and Entrepreneurship); gearing people up for the 'rat race' – the complete opposite of what Jimmy Reid advised.[51]

My friend Jim, who grew up in Shettleston and Easterhouse, said he'd heard people suggesting that following the second fire at Glasgow School of Art in 2018, the building should have been rebuilt out in Easterhouse rather than in the city centre. That is the sort of action that we need to decentralise wealth, resources and opportunity in this city to the extent that it necessary – so that 'social mobility' no longer requires literal mobility and we can stop the 'brain drain'. We need radically inclusive art schools which, like Joseph Beuys' class at the Düsseldorf Academy of Fine Arts, anyone who wants to can attend. In the foreword to Paulo Freire's book, *Pedagogy of the Oppressed*, Richard Shaull writes:

> Education either functions as an instrument which is used to facilitate the integration of the younger generation into the logic of the present system and bring about conformity to it, or it becomes 'the practice of freedom', the means by which men and women deal critically and creatively with reality and discover how to participate in the transformation of their world.[52]

It's only a creative education – that focuses specifically on developing those 'critical' and 'creative' skills – that can help everyone to 'discover how to participate in the transformation of their world',[53] so they can create a 'sense of purpose and meaning in life'. We need everyone to discover that 'creative outlet', where they can channel their alienation, anger and frustration in productive ways, thereby starting to overcome them. We must reclaim creativity from the neoliberals, those who made it into an 'industry', and a 'middle class job', so that it can once again become a central part of all our lives. As Joseph Beuys once said: 'Away with money, then art can play the most important role'.[54] Rather than 'releasing the inner entrepreneur', Oli Mould shows that creative work can, and should, 'release the inner revolutionary'.[55] He writes:

> Creativity could be – indeed, should be – thought of more as an emancipatory force of societal change.[56]

Not only does exercising our imagination enable us to 'travel without moving' – to live the 'low-carbon lifestyle of the future', rather than resorting to 'escapism or aggression'[57] and fixating on our next 'holiday' – but exercising our imagination is also an essential step for any social change. As Ada Colau, mayor of Barcelona, says:

> We're living in extraordinary times that demand brave and creative solutions. If we're able to imagine a different city, we'll have the power to transform it.[58]

It is only municipalism – the people-led global movement that truly manages to 'Think Global, Act Local' – which American activist Debbie Bookchin says 'returns politics to its original definition – a moral calling based on rationality, community, creativity, free association and freedom. It is a richly articulated vision of a decentralised democracy in which people act together to chart a radical future'.[59] It is only by rekindling the tradition of radical municipalism in towns and cities across Britain, that we can create the social conditions where 'every human being [actually can be] an artist',[60] where the distinctions between artist and non-artist dissolve and where art and politics become one.

Rekindling our radical past

The worst aspect of Glasgow's brutal post-war redevelopments was that not only were communities left totally disorientated and disempowered by all the upheaval (destroying what had once been a strong sense of community and affiliation), but collective memory was also erased. The history of radical socialism in the city was demolished along with all those inner-city tenement blocks, so that it became nothing more than the 'rumour'[61] which Alasdair Gray describes. A cynic might say that it was part of the master plan all along to destroy this radical history: another 'safety valve', as emigration was seen, 'relieving pressure for radical reforms'.[62]

But in this time of accelerating global social, environmental and economic crises, it is this radical past which must urgently be rekindled. It's not just our city's highway 'scar' that must be healed, but so many psychological ones as well.[63] Debbie Bookchin's description of politics above 'a moral calling based on rationality, community, creativity, free association and freedom'[64] is what was once common in Glasgow, before the Second World War. As Carol Craig writes:

> There was a flowering in Glasgow of a whole range of specifically socialist organisations – cycling clubs, rambling associations, Sunday schools, drama groups, youth groups and choirs. Education too was a major motivator and socialist discussion groups, reading groups, lectures and evening classes were common. Many of the movement's leaders had been teachers. Glasgow's ILP [Independent Labour Party] was not a centralised organisation but a loose federation of branches as well as women's organisations and other radical groups. Its main means of communication was via its newspaper *Forward*. The movement's vision was often expressed in terms of 'well-being' and 'happiness' and also emphasised the importance of 'spiritual emancipation'.[65]

Part of our task moving forward is to unearth this history and make it visible again so that we can now build our collective future by learning from the past. Writing about his experience of attending a Socialist Fellowship School at Pollok Community Centre, Brian Moore says 'it was immediately apparent that it was a place where divisions invoked by religion and sport were wonderfully absent'.[66] In our deeply divided society – where the 'piss back aff' attitude is rife – we need more places like this where we can come together in 'real life'. We need places where we can forge real social networks (not online ones) and find real solidarity (not the illusion of connection that technology provides). We need places, like the Socialist Sunday Schools, where we are all encouraged to 'Love learning, which is the food of the mind',[67] and to 'Observe and think in order to discover the truth'.[68] In Glasgow, it's essential we allow ourselves the space and the time to dream that it is still possible to create 'a New Society, with Justice as its foundation and Love its Law'.[69]

Love-hate relationship with the city

> Thou shalt not be a patriot… Your duty to yourself and your class demands that you be a citizen of the world.[70]

Those words, written in 1918 by Tom Anderson, founder of Glasgow's first Socialist Sunday School (first quoted in Chapter 5), stand in complete contrast to what former Prime Minister Theresa May told the Tory Party Conference nearly a century later in 2016:

> If you believe you are a citizen of the world, you are a citizen of nowhere. You don't understand what citizenship means.[71]

And so we come back to that *very fine line* between localism and protectionism, and bigotry and xenophobia – which now appears to me as one of the most pertinent issues of our times. I think that the only way forward in this time of increasing hatred and division is that we all do as the 'Think Global, Act Local' mantra demands. That is that we are both 'citizens of the world', in that we strive to understand the global context and have international solidarity with all fellow human beings, but that we are also 'citizens of somewhere', in that we commit to and take an interest in improving the places – towns and cities – where we live. This is a balance that the people's platform of Barcelona en Comú have struck perfectly. They are outward-looking 'internationalists', at the same time as delivering 'concrete actions and policies' for the local people they serve. This is what the people of Glasgow must now try to replicate – to 'win back our city'[72] and transform our broken political systems from within – rather than pinning all our hopes on a divisive nationalism. It's like Glaswegian poet Tom Leonard (1944–2018) says:

The local is the international. The national is the parochial.[73]

So I'm back to where I started in my Preface and 'settling down' or 'making ourselves at home'[74] now presents itself as part of the answer to building 'the sustainable city of the future' that we so urgently need. I am the one who's been 'footloose' after all. But there's one more fundamental contradiction to consider. That is if you get 'too comfortable', then there is no agitation, no motivation to fight for social, environmental or economic change. And that's where that love-hate relationship with the place you live becomes important. You have to be angry enough to fight to make it better (not too comfortable), but you also have to be compassionate enough to care and optimistic and hopeful enough to believe another city is possible. And it is change in Glasgow that we desperately need. Otherwise, as Cathy McCormack said back in the '90s, 'the only thing that remains sustainable… is poverty'.[75]

As I come to the end of this book, I must thank the three amazing women I lost in 2017, without whom it wouldn't exist: Diane Torr, who taught me to be myself and resist conforming to social norms; Roanne Dods, who taught me patience and empathy, and that 'small is beautiful'; and, finally, my beautiful mum, who taught me to love learning, to laugh and always look on the bright side – as well as nearly everything else I know. Sending this book out into the world, I might once again find myself falling into what Edward Said calls the paradox of the 'sensitive émigré' – whose 'attempts to *impress* others compound, rather than reduce, their original sense of isolation'.[76] But now that this epic undertaking is finally complete, I am going back out into the city to put all this reflection into action in the local projects and campaigns I began work on in 2016, and to get on with the rest of my life.

My message to anyone still reading is: channel your anger. Let's get educated, get disciplined and get organised!

Acknowledgements

IT SHOULD BE quite clear from the book itself who I owe most gratitude to for coming to my rescue when I needed them the most: my dear mum, of course, to whom the whole book is dedicated, my lovely old dad, my sister and her family and all my long-suffering friends: Neil, Sarah, Emily, Oliver, Alberta, Deniz, Tom and many more.

I would also like to offer a special thank you to: Martina Meijer and her family for looking after me in 2017; Dawn Irving for her support in 2017 and for her research into the Social Mobility Index used in Chapter 1; Gerrie van Noord, my old tutor at Glasgow School of Art, for her initial support and motivation in June 2018 as I geared up to find a publisher; Gavin MacDougall at Luath Press for coming to my talk at the Glasgow Film Theatre on 7 January 2017 (unbeknown to me) and for agreeing to publish the book when I approached him more than a year later on 27 June 2018; and my uncle, aunty and cousin for their help getting started with writing that July.

But my greatest thanks go to Anna McLauchlan. Without Anna, this book would never have been finished. She came to rescue me in my flat on 5 January 2019, amidst a sea of books, papers and dust, and set me targets to keep me focused through the bleak winter months. She read two full drafts (in February and in June) and advised on which bits needed more work. She checked in regularly and helped create the 'motivational structure' that I needed to see me right through to the final, final deadline on 8 July 2019.

Anna also helped organise additional readers to comment on a draft in June: Lee Gardner and Ian Macbeth. Thanks to both of them for all their time, help, support and thoughtful and challenging comments, and to Ian for his research on the Clyde shipyards referenced in Chapter 4.

I'd also like to give another special thanks to my lovely old dad, who proofread nearly the whole thing in June 2019 and didn't seem at all annoyed at some of the less than complimentary stuff I might have written. I love you, dad.

There are a few other people who I would like to thank for their support in other ways: Avril Stevenson for her kindness in offering me a home-from-home in Dundee; Eddie Summerton for being the best boss ever and all my other brilliant colleagues at Duncan of Jordanstone College of Art & Design; my second year students of 2018–19 who put up with me wittering on about this book all year and all my lovely new colleagues in the local projects and campaigns I'm now working on in Glasgow.

Finally, I would like to thank Maia Gentle and Jennie Renton at Luath Press for their help and support with editing and typesetting. Jennie went way beyond the call of duty, putting me up in her flat in Edinburgh the whole weekend before we went to print! And I guess I should also thank Creative Scotland again, for taking a risk on *Think Global, Act Local!* in 2015, even though I have funded the year needed to write this book completely off my 'own back'.

Endnotes

Resources marked with ↗ are available online.

Introduction

1 McGarvey, D, 2017, 'Neoliberalism', *Poverty Safari Live* EP, YouTube ↗
2 Scottish Government, 2017, 'Scottish Vacant & Derelict Land Survey 2016', Scottish Government ↗
3 Yates, A, 1998, 'Deep Pan Pizza chain to disappear', *The Independent*
4 Lowndes, S, 2010, *Social Sculpture: The Rise of the Glasgow Art Scene*, Luath Press, p.294
5 Quoted in Hassan, G, 2016, 'The Real Glasgow Effect on all of us', Gerry Hassan ↗
6 McCartney, G et al, 2011, *Accounting for Scotland's Excess Mortality: Towards a Synthesis*, Glasgow Centre for Population Health, p.4
7 Ibid, p.73
8 Walsh, D et al, 2016, *History, Politics & Vulnerability: Explaining Excess Mortality in Scotland & Glasgow*, Glasgow Centre for Population Health
9 Harrison, E, 2015, *Open Project Funding Application Form*, Creative Scotland, p.7
10 I described my motivations for choosing the title *The Glasgow Effect* in the interview I gave to CommonSpace, published online on 7 January 2016 (discussed in Chapter 5). See: Haggerty, A, 2016, 'Artists need to stick their necks out: Exclusive interview with The Glasgow Effect's Ellie Harrison', CommonSpace ↗
11 Foote, C, 2018, 'One in three Glasgow children "growing up in poverty"', STV News ↗
12 Dalziel, M, 2018, 'Urgent appeal as Glasgow food bank demand reaches record high', Glasgow Live ↗
13 McGarvey, D, 2017, *Poverty Safari*, Luath Press, p.210
14 Friedman, S, Laurison, D, 2019, '10 ways to break the class ceiling', Policy Press ↗

Chapter 1

1 Chambers, V, Chambers, F, 2013, 'Perivale Maternity Hospital', Lost Hospitals of London ↗
2 Libcom, 2007, '1978–9: Winter of Discontent', Libcom ↗
3 The phrase 'own back and front door' is used by Carol Craig to describe the aspirations of working class Glaswegian women for better quality housing. However, she later explains that this aspiration itself may have resulted from cultural 'colonisation' by the English. See:

Craig, C, 2010, *The Tears That Made the Clyde: Well-being in Glasgow*, Argyll Publishing, p.223
4 Both my parents were members of the National Association of Teachers in Further & Higher Education (NATFHE), which is now part of the University & College Union (UCU) to which I now belong.
5 CASE, 2017, 'Who we are', Campaign for the Advancement of State Education ↗
6 Harry Greenway was 'accused of accepting gifts, including foreign holidays, in return for using parliamentary influence to help gain British Rail contracts for the Plasser Railway Machinery company'. The case was eventually dropped due to 'insufficient evidence'. See: Mills, H, 1992, 'MP will not face bribery charges', *The Independent*
7 Weil, L, 1977, *Gertie & Gus*, Parents' Magazine Press
8 This title is borrowed from my friends at the Solidarity Against Neoliberal Extremism (SANE) collective who run a 'popular education' course in Glasgow called 'WTF is Neoliberalism?'
9 Jelinek, A, 2013, *This is Not Art: Activism & Other 'Not-Art'*, IB Tauris, p.19
10 This is the definition of socialism used by Glasgow's Socialist Sunday Schools, which operated in communities across the city from 1896 until 1980 in various guises, including the Proletarian Schools (1918–39), the Socialist Sunday Schools (1896–1965) and the Socialist Fellowship Schools (1965–80). See: Ewan, R et al, 2012, *The Glasgow Schools*, The Common Guild, p.12
11 Water privatisation was stopped in Scotland when Strathclyde Regional Council held a referendum in March 1994 which showed 97 per cent of people were against the policy (see Chapter 10 for further details). See: Arlidge, J, 1994, 'Scottish Tories urge rethink of water plan', *The Independent*
12 French economist Thomas Piketty has shown that inequality is not an accident, but rather a feature of capitalism, and that it can only be reversed through state intervention. See: Piketty, T, 2014, *Capital in the Twenty-First Century*, Harvard University Press
13 Cumbers, A et al, 2013, *Repossessing the Future: A Common Weal Strategy for Community & Democratic Ownership of Scotland's Energy Resources*, The Jimmy Reid Foundation, p.5
14 Sommerlad, J, 2018, 'This is how Section 28 affected the lives of LGBT+ people for 30 years', *The Independent*
15 Jelinek, A, 2013, *This is Not Art: Activism & Other 'Not-Art'*, IB Tauris
16 Robertson, J, 2016, 'How the Big Bang changed the City of London', BBC News ↗

CHAPTER 1 CONTINUED

17 Thames Water, 2018, 'Ownership structure', Thames Water ↗

18 WTO, 1995, 'History of the multilateral trading system', World Trade Organisation ↗

19 Klein, N, 2014, 'Trade & Climate: Two Solitudes' in *This Changes Everything: Capitalism vs the Climate*, Penguin, p.75–80

20 Thatcher, M, 1989, 'Global Environment Speech' at United Nations General Assembly, YouTube ↗

21 WTO, 1995, 'History of the multilateral trading system', World Trade Organisation ↗

22 Klein, N, 2014, *This Changes Everything: Capitalism vs the Climate*, Penguin, p.76

23 Frumhoff, P, 2014, 'Global Warming Fact: More than Half of All Industrial CO_2 Pollution Has Been Emitted Since 1988', Union of Concerned Scientists ↗

24 Cox, L, 2015, 'The New Economics' at The RRAAF Debate, Beaconsfield Gallery, Vimeo ↗

25 Cahn, E et al, 2008, *Co-production: A Manifesto for Growing the Core Economy*, New Economics Foundation, p.8

26 Harvey, D, 2007, *A Brief History of Neoliberalism*, Oxford University Press, p.119

27 Poet Simon Murray devised an alternative to the 'there is no alternative' (TINA) mantra to inspire people with the optimism and hope necessary to fight back. Instead he says 'there are billions of options' (TABOO). See: Murray, S, 2009, 'There Are Billions Of Options', Sai Murai ↗

28 Jelinek, A, 2013, *This is Not Art: Activism & Other 'Not-Art'*, IB Tauris, p.20–1

29 Lawson, N, 1991, 'Back seat for education', *Evening Standard*

30 Kingston, P, 2001, 'Are agencies doomed?', *The Guardian*

31 McGarvey, D, 2017, *Poverty Safari*, Luath Press, p.50

32 O'Brien, D et al, 2018, *Panic! Social Class, Taste & Inequalities in the Creative Industries*, Create, p.39

33 Friedman, S, Laurison, D, 2019, '10 ways to break the class ceiling', Policy Press ↗

34 Walsh, D et al, 2016, *History, Politics & Vulnerability: Explaining Excess Mortality in Scotland & Glasgow*, Glasgow Centre for Population Health, p.29

35 McGarvey, D, 2018, 'Poverty Safari Live' at The Stand's New Town Theatre, Edinburgh Festival Fringe

36 O'Brien, D et al, 2018, *Panic! Social Class, Taste & Inequalities in the Creative Industries*, Create, p.40

37 Alston, P, 2018, 'Press Conference of the UN Special Rapporteur on Extreme Poverty & Human Rights', Facebook ↗

38 Social Mobility Commission, 2017, 'Social Mobility Index', UK Government ↗

39 ONS, 2016, 'The National Statistics Socio-Economic Classification (NS-SEC)', Office for National Statistics ↗

40 Social Mobility Commission, 2017, 'Social Mobility Index', UK Government ↗

41 Bell, S, 1992, 'We Rule You, We Fool You...', *The Guardian*

42 Fogg, A, 2009, 'The Prophecy of 1994', *The Guardian*

43 Smith, FM, 2014, 'How UK Ravers Raged Against the Ban', *Vice*

44 Penman, D, 1994, 'The Park Lane Riot: How Park Lane was turned into a battlefield', *The Independent*

Chapter 2

1 When tuition fees were first introduced in 1998 the rate was £1,000 per year. Small maintenance grants were still available from Local Education Authorities and I qualified for one because my dad was a pensioner. I took out a loan each year from the newly created Student Loans Company, but my mum still paid my housing costs. I still owe them £12,747.85 (as of 31 January 2019), but I'm never asked to pay any back as I rarely earn over the threshold.

2 Part of the *Eat* 22 project involved uploading the photos to the project website on a weekly basis, so that people could 'follow' as it developed over the course of the year. People from all over the world tuned in. You could say the project 'went viral' at a time when the internet was still a friendly and supportive place, before it was colonised by the global tech giants. Articles about *Eat* 22 were published in magazines and newspapers in India, Taiwan, Czech Republic, France, Sweden, Germany, America and beyond.

3 Miller, J, 2006, 'Double or nothing: John Miller on the art of Douglas Huebler', *Artforum International*, p.220–7

4 Weiner, L, 1969, 'Statements' in *Art in Theory 1900–2000: An Anthology of Changing Ideas*, Blackwell, p.893–4

5 Jelinek, A, 2013, 'Privatisation & the Knowledge Economy' in *This is Not Art: Activism & Other 'Not-Art'*, IB Tauris, p.34–7

6 WTO, 1994, 'Intellectual Property: Protection & Enforcement', World Trade Organisation ↗

7 Bishop, C, 2012, *Artificial Hells: Participatory Art & the Politics of Spectatorship*, Verso, p.13

8 Quaintance, M, 2017, 'The New Conservatism: Complicity & the UK Art World's Performance of Progression', e-flux conversations ↗

9 ONS, 2018, 'Trends in self-employment in the UK', Office for National Statistics ↗

10 Harrison, E, 2011, 'Press Release: Work-a-thon for the Self-Employed', Ellie Harrison ↗
11 Harrison, E et al, 2005, *Day-to-Day Data*, Angel Row Gallery
12 Higgins, C, 2007, 'Blair reminisces about Labour's "golden age" of the arts', *The Guardian*
13 I attempted to visualise the history of public spending on the arts, and this 'end of an era' moment in my installation *The Redistribution of Wealth*, shown at Late at Tate Britain in October 2012. I created a lightshow projected onto three small stages (representing England, Scotland and Wales) to show the relative scale and distribution of UK government spending on the arts from the birth of the Council for the Encouragement of Music & the Arts in 1940, right up to the present day climate of cuts.
14 This title is borrowed from the final uncompleted book by French philosopher Michel Foucault (1926–84). He had aimed to trace the history of the concept of mental and physical self-improvement back to the ancient Greeks. See: Foucault, M, 1988, *Technologies of the Self*, University of Massachusetts Press
15 SparkNotes, 2018, 'The Apology: 35e–38b', SparkNotes ↗
16 Zaltzman, A, 2018, *My Life as a… Cynic*, BBC Radio 4
17 Hobbes, T, 1651, *Leviathan: the Matter, Forme, and Power of a Commonwealth, Ecclesiasticall and Civil*
18 Gonzalez, M, 2006, *A Rebel's Guide to Marx*, Bookmarks
19 Marx, K, Engels, F, 1848, *The Communist Manifesto*
20 Marx, K, 1845, *Theses on Feuerbach*
21 Quoted in Davies, W, 2018, 'What are they after?', *London Review of Books*, p.3–5

Chapter 3

1 Farndale, N, 2008, 'Lehman Brothers collapse: How the worst economic crisis in living memory began', *The Telegraph*
2 Grice, A, 2009, '£850bn: official cost of the bank bailout', *The Independent*
3 HM Treasury, 2009, *Budget 2009: Building Britain's Future*, Stationery Office, p.241
4 DCMS, 2009, *Annual Report & Accounts 2009*, Stationery Office, p.6
5 That particular bike I've had since my 19th birthday in March 1998 and still use today.
6 Hanley, L, 2018, 'It's not only Londoners who rely on buses and trains', *The Guardian*
7 GCPH, 2012, 'Scottish Cities: Cycling', The Glasgow Indicators Project ↗
8 CityMetric, 2016, 'Which English cities have the most cycling commuters?', CityMetric ↗
9 Rachels, J, 2006, 'Killing & Starving to

Death' in *The Legacy of Socrates: Essays in Moral Philosophy*, Columbia University Press, p.60–82
10 Harrison, E, 2008, *How Can We Continue Making Art?*, Ellie Harrison ↗
11 Abbing, H, 2002, *Why Are Artists Poor? The Exceptional Economy of the Arts*, Amsterdam University Press, p.95
12 Harrison, E, 2010, *Trajectories: How to Reconcile the Careerist Mentality with Our Impending Doom*, Ellie Harrison ↗
13 GSA, 2010, 'Ellie Harrison', Master of Fine Art Online Archive, Glasgow School of Art ↗
14 This distance is calculated using the online Rail Miles Engine, used to calculate the distances of all the UK train journeys I took from 2004–18 for my Carbon Graph on p.138–9.
15 Curtis, A, 2002, *The Century of the Self*, BBC 4
16 BBC News, 2014, 'Stagecoach boss donates £1m to SNP', BBC News ↗
17 Harrison, E et al, 2009, *Confessions of a Recovering Data Collector*, Plymouth College of Art
18 The No.10 website was promptly taken down when David Cameron's coalition took over government in May 2010.
19 Maybe my letter of complaint to National Express had been the last straw!
20 BBC Radio 5 Live, 2009, 'Richard Bacon Show', BBC Radio 5 Live, Mixcloud ↗
21 The Green Party, 2010, 'Our Environment', Manifesto 2010 ↗
22 BBC News, 2015, 'Labour backs railway nationalisation', BBC News ↗
23 Harrison, E, Lewis, O, 2016, *A Better Railway for Britain: Reunifying our railways under public ownership*, Bring Back British Rail
24 Labour4Clause4, 2019, 'Help Restore Clause 4: Labour's Socialist Pledge', Labour4Clause4 Newsletter, p.1
25 The railway network, electricity grid and our water infrastructure are referred to as 'natural monopolies' because it only makes sense for a society to have one efficient system. Forcing a market onto a 'natural monopoly' is therefore always an absurd idea, because there is no real 'choice' about which electricity grid we connect to, which railway line we get between two places or which water we get out of our tap.
26 Castella, T, 2013, 'Have train fares gone up or down since British Rail?', BBC News ↗
27 Harrison, E, 2014, 'Power For The People!', The Ecologist ↗
28 Zee, B, 2011, 'Climate Camp disbanded', *The Guardian*
29 Pooley, C, Turnbull, J, 2000, 'Commuting, Transport & Urban Form: Manchester & Glasgow in the mid-twentieth century', *Urban*

CHAPTER 3 CONTINUED

History, p.371
30 Craig, C, 2010, *The Tears that Made the Clyde: Well-being in Glasgow*, Argyll Publishing, p.115–6
31 Harrison, E, 2010, 'Hedonism vs Asceticism: A control freak's guide to the MFA' at CCA: Centre for Contemporary Arts, Vimeo ↗
32 Lowndes, S, 2010, *Social Sculpture: The Rise of the Glasgow Art Scene*, Luath Press, p.17
33 King, E, 1993, *The Hidden History of Glasgow's Women: The Thenew Factor*, Mainstream Publishing, p.123
34 Cameron, D, 2009, 'The Age of Austerity', The Conservative Party ↗
35 Ibid
36 BBC News, 2011, 'Osborne and Balls clash over GDP', BBC News ↗
37 Lanchester, J, 2018, 'After the Fall', *London Review of Books*, p.6
38 Keegan, M, 2018, 'Shenzhen's silent revolution: world's first fully electric bus fleet quietens Chinese megacity', *The Guardian*
39 Lanchester, J, 2018, 'After the Fall', *London Review of Books*, p.6
40 BBC News, 2010, 'Election 2010 Results', BBC News ↗
41 Gray, A, 2012, 'Settlers and Colonists' in *Unstated: Writers on Scottish Independence*, Word Power Books, p.109
42 I found Alasdair Gray and Adam Tompkin's book *How We Should Rule Ourselves* very useful when researching for my *Personal Political Broadcast*, especially in terms of their republican stance, which makes the SNP's proposal to keep the Queen in an independent Scotland look absurd. See: Gray, A, Tomkins, A, 2005, *How We Should Rule Ourselves*, Canongate
43 Craig, C, 2010, *The Tears That Made the Clyde*, Argyll Publishing, p.255
44 Cahn, E et al, 2008, *Co-production: A Manifesto for Growing the Core Economy*, New Economics Foundation, p.1–3
45 Looking at the picture on p.14 of me disguised as a small boy in the '80s, it seems crazy it took so long.
46 I discussed my new hobbies in my talk at Glasgow Women's Library on 22 March 2012 and how these had given rise to the *National Museum of Roller Derby* project. See: Harrison, E, 2012, '21 Revolutions' at Glasgow Women's Library, Vimeo ↗
47 Harvey, D, 2008, 'Reading Marx's Capital with David Harvey', David Harvey ↗
48 Marx, K, 1867, *Capital: Critique of Political Economy Volume I*, p.225
49 Harrison, E, Braid, O, 2012, 'Isolation', Ellie

& Oliver Show, Mixcloud ↗
50 This project with Glasgow Women's Library was the third local institution I'd worked with since graduating from the MFA, following my exhibition *The History of Financial Crises* at Market Gallery on Duke Street in November 2010 and my project *Fair Game* at The Briggait in September 2011.
51 Beaver, TD, 2012, '"By the Skaters, for the Skaters" The DIY Ethos of the Roller Derby Revival', *Journal of Sport & Social Issues*, p.25–49
52 Harrison, E, 2015, 'Ethics: Extremism & Compromise' at Camden Arts Centre, Vimeo ↗
53 Coughlan, S, 2010, 'Tuition fees increasing to £9,000', BBC News ↗
54 Hyslop, F, 2008, 'Graduate Endowment Scrapped', Scottish Government ↗
55 This was what I said during my 20-minute presentation at the interview for my permanent lecturing post on 13 December 2012.

Chapter 4

1 Weingartner, H, 2005, *The Edukators*, Y3 film & Coop99
2 Akehurst, S, 2017, 'Housing and the 2017 election: what the numbers say', Shelter ↗
3 Daniel, O, 2014, 'You're funnier when you're angry', *Performance Research*, p.114
4 Wallace, RJ, 2010, *The Bourgeois Predicament*, University of California, p.7
5 Macdonell, H, 2018, 'Big rise in single-person households', *The Times*
6 Campaign to End Loneliness, 2018, 'Threat to Health', Campaign to End Loneliness ↗
7 Marx, K, 1867, *Capital: Critique of Political Economy Volume I*
8 Lefebvre, H, 1947, *Critique of Everyday Life Volume I*, Verso, p.229
9 The Independent, 2010, 'Still irresistible, a working-class hero's finest speech', *The Independent*
10 Reid, J, 1972, 'Alienation Speech' at University of Glasgow, YouTube ↗
11 As a result of trade union activism led by Jimmy Reid, Edward Heath's Tory government invested £35 million into shipyards at Scotstoun, Linthouse and Govan – later nationalised as part of British Shipbuilders in 1977. It wasn't until 1988 that they were eventually privatised by Margaret Thatcher.
12 Harrison, E, 2015, 'Ethics: Extremism & Compromise' at Camden Arts Centre, Vimeo ↗
13 Lefebvre, H, 1947, *Critique of Everyday Life Volume I*, Verso, p.150
14 My withdrawal from social media, was a 'lifestyle choice' I discussed in my talks at the Time & Motion conference at the Royal College of Art on 6 March 2014 and at Transmediale

in Berlin on 31 January 2015 with the motto 'stupid people have smart phones'. See: Harrison, E, 2014, 'Time & Motion' at Royal College of Art, Vimeo ↗

Harrison, E, 2015, 'Confessions of a Recovering Data Collector' at Capture All, Transmediale, Vimeo ↗

15 O'Shea, J, 2012, The Meat Licence, John O'Shea ↗

16 Steinfeld, H et al, 2006, Livestock's Long Shadow, United Nations Food & Agriculture Organisation

17 Trebeck, K, Williams, J, 2019, The Economics of Arrival: Ideas for a Grown Up Economy, Policy Press, p.20

18 Harrison, E, 2013, 'Eat 22: The Personal is Political' in Playing for Time: Making Art as If the World Mattered, Oberon Books

19 Harrison, E, 2013, Anti-Capitalist Aerobics, Ellie Harrison, Vimeo ↗

20 Quoted in Trebeck, K, Williams, J, 2019, The Economics of Arrival: Ideas for a Grown Up Economy, Policy Press, p.8

21 Coote, A, Franklin, J, 2009, Green Well Fair: Three Economies for Social Justice, New Economics Foundation, p.3

22 McCormack, C, 2009, The Wee Yellow Butterfly, Argyll Publishing, p.247

23 Coote, A, Franklin, J, 2009, Green Well Fair: Three Economies for Social Justice, New Economics Foundation, p.15

24 Alinsky, SD, 1971, Rules for Radicals, Vintage Books

25 Branch, T, 1988, Parting the Waters: America in the King Years, Pocket Books

26 King, ML, 1968, 'The "Drum Major Instinct" sermon' at Ebenezer Baptist Church, YouTube ↗

27 PSU, 'The Privilege Walk', Pennsylvania State University ↗

28 Orton, B, 1999, At the Sharp End of the Knife, British Film Institute

29 McGarvey, D, 2018, 'Poverty Safari Live' at The Stand's New Town Theatre, Edinburgh Festival Fringe

30 McNicoll, T, 2018, '"End of the world" vs "end of the month": Macron walks tightrope amid fuel tax protests', France 24 ↗

31 Coleman, L, 2017, 'Slow Violence' at Slow Violence Symposium, University of Hertfordshire

32 Critchley, S, 2007, Infinitely Demanding: Ethics of Commitment, Politics of Resistance, Verso

33 Harrison, E, 2018, Power & Privilege, Ellie Harrison ↗

34 'Asceticism and the spirit of capitalism' is the title of a chapter in a book by German philosopher Max Weber (1864–1920), in which he identifies ascetic values as a key driving force of capitalism (see further details in Chapter 7). See: Weber, M, 1905, The Protestant Ethic & the Spirit of Capitalism

35 The phrase 'socialism in one person' is taken from Adam Curtis's documentary The Century of the Self to highlight the inevitably absurd consequence of an individualistic society such as our own. 'Socialism in one person' essentially means capitalism. See: Curtis, A, 2002, The Century of the Self, BBC 4

36 When Oliver moved out, I realised I'd only watched one TV programme in the previous two years and that was Adam Curtis's new documentary All Watched Over by Machines of Loving Grace screened on BBC 2 in summer 2011. See: Curtis, A, 2011, All Watched Over by Machines of Loving Grace, BBC 2

37 Harrison, E, 2015, 'Ethics: Extremism & Compromise' at Camden Arts Centre, Vimeo ↗

38 When I lived in Nottingham where water is privatised, I also, of course, had to pay a water bill to the private company Severn Trent Water.

39 Harrison, E, 2012, Counter Hegemonic Propaganda Machine, Ellie Harrison ↗

40 Kasser, T, 2013, 'Provocation' in The Art of Life: Understanding How Participation in Arts & Culture Can Affect Our Values, Mission Models Money & Common Cause, p.8–12

41 Crompton, T, 2010, The Common Cause Handbook, Public Interest Research Centre, p.5

42 Kasser, T, 2013, 'Provocation' in The Art of Life: Understanding How Participation in Arts & Culture Can Affect Our Values, Mission Models Money & Common Cause, p.9

43 Ibid, p.9

44 CASE, 2012, 'Probation Criteria', College of Art, Science & Engineering, University of Dundee

45 It's for this reason that I'm writing what I hope is a relatively accessible book, rather than making all this research into a PhD.

46 Harrison, E, 2014, The Art School Handbook, Ellie Harrison ↗

47 Bishop, C, 2012, 'Academic Capitalism' in Artificial Hells: Participatory Art & the Politics of Spectatorship, Verso, p.268–71

48 David Graeber describes 'bullshit jobs' as those where people 'spend their entire working lives performing tasks they secretly believe do not really need to be performed. The moral and spiritual damage that comes from this situation is profound'. See: Graeber, D, 2013, 'On the Phenomenon of Bullshit Jobs', Strike! magazine

49 In May 1968, students and staff at Hornsey College of Art began their six-week occupation of the college, in which they attempted to reassert 'the age old ideal of the university as a community of learning'. They were motivated by the belief 'that education should be relevant and joyous' and not 'insulated from the

CHAPTER 4 CONTINUED

problems of the contemporary world'. See: Hornsey College of Art, 1969, *The Hornsey Affair*, Penguin

50 Reid, J, 1972, 'Alienation Speech' at University of Glasgow, YouTube ↗

51 Harvey, D, 2014, 'The 17 Contradictions of Capitalism' at London School of Economics, YouTube ↗

52 ONS, 2014, 'Commuters travel further to work than 10 years ago', Office for National Statistics ↗

53 Marshalls ELearning, 'Stress in the Workplace', University of Dundee ↗

54 Gray, A, 2012, 'Settlers and Colonists' in *Unstated: Writers on Scottish Independence*, Word Power Books, p.103

55 Bradley, S, 2015, 'Railways: Nation, Network & People', *Book of the Week*, BBC Radio 4

56 Said, EW, 1984, 'Reflections on Exile' in *Reflections on Exile: And Other Literary & Cultural Essays*, Granta, p.174

57 Cobo-Guevara, P et al, 2018, *Situating Ourselves in Displacement: Conditions, Experiences & Subjectivity across Neoliberalism & Precarity*, Minor Compositions

58 Craig, C, 2010, *The Tears that Made the Clyde: Well-being in Glasgow*, Argyll Publishing, p.21

59 Dods, R et al, 2016, 'The Glasgow Effect: A Discussion' at Glad Café, Mixcloud ↗

60 Simms, A, 2007, *Tescopoly: How One Shop Came Out on Top & Why it Matters*, Constable, p.24

61 Coote, A, Franklin, J, 2009, *Green Well Fair: Three Economies for Social Justice*, New Economics Foundation, p.21

62 Simms, A, 2007, *Tescopoly: How One Shop Came Out on Top & Why it Matters*, Constable

63 Ward, B, Lewis, J, 2002, *Plugging the Leaks: Making the most of every pound that enters your local economy*, New Economics Foundation

64 Simms, A, 2007, *Clone Town Britain*, New Economics Foundation

65 I attempted to visualise the extent of the destruction to high streets across Britain in my 2015 performance/event *High Street Casualties: Ellie Harrison's Zombie Walk*, for which a huge crowd of 'zombie employees' revisited the vacant sites of their former shops across Birmingham and then swarmed together to trudge through the city centre en masse.

66 Boyle, D, 2009, *Localism: Unravelling the Supplicant State*, New Economics Foundation, p.18

67 Craig, C, 2010, *The Tears That Made the Clyde,* Argyll Publishing, p.217

68 Ibid, p.225

69 Riddoch, L, 2018, *Blossom: What Scotland Needs to Flourish*, Luath Press, p.223

70 Patrick Geddes was born in Ballater in Aberdeenshire in 1854. From 1888 to 1919 he held the Chair of Botany at University College Dundee (now the University of Dundee).

71 Glasgow City Council, 2016, 'Glasgow Disability Alliance: Making funding go further to serve Glasgow's disabled citizens', Glasgow City Council ↗

72 Walsh, D et al, 2016, *History, Politics & Vulnerability: Explaining Excess Mortality in Scotland & Glasgow*, Glasgow Centre for Population Health, p.57

73 McGarvey, D, 2016, 'Dear Creative Scotland: Ellie Harrison is a Class Act', Bella Caledonia ↗

74 Mullen, S, 2009, *It Wisnae Us: The Truth about Glasgow & Slavery*, The Royal Incorporation of Architects in Scotland

75 Just imagine if the four years' solid airtime which has so far been afforded to Brexit (2016–9), had been devoted to the urgent task of decarbonising our economy – we might have made a bit more progress! For trolls see: Nagle, A, 2017, *Kill All Normies: Online culture wars from 4chan & Tumblr to Trump & the alt-right*, Zero Books

76 Scottish Parliament, 2016, 'Your MSPs: Glasgow', Scottish Parliament ↗

77 Scottish Government, 2017, 'Growing the Economy: Inclusive Growth', Scottish Government ↗

78 MacLeod, D, 2002, 'Glasgow "posher" than Oxbridge', *The Guardian*

79 McGarvey, D, 2017, 'Garnethill' in *Poverty Safari*, Luath Press, p.144–9

80 Limond, B, 2014, 'Glasgow School of Art Tragedy: My Thoughts', *Limmy*, YouTube ↗

81 Davies, C, 2014, 'Glasgow counts down for Games with worst life expectancy in UK', *The Guardian*

82 Gray, A, 2012, 'Settlers and Colonists' in *Unstated: Writers on Scottish Independence*, Word Power Books, p.106

83 Ibid, p.104

84 Harrison, E, 2015, 'Ethics: Extremism & Compromise' at Camden Arts Centre, Vimeo ↗

85 Cobo-Guevara, P et al, 2018, *Situating Ourselves in Displacement: Conditions, Experiences & Subjectivity across Neoliberalism & Precarity*, Minor Compositions

86 I went to Venice Biennale in May 2015. My thinking for the *Think Global, Act Local!* proposal was cemented when I witnessed the decadence of the global artworld first hand. See: Harrison, E, 2015, *Venice Biennale: Think*

Local, Act Global!, Ellie Harrison ↗

87 £15,000 was the upper limit for the 'small grants', which have a quicker turnaround of eight weeks before you find out whether it's been awarded. If you apply for more than £15,000 (up to £150,000), then there is a more complex application process and a 12-week turnaround.

88 Reid, J, 1972, 'Alienation Speech' at University of Glasgow, YouTube ↗

89 Moorhouse, P, 2005, 'And the word was made art', Tate Etc

90 Veiel, A, 2017, Beuys, Zero One Film

91 If you look carefully at the video of the talk I gave at the Data Traces conference in Basel, you can see the homemade bandage on my right arm. See: Harrison, E, 2015, 'Confessions of a Recovering Data Collector' at Data Traces, Academy of Art & Design FHNW, Vimeo ↗

92 I described the This Is What Democracy Looks Like! project as 'playing with the aesthetic of a misjudged publicity stunt (see Labour's 2015 "pink bus")'. I had hired a seven-seater 'Conference Bike' and invited local politicians to join me on it for a series of roving surgeries where passers-by were encouraged to hop on for a chat. I took a series of photo-portraits featuring me shaking hands with the 14 politicians who agreed to participate. All of these are done with my left hand, with the bandaged right one tucked behind my back.

93 This particular piece of cycle lane on Cambridge Street was highlighted in the 2018 Connectivity Commission report as an example of one of Glasgow's many 'poorly-designed, unsafe routes with poor segregation'. See: Begg, D et al, 2018, Connecting Glasgow: Creating an Inclusive, Thriving, Liveable City, Glasgow City Council, p.14

94 Neil is married to Laura González, my former tutor at Glasgow School of Art who led the 'Psychoanalysis in Art & Culture' elective I attended in 2009.

95 I later compiled all the correspondence with the university from 29 June 2015 to 20 May 2016 into a 134-page redacted document called An Anatomy of a Research Project.

96 BEIS, 2018, 'Government emission conversion factors for greenhouse gas company reporting', Department for Business, Energy & Industrial Strategy ↗

Chapter 5

1 On New Year's Eve 2015, I left London Euston on the 7.30am train to travel back to Glasgow to begin the project. For Oliver's Christmas present, I'd bought us both tickets to see the Peter Pan panto at the SEC starring The Krankies and David Hasselhoff, which was due to start at 1pm. On the way up, Network Rail discovered that Storm Frank had 'severely weakened the viaduct in South Lanarkshire', which carried the railway line to Glasgow. We were all booted off the train at Carlisle and had to wait for rail replacement bus services, which delayed us by more than two hours. We missed the panto and the railway line then remained shut for seven weeks. That evening I went to three Hogmanay parties, firstly one at my friend Sarah's house, then to see my friends Ivor and Rosana perform at The Old Hairdresser's for the bells and finally to an unknown house party in the West End with my friend Anna. All the while I was wearing my new 'French Fries Invasion Sweater' made with a chip-smothered fabric, which I had bought to celebrate being awarded the grant. That was to be my last big blowout. For Storm Frank see: BBC News, 2016, 'West Coast Mainline to reopen next week', BBC News ↗

2 Jeffrey, M, 2010, 'The Finished Article', The Scotsman, Review Section, p.7

3 I noted this quote down in Notebook 36 on the train heading back from Edinburgh on 30 November 2015.

4 Craig Tannock is the man almost single-handedly responsible for transforming Glasgow into 'the vegan capital of Britain' by setting up The 13th Note, Mono, Stereo, The 78 and inspiring Saramago at the CCA and many other cafes and restaurants around the city, which do not use animal products in any of their food. See: Saner, E, 2013, 'Glasgow: the vegan capital of Britain?', The Guardian

5 Harrison, E, 2015, Open Project Funding Application Form, Creative Scotland, p.7–8

6 I found this note scrawled in biro in Notebook 36.

7 In the end I decided that all I would post on the Tumblr blog would be my original funding application to Creative Scotland, which is still available to read at: glasgoweffect.tumblr.com ↗

8 TUC, 2015, 'Rail fares have risen by 25 per cent since 2010', Action for Rail ↗

9 Crary, J, 2014, 24/7: Late Capitalism & the Ends of Sleep, Verso

10 This title is paraphrased from the 1974 performance I Like America and America Likes Me by German artist Joseph Beuys, for which he visited America for three days to be locked in a room with a coyote – the only living being he had contact with – before returning back to Europe.

11 BBC Radio 5 Live, 2016, '5 Live Drive Show', BBC Radio 5 Live, Mixcloud ↗

12 Sloman, L, Taylor, I, 2012, Rebuilding Rail, Transport for Quality of Life, p.7

13 TUC, 2016, 'UK commuters spend up to six

CHAPTER 5 CONTINUED

times as much of their salary on rail fares as other European passengers', Action for Rail ↗
14 Kerr, A, 2016, 'London artist paid £15k public money to live in Glasgow for a year', *Daily Record*
15 Ibid
16 Harrison, E, Braid, O, 2011, 'Ellie Harrison & Oliver Braid', Wunderbar Radio, Mixcloud ↗
17 Braid, O, 2016, 'Speaks & Loki: 2016 Scottish Mini-Tour', Oliver Braid's Tiny Tumble ↗
18 BBC Radio Scotland, 2016, 'The Glasgow Effect phone-in', *Kaye Adams Programme*, BBC Radio Scotland, Mixcloud ↗
19 Somebody had drawn attention to *The Guide to Open Project Funding* 2014–5 by Creative Scotland, which states on p.11 that: 'Academics or other education professionals seeking funding related to their educational role cannot apply'. This meant I could no longer 'donate' the funding to the university as had been planned during the negotiations before Christmas. Instead, I would now have to attempt to negotiate 'Special Leave (without pay)' in order to complete the project and then carefully plan how to spend the £15,000 myself (see details at the end of Chapter 6).
20 Alinsky, SD, 1971, *Rules for Radicals*, Vintage Books
21 Haggerty, A, 2016, 'Artists need to stick their necks out: Exclusive interview with The Glasgow Effect's Ellie Harrison', CommonSpace ↗
22 Beadie, B, 2016, 'A tumultuous week in Glasgow', Kiltr ↗
23 McGarvey, D, 2016, 'Dear Creative Scotland: Ellie Harrison is a Class Act', Bella Caledonia ↗
24 Round, K, 2016, *The Divide*, Dartmouth Films & Literally Films
25 Wilkinson, R, Pickett, K, 2009, *The Spirit Level: Why Equality is Better for Everyone*, Penguin
26 Wilkinson, R, Pickett, K, 2009, 'What's trust got to do with it?' in *The Spirit Level: Why Equality is Better for Everyone*, Penguin, p.51–4
27 McCormack, C, 2009, *The Wee Yellow Butterfly*, Argyll Publishing, p.172
28 Quoted in Ibid, p.11
29 Once I had read *Small is Beautiful: A Study of Economics as if People Mattered* and understood the relevance of Schumacher's ideas to *The Glasgow Effect*, his book became the central reference for me (a 'bible' if you like) when writing the rest of this book.
30 Schumacher, EF, 1973, *Small is Beautiful*, Abacus, p.49
31 'Deeds not words' was the guiding principle of the women whose long fight for justice won us all the right to vote.
32 Harrison, E, 2016, 'Statement', *The Glasgow Effect*, Facebook ↗
33 McGarvey, D, 2016, 'Dear Creative Scotland: Ellie Harrison is a Class Act', Bella Caledonia ↗
34 Gallogly-Swan, K, 2016, 'Cos I Can', Bella Caledonia ↗
35 McGarvey, D, 2017, 'The Changeling' in *Poverty Safari*, Luath Press, p.201–13
36 Dods, R et al, 2016, 'The Glasgow Effect: A Discussion' at Glad Café, Mixcloud ↗
37 Walsh, D et al, 2010, *Investigating a 'Glasgow Effect': Why do equally deprived UK cities experience different health outcomes?*, Glasgow Centre for Population Health, p.8
38 Walsh, D et al, 2016, *History, Politics & Vulnerability: Explaining Excess Mortality in Scotland & Glasgow*, Glasgow Centre for Population Health
39 In the emails I received from GCPH, they also explained that their 2016 report had in fact been 'delayed' until after the Scottish Parliamentary elections in May (I wonder why?) and informed me that 'obviously' *The Glasgow Effect* project I was undertaking was 'about something completely different' to population health.
40 Wilson, J, 2016, 'Is the Glasgow Effect already a failure?', *The Scotsman*
41 Dods, R, 2012, 'Diversity' at LAB, Vimeo ↗
42 Craig, C, 2010, *The Tears that Made the Clyde: Well-being in Glasgow*, Argyll Publishing, p.263
43 Ibid, p.263
44 Haggerty, A, 2016, 'Artists need to stick their necks out: Exclusive interview with The Glasgow Effect's Ellie Harrison', CommonSpace ↗
45 Craig, C, 2010, *The Tears That Made the Clyde*, Argyll Publishing, p.100
46 Ibid, p.254
47 Lamont, G, 2016, 'Glasgow is much more than a plate of chips', *Evening Times*
48 Quoted in Ewan, R et al, 2012, *The Glasgow Schools*, The Common Guild, p.23
49 Perhaps this is the difference between being from Ealing and being from Glasgow – a place which has been labelled as 'deprived' by politicians and public health 'professionals' for decades. It makes you much more defensive.
50 Marx, K, 1844, 'Letter from Marx to Arnold Ruge' in *Deutsch-Französische Jahrbücher*
51 Quoted in Craig, C, 2010, *The Tears That Made the Clyde*, Argyll Publishing, p.279
52 Gardiner, K, 2016, 'Is it the end of the Scottish Cringe?', BBC News ↗

53 Colau, A, 2019, 'Transforming fear into hope' in *Fearless Cities: A Guide to the Global Municipalist Movement*, New Internationalist, p.145–8

54 Harrison, E, 2015, *Open Project Funding Application Form*, Creative Scotland, p.8

55 Said, EW, 1984, 'Reflections on Exile' in *Reflections on Exile: And Other Literary & Cultural Essays*, Granta, p.176

56 Ibid, p.177

57 Ibid, p.180

Chapter 6

1 Godfrey, T, 1998, *Conceptual Art*, Art & Ideas, Phaidon, p.4

2 Limond, B, 2014, 'Glasgow School of Art Tragedy: My Thoughts', *Limmy*, YouTube ↗

3 Jelinek, A, 2013, *This is Not Art: Activism & Other 'Not-Art'*, IB Tauris, back cover

4 Harrison, E, 2016, 'Statement', *The Glasgow Effect*, Facebook ↗

5 Bishop, C, 2012, *Artificial Hells: Participatory Art & the Politics of Spectatorship*, Verso, p.45

6 Cameron, K, 2013, *Alasdair Gray: A Life in Progress*, Hopscotch Films

7 Quoted in Han, BC, 2017, *Psychopolitics: Neoliberalism & New Technologies of Power*, Verso, p.81

8 Quaintance, M, 2016, 'Rules of Engagement', *Art Monthly*, p.7–10

9 O'Brien, D et al, 2018, *Panic! Social Class, Taste & Inequalities in the Creative Industries*, Create

10 Quaintance, M, 2017, 'The New Conservatism: Complicity & the UK Art World's Performance of Progression', e-flux conversations ↗

11 Ibid

12 Jelinek, A, 2013, *This is Not Art: Activism & Other 'Not-Art'*, IB Tauris, p.17

13 Fox, D, 2009, 'A Serious Business', *Frieze*

14 Abbing, H, 2002, *Why Are Artists Poor? The Exceptional Economy of the Arts*, Amsterdam University Press

15 Harrison, E, 2013, *Anti-Capitalist Aerobics*, Ellie Harrison, Vimeo ↗

16 Simms, A, Potts, R, 2012, *The New Materialism: How our relationship with the material world can change for the better*, Bread, Print & Roses

17 Harrison, E, 2013, 'For The Love of It' at Artquest, Mixcloud ↗

18 Four-fifths of artists earn less than £10k a year from their work. See: McCalden, J, 2016, 'The ethics and effects of funding artists: a response to the Ellie Harrison affair', a-n The Artists Information Company ↗

19 Thangavelu, P, 2018, 'How to Calculate the GDP of a Country', Investopedia ↗

20 Waring, M, 1990, *If Women Counted: A New Feminist Economics*, Harper Collins

21 Nash, T, 1995, *Who's Counting? Marilyn Waring on Sex, Lies & Global Economics*, National Film Board of Canada

22 Broderick, R, 2016, 'Everyone's Taking The Piss Out Of An Artist Who Got £15,000 To Live In Glasgow For A Year', BuzzFeed ↗

23 Craig, C, 2010, *The Tears that Made the Clyde: Well-being in Glasgow*, Argyll Publishing, p.331

24 Ibid, p.127

25 Schumacher, EF, 1973, *Small is Beautiful: A Study of Economics as if People Mattered*, Abacus, p.148

26 Peterson, J, 2017, 'Marxism is Ignorant of the Pareto Principle', *Joe Rogan Experience*, YouTube ↗

27 Coughlan, S, 2018, 'University pension boss's £82,000 pay rise', BBC News ↗

28 I joined ShareAction's 'Ethics for USS' campaign (formerly 'Listen to USS!'), soon after I began work at the university in 2013. In 2014, I wrote an article called 'How the "Them" became "Us"' explaining why I'd come to see 'divestment' – that is the conscious moving of investments away from fossil fuels into renewable technologies – as one of the key fights of our time. See: Harrison, E, 2014, 'How the "Them" became "Us"', ShareAction ↗

29 Matthew 13:11–2

30 Marx, K, 1875, *Critique of the Gotha Programme*

31 Watt, G, Burns, R, 2019, 'Money advice embedded in the Deep End' at Money, Debt & Health, Glasgow's Healthier Future Forum 23

32 Dods, R et al, 2012, 'Open letter to Creative Scotland', BBC News ↗

33 French, L, Logan, O, 2013, 'Creative Scotland: A Timeline', Creative Scotland Campaign ↗

34 Harrison, E, 2015, *Open Project Funding Application Form*, Creative Scotland, p.11

35 Wee D, 2016, 'The Glasgow Effect', *Powercut Productions*, YouTube ↗

36 Harrison, E, 2016, 'The Glasgow Effect' at Cross-Party Group on Culture, Scottish Parliament, Vimeo ↗

37 Kerr, A, 2016, 'London artist paid £15k public money to live in Glasgow for a year', *Daily Record*

38 It was telling how much of the 'emotional abuse' on *The Glasgow Effect* event page came from other artists – 'proper Glasgow-based artists' as Kendal Orr described them (see Facebook comments in Chapter 5).

39 Quaintance, M, 2016, 'Illiquid Assets', *Art Monthly*, p.6–8

40 This was something American artist Joanna Spitzner attempted to highlight with her

CHAPTER 6 CONTINUED

playful work *Joanna Spitzner Foundation, Inc.* (described in Chapter 2).
41 Tate à Tate, 2012, 'Oil Sponsorship of the Arts', Tate à Tate ↗
42 Evans, M, 2015, *Artwash: Big Oil & the Arts*, Pluto Press
43 Dods, R et al, 2016, 'The Glasgow Effect: A Discussion' at Glad Café, Mixcloud ↗
44 Metzger, G, 1974, 'Art Strike 1977–80', The Seven by Nine Squares ↗
45 Jelinek, A, 2013, *This is Not Art: Activism & Other 'Not-Art'*, IB Tauris, back cover
46 Schumacher, EF, 1973, *Small is Beautiful*, Abacus, p.55
47 Black, I et al, 2015, *From 'I' to 'We': Changing the narrative in Scotland's relationship with consumption*, Common Weal, p.4
48 The *ArcelorMittal Orbit* is installed close to where I staged my *This Is What Democracy Looks Like!* project in July 2015 (described in Chapter 4), so during that time I had the pleasure of seeing it every day.
49 Jones, J et al, 2012, 'London 2012 Olympics: first view from the ArcelorMittal Orbit tower', *The Guardian*
50 Although people felt that the initial premise of *The Glasgow Effect* was a symbol of a similarly over-budgeted, aloof and meaningless kind of art, the irony was that it clearly *did resonate* with the reality of their lives. It was connected to the human scale and to their specific locality – Glasgow – and that's why they felt so compelled to plunge into the debate. This is something that never would have happened had I kept the more generic title *Think Global, Act Global!* As German artist Joseph Beuys once said: 'a provocation always causes something to come alive'.
51 Beuys, J, 1973, 'I Am Searching for Field Character' in *Joseph Beuys in America: Energy Plan for the Western Man*, Da Capo Press, p.22
52 Kuoni, C, 1993, *Joseph Beuys in America: Energy Plan for the Western Man*, Da Capo Press, p.13
53 Mould, O, 2018, *Against Creativity: Everything you have been told about creativity is wrong*, Verso, p.8
54 Veiel, A, 2017, *Beuys*, Zero One Film
55 I explored the absurd consequences of 'relentless innovation' cycles in my 2014 performance/event *The Global Race* in Berlin. I used the Segway as a symbol of human stupidity. Through a series of Segway races around a local athletics track, I created a vision of an Olympic games of the future, where humans no longer need to break a sweat.
56 Schumacher, EF, 1973, *Small is Beautiful*,

Abacus, p.71
57 Burns, H, 2016, 'The Digital Society & Health Inequalities' at Net Benefits: Is the Digital Society Good for Us?, University of Strathclyde
58 McGarvey, D, 2017, *Poverty Safari*, Luath Press, p.38
59 Craig, C, 2010, *The Tears That Made the Clyde*, Argyll Publishing, p.245
60 Ibid, p.252
61 This title is taken from the 1976 film *Network* which I saw on 22 February 2016 at the Glasgow Film Festival, where I was working as a volunteer usher. I saw similarities between the response to *The Glasgow Effect* and the famous scene where the film's main character, TV newsreader Howard Beal, goes off script encouraging people all over America to go to their windows and scream out: 'I'm as mad as hell, and I'm not going to take this anymore!'
62 Wallace-Wells, D, 2019, *The Uninhabitable Earth: A Story of the Future*, Allen Lane, p.148
63 Coote, A, Franklin, J, 2009, *Green Well Fair: Three Economies for Social Justice*, New Economics Foundation, p.3
64 I was interested in the comments posted on the Facebook event page, which suggested that I too was a troll. By using the photo of the chips and the title *The Glasgow Effect*, was I was giving as good as I got? Social media trolls are said to have taken on the role of modern-day satirists – those who use ridicule and irony to 'expose folly and vice'. Just as the great Irish satirist Jonathan Swift (1667–1745) had made the controversial suggestion that the 1729 famine in Ireland could be solved with 'a modest proposal' of eating babies, so too was *The Glasgow Effect* drawing attention of the failure of policy makers and academics to actually deliver social justice in Glasgow and beyond. Could 'the Glasgow effect' really be solved by withholding chips from the masses? One of the audience members at our *The Glasgow Effect: A Discussion* event said: 'I didn't have a problem with you calling it *The Glasgow Effect* at all, I think the more focus on that the better'. Another audience member said: 'I think it's a reasonable enough pun… [having the chips up there] I think it's very much the kind of joke that we make in Glasgow, and you've graduated from Glasgow School of Art, you're a "Glasgow artist", you're a Glaswegian as far as I'm concerned'. I also found that fascinating. My work had changed quite dramatically in the time I'd been in Glasgow – become more political, but also much darker and more misanthropic. After all those years of social isolation and the lack of 'local love' that I've endured here in this city, perhaps I had also come to embody the Glaswegian humour

of 'putting folk down' (described in Chapter 5).
For trolls see: Nagle, A, 2017, *Kill All Normies: Online culture wars from 4chan & Tumblr to Trump & the alt-right*, Zero Books

65 Chakrabortty, A, 2018, 'Yes, there is an alternative. These people have shown how to "take back control"', *The Guardian*

66 Walsh, D et al, 2016, *History, Politics & Vulnerability: Explaining Excess Mortality in Scotland & Glasgow*, Glasgow Centre for Population Health, p.29

67 Nash, T, 1995, *Who's Counting? Marilyn Waring on Sex, Lies & Global Economics*, National Film Board of Canada

68 Without EU Directives 2009/72/EC and 2012/34/EU imposing markets on our key services and infrastructure which are 'natural monopolies', there are less obstacles to returning our energy and railways to public ownership.

69 Steinfort, L, 2018, 'The Almighty Investor' in *Public Ownership Rises Again*, New Internationalist, p.24–5

70 At Glasgow's State of the City Economy conference in 2018, I heard Professor Graeme Roy talking about the impact of Brexit on the city and the opportunities it presented for localising supply chains. He illustrated the absurdity of international trade, saying: 'If you're importing milk from Australia, you're importing cheese!' See: Roy, G, 2018, 'Brexit & the Glasgow City Region' at State of the City Economy, Glasgow City Council

71 Harrison, E, 2015, *Venice Biennale: Think Local, Act Global!*, Ellie Harrison ↗

72 Steyerl, H, 2017, *Duty Free Art: Art in the Age of Planetary Civil War*, Verso, p.184

73 Schumacher, EF, 1973, *Small is Beautiful*, Abacus, p.64

74 Ibid, p.140

75 McAlpine, R, 2019, 'The End of "Stuff"' at Ecopedagogy, Locavore

76 Quoted in Bishop, C, 2012, *Artificial Hells: Participatory Art & the Politics of Spectatorship*, Verso, p.297

77 Ibid, p.297

78 Schumpeter, J, 1942, 'Can Capitalism Survive?' in *Capitalism, Socialism & Democracy*

79 Homersham, L, 2017, 'Profile: Hannah Black', *Art Monthly*, p.19

80 Kasser, T, 2013, 'Provocation' in *The Art of Life: Understanding How Participation in Arts & Culture Can Affect Our Values*, Mission Models Money & Common Cause, p.9

81 Harrison, E, 2017, *End of project monitoring report*, Creative Scotland, p.7

82 Harrison, E, 2015, *Open Project Funding Application Form*, Creative Scotland, p.7

83 Schumacher, EF, 1973, *Small is Beautiful*, Abacus, p.81

84 Ibid, p.173

85 Ibid, p.174

86 Ibid, p.174

87 Though slashing my carbon footprint for transport to zero did also, of course, have real tangible benefits which you can see in my Carbon Graph on p.138–9.

88 Herbert, M, 2016, *Tell Them I Said No*, Sternberg Press, p.44

89 Wray, B, 2016, 'The Glasgow Effect: Activism as a public health issue', CommonSpace ↗

90 McCormack, C, 2009, *The Wee Yellow Butterfly*, Argyll Publishing, p.78

91 Ibid, p.93

92 Black, I et al, 2015, *From 'I' to 'We': Changing the narrative in Scotland's relationship with consumption*, Common Weal, p.4

93 Schumacher, EF, 1973, *Small is Beautiful*, Abacus, p.174

94 Committee on Standards in Public Life, 1995, 'The 7 Principles of Public Life', UK Government ↗

95 Ibid

96 The FinWell 'research project' received £211,197 of public money (14 times more than *The Glasgow Effect*). How's that for a 'poverty safari'? See: CSO, 2015, 'Health Services & Population Health Research Committee', Chief Scientist Office ↗

97 Morduch, J, 2016, 'The hidden financial lives of low-income households' at Glasgow Centre for Population Health

98 The One Show, 2010, 'Jimmy Reid', BBC 1, YouTube ↗

Chapter 7

1 Lanchester, J, 2018, 'After the Fall', *London Review of Books*, p.8

2 Coote, A et al, 2010, *21 Hours: Why a shorter working week can help us all to flourish in the 21st century*, New Economics Foundation, p.3

3 Ibid, p.2

4 Stephen, W, 2004, *Think Global, Act Local: The Life and Legacy of Patrick Geddes*, Luath Press, p.71

5 Holm, L, 2018, *Politics of Small Places: Paul Noble & Patrick Geddes*, Cooper Gallery, p.17

6 Schumacher, EF, 1973, *Small is Beautiful: A Study of Economics as if People Mattered*, Abacus, p.140

7 Taleb, F, 2019, 'Standing up to the far right' in *Fearless Cities: A Guide to the Global Municipalist Movement*, New Internationalist, p.27–9

8 Schumacher, EF, 1973, *Small is Beautiful*,

CHAPTER 7 CONTINUED

Abacus, p.140

9 Burns, H, 2016, 'The Digital Society & Health Inequalities' at Net Benefits: Is the Digital Society Good for Us?, University of Strathclyde

10 Ibid

11 Kane, P, 2016, 'Is the "Glasgow Effect" within control of those with no power in their lives?', *The National*, p.14–5

12 Ibid

13 Pedersen, S, 2018, 'One-Man Ministry', *London Review of Books*, p.3–6

14 Copestake, AM, Jackson, I, Lowther, R, 2015, *For Ailsa*, Vimeo ↗

15 Miller, AG, 2017, *What is a Basic Income?*, Briefing Paper 1, p.4

16 Walsh, D et al, 2016, *History, Politics & Vulnerability: Explaining Excess Mortality in Scotland & Glasgow*, Glasgow Centre for Population Health, p.57

17 Hanlon, P et al, 2011, 'Making the case for a "fifth wave" in public health', *Public Health*, p.30

18 Ibid

19 Walsh, D et al, 2016, *History, Politics & Vulnerability*, Glasgow Centre for Population Health, p.202

20 OAS, 2017, 'About Obesity and Scotland's biggest killer', Obesity Action Scotland ↗

21 GCPH, 2017, 'Obesity across Scotland', The Glasgow Indicators Project ↗

22 Craig, C, 2010, *The Tears that Made the Clyde: Well-being in Glasgow*, Argyll Publishing, p.29

23 Bauman, Z, 2013, *Society under Siege*, John Wiley & Sons, p.68

24 Fisher, M, 2009, *Capitalist Realism: Is There No Alternative?*, Zero Books, p.19

25 O'Neill, M, 2017, 'Museums and Public Health in Glasgow: The Lessons of History' at GCPH Seminar Series 14

26 Baba, C, 2014, 'A "Fifth Wave" in Public Health: Where Do We Start?', Institute of Health & Wellbeing ↗

27 McCormack, C, 2009, *The Wee Yellow Butterfly*, Argyll Publishing, p.79

28 Ibid, p.76

29 Ibid, p.84

30 Walsh, D et al, 2016, *History, Politics & Vulnerability*, Glasgow Centre for Population Health, p.9

31 Ibid, p.6

32 Ibid, p.100

33 Craig, C, 2010, *The Tears That Made the Clyde*, Argyll Publishing, p.33

34 Walsh, D et al, 2016, *History, Politics & Vulnerability*, Glasgow Centre for Population Health, p.70

35 Ibid, p.44

36 Pooley, C, Turnbull, J, 2000, 'Commuting, Transport & Urban Form: Manchester & Glasgow in the mid-twentieth century', *Urban History*, p.371

37 Walsh, D et al, 2016, *History, Politics & Vulnerability*, Glasgow Centre for Population Health, p.41

38 Ibid, p.43

39 Ibid, p.48–9

40 Ibid, p.8

41 Ibid, p.68

42 Cahn, E et al, 2008, *Co-production: A Manifesto for Growing the Core Economy*, New Economics Foundation, p.8

43 Marmot, M et al, 2008, 'Inequities are killing people on grand scale, reports WHO's Commission', World Health Organisation ↗

44 Reid, M, 2011, 'Behind the "Glasgow effect"', *Bulletin of the World Health Organisation*, p.706

45 Quoted in Craig, C, 2010, *The Tears That Made the Clyde*, Argyll Publishing, p.220

46 Walsh, D et al, 2016, *History, Politics & Vulnerability*, Glasgow Centre for Population Health, p.8

47 Ibid, p.57

48 Ibid, p.9

49 Ibid, p.56

50 Ibid, p.67–8

51 McCormack, C, 2009, *The Wee Yellow Butterfly*, Argyll Publishing, p.108

52 Schumacher, EF, 1973, *Small is Beautiful*, Abacus, p.83

53 BBC 4, 2013, 'The Bruce Plan for Glasgow', *Dreaming the Impossible: Unbuilt Britain*, BBC 4

54 Begg, D et al, 2018, *Connecting Glasgow: Creating an Inclusive, Thriving, Liveable City*, Glasgow City Council, p.8

55 Ibid, p.4

56 McGarvey, D, 2016, 'Dear Creative Scotland: Ellie Harrison is a Class Act', Bella Caledonia ↗

57 McGarvey, D, 2017, *Poverty Safari*, Luath Press, p.209

58 Maas, W, 2017, 'Reinventing Glasgow: what the Dear Green Place can learn from Rotterdam', *The Herald*

59 I got this quote via Pat Baillie, who I met online through public transport campaigning in Glasgow.

60 Quoted in Craig, C, 2010, *The Tears That Made the Clyde*, Argyll Publishing, p.298

61 McCormack, C, 2009, *The Wee Yellow Butterfly*, Argyll Publishing, p. 257–8

62 Ibid, p.108

63 The Glasgow Story, 2004, 'Easterhouse Housing', The Glasgow Story ↗

64 McCormack, C, 1995, 'War Without

ENDNOTES

373

Bullets', *Cathy McCormack*, YouTube ↗
65 France 24, 2018, 'Colombia inaugurates first cable car in capital', France 24 ↗
66 Breen, M, 2017, *Options for Extending the Subway*, Strathclyde Partnership for Transport
67 Those like Darren Chadwick and others, who sought to undercover all the clues when I published my funding application for *The Glasgow Effect* online on 5 January 2016, may have noticed that it was no coincidence it was submitted with a map showing Scotland's seven Regional Transport Partnerships as defined in the Transport (Scotland) Act 2005 to depict the area within which I proposed to stay. My original proposal was to be geographically restricted to the area marked in dark blue as 'WEST RTP' – the region covered by Strathclyde Partnership for Transport. When I met that Labour councillor on 24 May 2016, the one useful thing he told me was that I could, as a member of the public, attend SPT's board meetings as an observer. I went along to every one of them for the remainder of the year to watch what they were doing and attempt to hold them to account. I was always the only member of the public there. For the remaining seven months of the year, I made myself unofficial 'undercover artist in residence' at SPT much to their horror and bemusement. I learnt a lot. It was so frustrating, firstly to see people who clearly did not rely on public transport (ie car owners) making decisions on a behalf of the majority of people in Glasgow who do. Where was the passion? And secondly, to see them fumbling around within a clearly dysfunctional system, in which they are totally powerless to control the private bus companies – trying to make the best of a bad situation – without stopping to demand that we transform the broken system and reclaim the powers for reregulation and public ownership that we so desperately need.
68 Scottish Government, 2018, 'Scottish Greenhouse Gas Emissions 2016', Scottish Government ↗
69 Stacey, T, Shaddock, L, 2015, *Taken For a Ride: How UK Public Transport Subsidies Entrench Inequality*, The Equality Trust, p.3
70 National Statistics, 2018, Scottish Transport Statistics, Transport Scotland, p.55
71 Marshalls ELearning, 'Stress in the Workplace', University of Dundee ↗
72 Crewe, T, 2016, 'The Strange Death of Municipal England', *London Review of Books*, p.6–10
73 Balkind, D, 2016, 'What is "Municipalism"?', Municipalist ↗
74 Hamilton, K, Potter, S, 1985, 'Beeching' in *Losing Track*, Routledge & Kegan Paul, p.55–65

75 BBC News, 2011, 'Trains are a "rich man's toy", says transport secretary', BBC News ↗
76 Gosling, T, 2014, '"Bomber Beeching": Undo vandalism & get Britain back on track', RT International ↗
77 Barclay, K, 2017, *Scottish Buses During Deregulation*, Amberley Publishing
78 Ward, R, 2004, 'Greater Glasgow Passenger Transport Executive', Photo Transport ↗
79 Tooher, P, 1996, 'Takeover nets bus workers £35,000', *The Independent*
80 Craig, C, 2010, *The Tears That Made the Clyde*, Argyll Publishing, p.331
81 McCall, C, 2016, 'Five bold Scottish transport projects that never took off', *The Scotsman*
82 Macdonald, M, 1999, 'The Significance of the Scottish Generalist Tradition' in *Popular Education and Social Movements in Scotland Today*, National Institute of Adult Continuing Education, p.83–91
83 Not to mention the fact that these decisions were all made by men and it is women that disproportionately rely on public transport, making up 61 per cent of bus users. See: Wray, B, 2018, 'Analysis: Scotland's bus regulations have barely changed since Thatcher – is it not time for public ownership?', CommonSpace ↗
84 Macdonald, M, 1999, 'The Significance of the Scottish Generalist Tradition' in *Popular Education and Social Movements in Scotland Today*, National Institute of Adult Continuing Education, p.87
85 Stephen, W, 2004, *Think Global, Act Local: The Life and Legacy of Patrick Geddes*, Luath Press, p.70
86 Stephen, W, 2004, *A Vigorous Institution: The Living Legacy of Patrick Geddes*, Luath Press, p.113
87 Holm, L, 2018, *Politics of Small Places: Paul Noble & Patrick Geddes*, Cooper Gallery, p.11
88 Schumacher, EF, 1973, *Small is Beautiful*, Abacus, p.141
89 Kite, L, 2018, 'Poverty Safari', The Orwell Prize ↗
90 Marx, K, 1845, *Theses on Feuerbach*
91 McAlpine, R, 2019, 'The End of "Stuff"' at Ecopedagogy, Locavore
92 Harrison, E, 2017, 'The Glasgow Effect: A Talk by Ellie Harrison' at Glasgow Film Theatre, Vimeo ↗
93 McCormack, C, 2009, *The Wee Yellow Butterfly*, Argyll Publishing, p.165
94 Ibid, p.184
95 Ibid, p.185
96 Ibid, p.170
97 Freire, P, 1972, *Pedagogy of the Oppressed*, Penguin, London, p.28
98 Ibid, p.65

CHAPTER 7 CONTINUED

99 Ibid, p.66
100 Walsh, D et al, 2016, *History, Politics & Vulnerability*, Glasgow Centre for Population Health, p.77
101 Dods, R et al, 2016, 'The Glasgow Effect: A Discussion' at Glad Café, Mixcloud ↗
102 Incite!, 2007, *The Revolution Will Not Be Funded: Beyond the Non-Profit Industrial Complex*, South End Press
103 Marmot, M et al, 2008, 'Inequities are killing people on grand scale, reports WHO's Commission', World Health Organisation ↗
104 Schumacher, EF, 1973, *Small is Beautiful*, Abacus, p.30
105 Ibid, p.31
106 McCormack, C, 2009, *The Wee Yellow Butterfly*, Argyll Publishing, p.225
107 Bradshaw, P, 2011, 'Ken Loach's Save the Children: the film that bit the hand that fed it', *The Guardian*
108 Loach, K, 1969, *The Save the Children Fund Film*, Save the Children
109 Pleyers, G, 2016, 'Mobilising Dissent: Social Activism in a Global Age', GCPH Seminar Series 12
110 McGarvey, D, 2017, 'A Tale of Two Cities' in *Poverty Safari*, Luath Press, p.86–91
111 Craig, C, 2010, *The Tears That Made the Clyde*, Argyll Publishing, p.123
112 Riddoch, L, 2018, *Blossom: What Scotland Needs to Flourish*, Luath Press, p.27
113 Reid, M, 2011, 'Behind the "Glasgow effect"', *Bulletin of the World Health Organisation*
114 Dods, R et al, 2016, 'The Glasgow Effect: A Discussion' at Glad Café, Mixcloud ↗
115 Paterson, S, 2019, 'Revealed: The Glasgow areas branded "food deserts" as hunger problem is targeted', *Evening Times*
116 Wray, B et al, 2019, 'Food, Poverty & Brexit' at CommonSpace Forum, Kinning Park Complex
117 Gallogly-Swan, K, 2016, 'Cos I Can', Bella Caledonia ↗
118 Dods, R et al, 2016, 'The Glasgow Effect: A Discussion' at Glad Café, Mixcloud ↗
119 McGarvey, D, 2017, *Poverty Safari*, Luath Press, p.209–10
120 Study.com, 2018, 'False Consciousness in Sociology: Definition & Examples', Study.com ↗
121 Renton, A, 2006, 'The Rot Starts Here', *The Observer*
122 Bike for Good founder Greg Kinsman-Chauvet gave me a copy of this sticker after I became obsessed with it and the simplicity and truth of its message.
123 Muirie, J, 2017, *Active travel in Glasgow: what we've learned so far*, Glasgow Centre for Population Health, p.25
124 ETA, 2017, 'Regular cycling for transport halves stress', ETA Services ↗
125 I have already witnessed a dramatic culture change in the decade I've been living in the city, and even in the short time I've been writing this book. There are now lots of brilliant organisations across the city, aimed at getting more people cycling, such as: my friends at Bike for Good, GoBike, Free Wheel North and SoulRiders, all listed on the new Aye Cycle Glasgow website launched in 2019.
126 The Glasgow Story, 2004, 'Govan Clarion Cycling Club', The Glasgow Story ↗
127 I got to go inside the Birmingham Municipal Bank in March 2013 when I showed my installation *The History of Financial Crises* in the building as part of the exhibition *Thrift Radiates Happiness* curated by Charlie Levine.
128 Quoted in Craig, C, 2012, 'Enlightenment in the Age of Materialism', *TEDxGlasgow*, YouTube ↗
129 Weber, M, 1905, *The Protestant Ethic & the Spirit of Capitalism*
130 Black, I et al, 2015, *From 'I' to 'We': Changing the narrative in Scotland's relationship with consumption*, Common Weal, p.4
131 Schumacher, EF, 1973, *Small is Beautiful*, Abacus, p.24
132 Frumkin, H, 2006, 'Urban vision and public health: designing and building wholesome places' at GCPH Seminar Series 2
133 I sang the praises of Scottish Water, when I took part in the launch of We Own It's campaign against water privatisation across England in July 2018. 'Since [moving to Scotland], I haven't had to worry about a water bill', I said in the campaign video on YouTube.
134 Craig, C, 2010, *The Tears That Made the Clyde*, Argyll Publishing, p.313
135 Parkinson, A, Buttrick, J, 2018, *Equality & Diversity within the Arts & Cultural Sector in England, 2013–6: Evidence Review*, Arts Council England, p.55
136 Craig, C, 2010, *The Tears That Made the Clyde*, Argyll Publishing, p.252
137 Ibid, p.246
138 McGarvey, D, 2017, 'Class & Intersectionality', *Poverty Safari Live* EP, YouTube ↗
139 The process of distancing myself from the character featured in the memes (pictured on p.169) was made easier by the fact that most used the profile picture from my Facebook page that was nearly ten years old. It was a still from a film made in 2005, featuring my *Daily Data Logger* character.
140 Quoted in Boyle, D, 2009, *Localism:*

Unravelling the Supplicant State, New Economics Foundation, p.23
141 Quoted in Peterson, J, 2017, 'On Middle Class Socialists', *Notes For Space Cadets*, YouTube ↗

Chapter 8

1 Harrison, E, 2015, *Open Project Funding Application Form*, Creative Scotland, p.14
2 McCall, C, 2016, 'Plans to put Sighthill back at the centre of Glasgow', *The Scotsman*
3 In 2016 used Notebooks 36–43 plus preparation in 2014–5 in Notebooks 28–35.
4 This figure of 3.48 tonnes of carbon for all my 'personal transportation' in 2015 was based on the initial carbon footprint calculation that I did during *The Glasgow Effect* and released on Facebook to celebrate World Car-free Day on 22 September 2016. It was calculated using the Conversion Factors taken from the 2007 Collin's Gem *Carbon Counter* book leant to me by my neighbour at the studios Pum Dunbar. In April and May 2019, I finally got round to calculating my full carbon footprint for all my 'personal transportation' dating back to when I first registered as self-employed in 2004 up to and including 2018 for my Carbon Graph on p.138–9 (that had been on my list of things to do for two and a half years). For the 2019 calculation, I used the most up-to-date carbon Conversion Factors taken from the UK Government's guidance for 'Company Reporting' issued in July 2018. The Conversion Factors had changed quite dramatically in those 11 years – with plane travel being seen as more damaging and train travel less so. The new calculation showed I only generated 1.76 tonnes of carbon in 2015 as opposed to the 3.48 tonnes I'd initially stated. In the new Carbon Graph, the years where I took long-haul plane journeys are now far more pronounced. See:
BEIS, 2018, 'Government emission conversion factors for greenhouse gas company reporting', Department for Business, Energy & Industrial Strategy ↗
5 In May 2019, Scotland's carbon reduction target was upgraded to a new target of net-zero emissions by 2045. See: Keane, K, 2019, 'Scotland to set "faster" climate change target', BBC News ↗
6 Trebeck, K, 2016, 'Policy & Practice', Oxfam Scotland ↗
7 BBC Radio Scotland, 2017, 'The Glasgow Effect interview', *Stephen Jardine Programme*, BBC Radio Scotland, Mixcloud ↗
8 Abbing, H, 2002, *Why Are Artists Poor? The Exceptional Economy of the Arts*, Amsterdam University Press
9 SPT, 2006, 'About SPT', Strathclyde

Partnership for Transport ↗
10 Wallace-Wells, D, 2019, *The Uninhabitable Earth: A Story of the Future*, Allen Lane, p.148
11 Schumacher, EF, 1973, *Small is Beautiful: A Study of Economics as if People Mattered*, Abacus, p.234
12 I noted this down on the train in Notebook 42.
13 Jackson, T, 2010, 'An Economic Reality Check', TED Talks, YouTube ↗
14 Cahn, E et al, 2008, *Co-production: A Manifesto for Growing the Core Economy*, New Economics Foundation, p.8
15 Craig, C, 2017, *Hiding in Plain Sight: Exploring Scotland's Ill Health*, CCWB Press
16 Said, EW, 1984, 'Reflections on Exile' in *Reflections on Exile: And Other Literary & Cultural Essays*, Granta, p.181
17 Schumacher, EF, 1973, *Small is Beautiful*, Abacus, p.202
18 This was Notebook 43.
19 Warchus, M, 2014, *Pride*, BBC Films
20 Quoted in Craig, C, 2010, *The Tears that Made the Clyde: Well-being in Glasgow*, Argyll Publishing, p.339

Chapter 9

1 This particular laptop I have had since before I moved to Glasgow in August 2008. I've taken great pride in using my practical skills to keep it going so long – I have replaced the hard drive four times and I reformat and reinstall the operating system every two years to keep it ticking over smoothly. Modern laptops have deliberately been designed to make self-maintenance more difficult, so you're forced to buy a new one if something goes wrong.
2 Carrington, D, 2019, 'Youth climate strikers: "We are going to change the fate of humanity"', *The Guardian*
3 IPCC, 2018, 'Summary for Policymakers of IPCC Special Report on Global Warming of 1.5°C approved by Governments', Intergovernmental Panel on Climate Change ↗
4 Watt, J, 2018, 'We have 12 years to limit climate change catastrophe, warns UN', *The Guardian*
5 Carrington, D, 2019, 'Youth climate strikers: "We are going to change the fate of humanity"', *The Guardian*
6 Carrell, S, 2018, 'Glasgow's major roads, railways and hospitals at risk from climate change', *The Guardian*
7 Climate Ready Clyde, 2018, *Towards a Climate Ready Clyde: Climate Risks & Opportunities for Glasgow City Region*, Climate Ready Clyde, p.3
8 Callan, I, 2018, 'New map shows how rising water levels could submerge Glasgow Airport',

CHAPTER 9 CONTINUED

Glasgow Live ↗
9 BBC News, 2019, 'Sturgeon declares "climate emergency"', BBC News ↗
10 CCC, 2019, 'Phase out greenhouse gas emissions by 2050 to end UK contribution to global warming', Committee on Climate Change ↗
11 BBC News, 2019, 'UK Parliament declares climate emergency', BBC News ↗
12 Laville, S, 2019, 'Top oil firms spending millions lobbying to block climate change policies, says report', The Guardian
13 We Own It, 2019, A model for public ownership in the 21st century, We Own It
14 McAlpine, R, 2019, 'The End of "Stuff"' at Ecopedagogy, Locavore
15 Wallace-Wells, D, 2019, The Uninhabitable Earth: A Story of the Future, Allen Lane, p.148
16 UNFCCC, 2016, 'Shipping Aviation and Paris', United Nations Framework Convention on Climate Change ↗
17 Monbiot, G, 2006, 'On the Flight Path to Global Meltdown', The Guardian
18 Monbiot, G, 2016, 'The Heathrow "hooligans" are our modern day freedom fighters', The Guardian
19 Gabbatiss, J, 2018, 'Tourism is responsible for nearly one tenth of the world's carbon emissions', The Independent
20 Harrison, E, 2017, 'The Glasgow Effect: A Talk by Ellie Harrison' at Glasgow Film Theatre, Vimeo ↗
21 Alston, P, 2018, 'Press Conference of the UN Special Rapporteur on Extreme Poverty & Human Rights', Facebook ↗
22 Bregman, R, 2019, 'Stop talking about philanthropy, start talking about taxes', Twitter ↗
23 Matthews, D, 2019, 'Meet the folk hero of Davos: the writer who told the rich to stop dodging taxes', Vox ↗
24 PWC, 2016, Five Megatrends & Their Implications for Global Defence & Security, PWC, London, p.2
25 Stefanova, K, 2016, 'The Rise And Fall Of The British Pound', Forbes ↗
26 Lanchester, J, 2018, 'After the Fall', London Review of Books, p.8
27 Barcelona en Comú et al, 2019, Fearless Cities: A Guide to the Global Municipalist Movement, New Internationalist, p.112
28 Trebeck, K, Williams, J, 2019, 'Sharing & Predistribution' in The Economics of Arrival: Ideas for a Grown Up Economy, Policy Press, p.95–100
29 Trebeck, K, Williams, J, 2019, 'What about decoupling' in The Economics of Arrival: Ideas for a Grown Up Economy, Policy Press, p.91–3

30 Jack, I, 2017, 'The derided railway sandwich, and its role in the privatisation of Britain', The Guardian
31 Rentoul, J, 2018, 'Theresa May and Jeremy Corbyn are both backward-looking politicians', The Independent
32 Klein, N, 2014, This Changes Everything: Capitalism vs the Climate, Penguin, p.91
33 The Equality Trust, 2017, 'How Has Inequality Changed?', The Equality Trust ↗
34 Evans, R, 2017, 'How Labour will take tax rates back to the 1970s', The Telegraph
35 Hutcheon, P, 2019, 'University principal received £40,000 "disruption allowance" for moving from England', The Herald
36 McAlpine, R, 2019, 'The End of "Stuff"' at Ecopedagogy, Locavore
37 Incite!, 2007, The Revolution Will Not Be Funded: Beyond the Non-Profit Industrial Complex, South End Press
38 Brook, P, 2017, 'This is what Britain's Gay Liberation Front movement looked like in the 1970s', Timeline ↗
39 Bishop, C, 2012, 'The Community Arts Movement' in Artificial Hells: Participatory Art & the Politics of Spectatorship, Verso, p.177–91
40 Walker, J, 2002, Left Shift: Radical Art in 1970s Britain, IB Tauris
41 Patience, J, 2017, 'Art and activism in Glasgow's Sauchiehall Street', The Herald
42 Norges Bank, 2019, 'About the fund', Norges Bank Investment Management ↗
43 Coote, A et al, 2010, 21 Hours: Why a shorter working week can help us all to flourish in the 21st century, New Economics Foundation, p.11
44 Coote, A, Franklin, J, 2013, Time on Our Side: Why We All Need a Shorter Working Week, New Economics Foundation
45 Coote, A et al, 2010, 21 Hours: Why a shorter working week can help us all to flourish in the 21st century, New Economics Foundation, p.2
46 Bray, J, 2018, 'Was Barbara Castle the best transport secretary Britain ever had?', CityMetric ↗
47 National Library of Scotland, 1979, 'Trans-Clyde Integrated Transport', Moving Image Archive ↗
48 FirstGroup plc, 2018, 'Results for the year to 31 March 2018', FirstGroup plc
49 Under privatisation, nearly fifty years after SELNEC's pioneering 'Silent Rider', Glasgow still only has two electric buses! See: Morris, S, 2018, Buses of Greater Manchester in the 1970s, Museum of Transport Greater Manchester, p.37
STV, 2018, 'New electric buses get motoring on Glasgow routes', STV News ↗

50 The Scottish Government's new Transport Bill, which has been going through the Parliament in 2018–9 as I have been writing this book, still tinkers the Transport Act 1985. In many places it refers to Thatcher's Act and make small amendments, rather than repealing the whole thing, which is what should have happened when the Scottish Parliament was first established in 1999. What's the point of having power over our public transport if we don't? It's only thanks to our ongoing campaign, begun in 2016, that we managed to win amendments to the Transport Bill which will allow everyone in Scotland to be served by a publicly-owned bus company like Edinburgh's Lothian Buses. Change is coming!

51 Schumacher, EF, 1973, *Small is Beautiful: A Study of Economics as if People Mattered*, Abacus, p.124

52 The 1972 book *A Blueprint for Survival*, begins with 'Indefinite growth of whatever type cannot be sustained by finite resources. This is the nub of the environmental predicament'. See: Goldsmith, E et al, 1972, *A Blueprint for Survival*, Penguin, p.17

53 Nash, T, 1995, *Who's Counting? Marilyn Waring on Sex, Lies & Global Economics*, National Film Board of Canada

54 Trebeck, K, Williams, J, 2019, *The Economics of Arrival: Ideas for a Grown Up Economy*, Policy Press, p.5

55 Ibid, p.xv

56 Ibid, p.5

57 Ibid, p.67

58 Ibid, p.6

59 Ibid, p.xv

60 Ibid, p.201

61 Dunlop, S et al, 2012, *Oxfam Humankind Index: The new measure of Scotland's Prosperity*, Oxfam Scotland, p.4

62 Ibid, p.4

63 McGarvey, D, 2016, 'Dear Creative Scotland: Ellie Harrison is a Class Act', Bella Caledonia ↗

64 Mahdawi, A, 2015, 'Can cities kick ads? Inside the global movement to ban urban billboards', *The Guardian*

65 In 2016 a huge new electric billboard was installed outside my studios on Alexandra Parade. Wasting energy on brainwashing our population, I watched through the window as this ridiculous thing – supposedly visible to cars on the M8 – was built with the urgency that our new sustainable local infrastructure (world-class public transport, renewable energy, insulated homes and decarbonised heating) must now be. For bus adverts see: Ward, R, 2004, 'Greater Glasgow Passenger Transport Executive', Photo Transport ↗

66 Black, I et al, 2015, *From 'I' to 'We':*

Changing the narrative in Scotland's relationship with consumption, Common Weal, p.15

67 Ibid, p.13

Chapter 10

1 Holm, L, 2018, *Politics of Small Places: Paul Noble & Patrick Geddes*, Cooper Gallery, p.16

2 Monbiot, G, 2018, 'Not capitalism, not communism: George Monbiot on why we need the commons', *We Own It*, YouTube ↗

3 Barcelona en Comú et al, 2019, *Fearless Cities: A Guide to the Global Municipalist Movement*, New Internationalist, p.12

4 Dods, R et al, 2016, 'The Glasgow Effect: A Discussion' at Glad Café, Mixcloud ↗

5 Barcelona en Comú, 2016, *How to Win Back the City En Comú: Guide to Building a Citizen Municipal Platform*, Barcelona en Comú, p.4

6 Morris, S, 2018, *Buses of Greater Manchester in the 1970s*, Museum of Transport Greater Manchester, p.9

7 Schumacher, EF, 1973, *Small is Beautiful: A Study of Economics as if People Mattered*, Abacus, p.55

8 Arlidge, J, 1994, 'Scottish Tories urge rethink of water plan', *The Independent*

9 Under the governance of the Strathclyde Regional Council, the Strathclyde Passenger Transport Executive opened several new railway lines: the Argyle Line connecting Partick to Rutherglen in 1979, Cumbernauld to Queen Street via Springburn and Stepps in 1989, Central to Paisley Canal in 1990, Queen Street to Maryhill in 1993, before developing plans for the Strathclyde Tram linking Drumchapel to Easterhouse via the city centre, which were thwarted by the private bus companies. See: Alderson, J, McDonald, I, 2017, *Britain's Growing Railway*, Railfuture

10 Riddoch, L, 2018, *Blossom: What Scotland Needs to Flourish*, Luath Press, p.207

11 Bort, E et al, 2012, *The Silent Crisis: Failure & Revival in Local Democracy in Scotland*, The Jimmy Reid Foundation, p.26

12 Sweeney, P, 2018, 'Nothing Short of Rank Socialism!' in *Shifting Power to the People*, Red Paper Collective, p.10–1

13 BBC News, 2017, 'Where is the best place to be a woman in Britain?', BBC News ↗

14 Alston, P, 2018, 'Press Conference of the UN Special Rapporteur on Extreme Poverty & Human Rights', Facebook ↗

15 Centre for Cities, 2019, *Cities Outlook 2019*, Centre for Cities, p.17

16 Whittaker, A, 2014, 'Scotland most "centralised" in Europe – Cosla', *The Scotsman*

17 Powers over the public ownership of Scotland's railways were granted in the

CHAPTER 10 CONTINUED

Scotland Act 2016 yet, on 14 November 2018, the SNP and Tories voted together against a motion in the Scottish Parliament to exercise a break clause in Abellio's contract to make way for a public sector operator in 2020. See: Bell, S, 2018, 'TSSA warns of "Winter of Discontent" as Holyrood votes against ScotRail nationalisation', CommonSpace ↗
18 Quoted in Riddoch, L, 2018, *Blossom: What Scotland Needs to Flourish*, Luath Press, p.219
19 In the last decade, Transport Scotland have spent £1 billion on the Aberdeen bypass (opened December 2018), £1.35 billion on the Queensferry Crossing (opened August 2017), £692 million on the M74 extension through Glasgow which opened in June 2011 despite the sustained Jam 74 campaign run by local people and against the recommendations of the public inquiry. This would have been more than enough funding to complete Glasgow's Subway extensions, which would facilitate low-carbon lifestyles for the most populous part of Scotland for centuries to come.
20 Quoted in Craig, C, 2010, *The Tears that Made the Clyde: Well-being in Glasgow*, Argyll Publishing, p.220
21 In its 21 years in operation, the Strathclyde Regional Council was only ever led by men from the Labour party.
22 Aitken, S, 2019, 'Message from the Chair of the Glasgow City Region Cabinet', Glasgow City Region ↗
23 Centre for Cities, 2019, *Cities Outlook 2019*, Centre for Cities, p.d
24 Riddoch, L, 2018, *Blossom: What Scotland Needs to Flourish*, Luath Press, p.212
25 Glasgow City Council, 2019, 'Community Councils: Funding & Finance Information', Glasgow City Council ↗
26 Riddoch, L, 2018, *Blossom: What Scotland Needs to Flourish*, Luath Press, p.213
27 Ibid, p.215
28 Black, I et al, 2015, *From 'I' to 'We': Changing the narrative in Scotland's relationship with consumption*, Common Weal, p.15
29 Barcelona en Comú et al, 2019, *Fearless Cities: A Guide to the Global Municipalist Movement*, New Internationalist, p.178
30 McCormack, C, 2009, *The Wee Yellow Butterfly*, Argyll Publishing, p.232
31 Ibid, p.252–3
32 Orton, B, 1999, *At the Sharp End of the Knife*, British Film Institute
33 Burgen, S, 2016, 'How to win back the city: the Barcelona en Comú guide to overthrowing the elite', *The Guardian*
34 Barcelona en Comú, 2016, *How to Win*

Back the City En Comú: Guide to Building a Citizen Municipal Platform, Barcelona en Comú, p.4
35 Barcelona en Comú et al, 2019, *Fearless Cities: A Guide to the Global Municipalist Movement*, New Internationalist, p.113–4
36 Cheeseman, GM, 2019, 'US Cities are Tackling Climate Change', PlanetWatch ↗
37 Los Angeles suffered an even worse fate than Glasgow with its public transport network. The film *Bikes vs Cars* documents how 'the greatest mass transit system ever built in the history of mankind' was gradually bought out by General Motors starting in the '40s and then purposefully destroyed (with its tramcars sunk to the bottom of the Pacific ocean!) all to 'create a market' cars. See: Gertten, F, 2015, *Bikes vs Cars*, WG Film
38 Karanth, S, 2019, 'Los Angeles Launched Its Own Green New Deal', Huffington Post ↗
39 Chakrabortty, A, 2018, 'How a small town reclaimed its grid and sparked a community revolution', *The Guardian*
40 Barcelona en Comú et al, 2019, *Fearless Cities: A Guide to the Global Municipalist Movement*, New Internationalist, p.121
41 The wind turbine which stands alone to the east on Cathkin Braes was initiated by a community group from Castlemilk and Carmunnock. However the council seized ownership of the project as a PR stunt for the Commonwealth Games because the site was used for the mountain bike competition. The turbine is now 50 per cent owned by private company SSE.
42 Cumbers, A et al, 2013, *Repossessing the Future: A Common Weal Strategy for Community & Democratic Ownership of Scotland's Energy Resources*, The Jimmy Reid Foundation, p.22
43 Reid, J, 1972, 'Alienation Speech' at University of Glasgow, YouTube ↗
44 Black, I et al, 2015, *From 'I' to 'We': Changing the narrative in Scotland's relationship with consumption*, Common Weal, p.13
45 Walsh, D et al, 2016, *History, Politics & Vulnerability: Explaining Excess Mortality in Scotland & Glasgow*, Glasgow Centre for Population Health, p.8
46 Ibid, p.63
47 McCormack, C, 2009, *The Wee Yellow Butterfly*, Argyll Publishing, p.259
48 Trebeck, K, Williams, J, 2019, *The Economics of Arrival: Ideas for a Grown Up Economy*, Policy Press, p.20
49 Triggle, N, 2012, 'Inactivity "as deadly as smoking"', BBC News ↗
50 BBC News, 2017, 'Glasgow "more polluted" than London', BBC News ↗

51 Scottish Government, 2018, 'Scottish Greenhouse Gas Emissions 2016', Scottish Government ↗

52 ETA, 2017, 'Regular cycling for transport halves stress', ETA Services ↗

53 Simms, A, 2018, *Climate & Rapid Behaviour Change: What do we know so far?*, Rapid Transition Alliance, p.7

54 In the 2018 Connectivity Commission report which was initiated by the new council administration in 2017 in response to our public transport campaign's demands, David Begg shows that if you are in the one third of Glasgow's population that does not live near a railway station or own a car (and are therefore relying entirely on buses), you are significantly disadvantaged and 'simply being left behind'. See: Begg, D et al, 2018, *Connecting Glasgow: Creating an Inclusive, Thriving, Liveable City*, Glasgow City Council, p.4

55 In the 2019 Connectivity Commission report, they recommend building a Glasgow Metro (tramway system) linking up Castlemilk, which must now become our city's priority.

56 There have been numerous proposals for Subway extensions: Robroyston in the north to King's Park in the south in 1937; an 'East End Circle' first proposed in 1944 which would have linked up Celtic Park (just as the West End Circle now links up Ibrox); Drumchapel to Castlemilk in 1954; Summerston to Kelvinbridge and a High Street to Gorbals loop in 1988; and most recently in 2010, an East End Circle again and an extension out to Cathcart and Maryhill. None of these has yet materialised. See: Breen, M, 2017, *Options for Extending the Subway*, Strathclyde Partnership for Transport

57 Both Cathy McCormack, who's lived her whole adult life in Easterhouse, and Darren McGarvey who grew up in Pollok, complain about the poor bus services in their respective books. Cathy McCormack writes: 'Looking back, I don't know what the worst aspects were of bringing up weans in poverty. But I hated the isolation. It was a half hour bus trip to the city centre and the bus fares were already expensive, now they had become luxury. We couldn't visit either Tony's family or mine and I missed that whole support network'. And Darren McGarvey writes: 'I started secondary school in 1996... In four years, I didn't venture too far from Pollok as it was a bit of a stoat into the city centre – around 40 minutes by bus – which was something the politicians hoped to remedy by green-lighting the new motorway, much to the anger of many locals'. See:
McCormack, C, 2009, *The Wee Yellow Butterfly*, Argyll Publishing, p.46
McGarvey, D, 2017, *Poverty Safari*, Luath Press, p.43

58 Docherty, I, 2017, 'Sustainable Transport & Glasgow's Peer Cities' at Glasgow's Transport Summit, Glasgow City Council, YouTube ↗

59 Schumacher, EF, 1973, *Small is Beautiful*, Abacus, p.140

60 GMCA, 2019, 'Our Network', Greater Manchester Combined Authority ↗

61 Electric buses are the answer! This is the most efficient way of moving people around the city with the lowest start-up costs. You can fit ten times as many people per lane per hour in a bus as you can in a car. Sharing is more sustainable. See: Begg, D et al, 2018, *Connecting Glasgow: Creating an Inclusive, Thriving, Liveable City*, Glasgow City Council, p.7

62 Sloman, L, Taylor, I, 2016, *Building a World-class Bus System for Britain*, Transport for Quality of Life, p.19

63 Bishop, J, 2018, 'Munich Named The Most Liveable City In The World', Forbes ↗

64 Walsh, D et al, 2016, *History, Politics & Vulnerability*, Glasgow Centre for Population Health, p.70

65 McCall, C, 2017, 'Why are so many people in Scotland living alone?' *The Scotsman*

66 Scottish Government, 2018, *A Connected Scotland: Tackling Social Isolation & Loneliness & Building Stronger Social Connections*, Scottish Government, p.6

67 Campaign to End Loneliness, 2018, 'Threat to Health', Campaign to End Loneliness ↗

68 Hammond, C, 2018, 'Episode 3', *The Anatomy of Loneliness*, BBC Radio 4

69 Craig, C, 2010, *The Tears That Made the Clyde*, Argyll Publishing, p.238

70 It's no wonder the inevitable conclusion of a capitalist system is to have everyone living in their own tiny little lonely bubbles. If everyone has to buy their own toaster, their own kettle etc, the market for consumer goods grows exponentially.

71 Theodoraki, T, Taulbut, M, 2016, 'Good work and health in Scotland: setting the scene' at Public Health Information Network for Scotland Seminar

72 Quoted in Foster, J, 2018, 'Ownership and Economic Development' in *Shifting Power to the People*, Red Paper Collective, p.14–6

73 The latter of these has never been an issue for most women performing traditional care work at home – cooking, cleaning, washing, bringing up the kids and more. As was highlighted by the 'Wages for Housework' campaign (described in Chapter 7), the problem has been that this labour is not acknowledged by the mainstream economy, making it harder for women to pay the costs of accessing 'those things we all need to live happily and well'.

CHAPTER 10 CONTINUED

74 Schumacher, EF, 1973, *Small is Beautiful*, Abacus, p.10
75 Ibid, p.61
76 Ibid, p.208
77 Ibid, p.127
78 Dunlop, S et al, 2012, *Oxfam Humankind Index: The new measure of Scotland's Prosperity*, Oxfam Scotland, p.4
79 There are three versions of the National Statistics Socio-economic Classification (NS-SEC) scale which allow for varying levels of detail – one broken down into eight classes (which was introduced in Chapter 1), one broken down into five classes, used in Carstairs Index of Deprivation, and a simplified version broken down into just three classes.
80 Schumacher, EF, 1973, *Small is Beautiful*, Abacus, p.30
81 Ibid, p.201
82 O'Brien, D et al, 2018, *Panic! Social Class, Taste & Inequalities in the Creative Industries*, Create, p.39
83 Burns, H, 2016, 'The Digital Society & Health Inequalities' at Net Benefits: Is the Digital Society Good for Us?, University of Strathclyde
84 Schumacher, EF, 1973, *Small is Beautiful*, Abacus, p.124
85 Black, I et al, 2015, *From 'I' to 'We': Changing the narrative in Scotland's relationship with consumption*, Common Weal, p.11
86 Coote, A, Franklin, J, 2013, *Time on Our Side: Why We All Need a Shorter Working Week*, New Economics Foundation
87 Marx, K, Engels, F, 1846, *The German Ideology*

Chapter 11

1 Timothy 6:10
2 Negative Income Tax is not too dissimilar to New Labour's Working Tax Credits which saw people on low incomes, like me at the time (2010–2), receive money back from HMRC.
3 Friedman, M, 1968, 'The Negative Income Tax', *LibertyPen*, YouTube ↗
4 Pearce, J, 1983, 'Job attitude and motivation differences between volunteers and employees from comparable organisations', *Journal of Applied Psychology*, p.646–52
5 Moore, HL et al, 2017, *Social Prosperity for the Future: A Proposal for Universal Basic Services*, Institute for Global Prosperity
6 Moore, HL 2017, Universal Basic Services, *Institute for Global Prosperity*, YouTube ↗
7 Bowcott, O et al, 2018, 'Revealed: legal aid cuts forcing parents to give up fight for

children', *The Guardian*
8 Rubel, M, Crump, J, 1987, *Non-Market Socialism in the Nineteenth and Twentieth Centuries*, Palgrave Macmillan, p.86
9 Harvey, D, 2017, 'Why We Need A Moneyless Society', *The Laura Flanders Show*, YouTube ↗
10 Kasser, T, 2013, 'Provocation' in *The Art of Life: Understanding How Participation in Arts & Culture Can Affect Our Values*, Mission Models Money & Common Cause, p.9
11 Quoted in Harvey, D, 2017, 'Why We Need A Moneyless Society', *The Laura Flanders Show*, YouTube ↗
12 Moore, HL et al, 2017, *Social Prosperity for the Future: A Proposal for Universal Basic Services*, Institute for Global Prosperity, p.55
13 Ibid, p.29
14 National Statistics, 2018, Scottish Transport Statistics, Transport Scotland, p.55
15 Sloman, L, Taylor, I, 2016, *Building a World-class Bus System for Britain*, Transport for Quality of Life, p.16
16 Kishimoto, S et al, 2017, *Reclaiming Public Services: How cities and citizens are turning back privatisation*, Transnational Institute
17 In May 2019 FirstGroup announced that they wanted to sell their UK bus businesses. We launched the 'Time to take back our buses!' campaign in June to put pressure on the council to seize this golden opportunity to remunicipalise Glasgow's buses. We're slowly inching closer to this become a reality. Activism works!
18 Sloman, L et al, 2018, *We need fare-free buses! It's time to raise our sights*, Transport for Quality of Life
19 Quoted in Craig, C, 2012, 'Enlightenment in the Age of Materialism', *TEDxGlasgow*, YouTube ↗
20 This was the huge oversight that occurred when Strathclyde's Buses became the private property of the employees and not the service users – it might have fared better as a passengers' co-operative than a workers' one.
21 Paterson, S, 2019, 'Revealed: The Glasgow areas branded "food deserts" as hunger problem is targeted', *Evening Times*
22 Castleton Primary, 2019, 'It's Just Not Fair', *Glasgow's Improvement Challenge*, YouTube ↗
23 Federici, S, 2019, 'Re-enchanting the World: Feminism and the Politics of the Commons', at Edinburgh Anarchist Feminist Bookfair
24 Quoted in Burns, H, 2016, 'The Digital Society & Health Inequalities' at Net Benefits: Is the Digital Society Good for Us?, University of Strathclyde
25 Dassanayake, D, 2015, 'Germaine Greer blasts the NHS as a "national illness service" on Question Time', *The Express*

26 Cahn, E et al, 2008, *Co-production: A Manifesto for Growing the Core Economy*, New Economics Foundation, p.8

27 Gallagher, J, 2015, '"Half of UK people" will get cancer', BBC News ↗

28 Pencheon, D, 2019, 'Health, climate change and sustainable development: engaging policy-makers, politicians, the public and private sectors' at GCPH Seminar Series 15

29 I attended Weight Watchers at Maryhill Central Hall on three occasions: in summer 2013, summer 2015 and again in January 2016.

30 Chakrabortty, A, 2018, 'Forget profit. It's love and fun that drive innovations like Parkrun', *The Guardian*

31 Hamilton, B, 2009, 'Welcome', Common Good Awareness Project ↗

32 Trebeck, K, Williams, J, 2019, *The Economics of Arrival: Ideas for a Grown Up Economy*, Policy Press, p.67

33 Dunlop, S et al, 2012, *Oxfam Humankind Index: The new measure of Scotland's Prosperity*, Oxfam Scotland, p.4

34 Hamlet, N, 2018, 'From Houses to Homes to Health', at Public Health Information Network for Scotland Seminar

35 McKinney, CJ, 2016, 'Spending on Housing Benefit', Full Fact ↗

36 Craig, C, 2010, *The Tears that Made the Clyde: Well-being*, Argyll Publishing, p.369

37 Gray, A, 2012, 'Settlers and Colonists' in *Unstated: Writers on Scottish Independence*, Word Power Books, p.109

38 Schumacher, EF, 1973, *Small is Beautiful: A Study of Economics as if People Mattered*, Abacus, p.220

39 Quoted in Ibid, p.220

40 Green, B, Shaheen, F, 2014, *Economic inequality and house prices in the UK*, New Economics Foundation

41 Living Rent, 2019, *The Rent Controls Scotland Needs*, Common Weal

42 Cahn, E et al, 2008, *Co-production: A Manifesto for Growing the Core Economy*, New Economics Foundation, p.9

43 Burns, H, 2016, 'The Digital Society & Health Inequalities' at Net Benefits: Is the Digital Society Good for Us?, University of Strathclyde

44 Cahn, E et al, 2008, *Co-production: A Manifesto for Growing the Core Economy*, New Economics Foundation, p.8

45 Ibid, p.22–3

46 Coote, A, Franklin, J, 2009, *Green Well Fair: Three Economies for Social Justice*, New Economics Foundation, p.13

47 Ibid, p.10

48 Watson, D, 2018, 'Making the Local, Local' in *Shifting Power to the People*, Red Paper Collective, p.5–7

49 McAlpine, R, 2019, 'The End of "Stuff"' at Ecopedagogy, Locavore

50 Craig, C, 2010, *The Tears That Made the Clyde*, Argyll Publishing, p.55–6

51 Ibid, p.95

52 Ibid, p.191

53 Schumacher, EF, 1973, *Small is Beautiful*, Abacus, p.45

54 Monbiot, G, 2018, 'Not capitalism, not communism: George Monbiot on why we need the commons', *We Own It*, YouTube ↗

55 Schumacher, EF, 1973, *Small is Beautiful*, Abacus, p.133

56 Ibid, p.231

57 Ibid, p.239

58 Davies, R, 2018, 'Barclays customers in switch threat over tar sands investment', *The Guardian*

59 Foster, J, 2018, 'Ownership and Economic Development' in *Shifting Power to the People*, Red Paper Collective, p.14–6

60 Barcelona en Comú et al, 2019, *Fearless Cities: A Guide to the Global Municipalist Movement*, New Internationalist, p.131

61 Cato, MS et al, 2015, *People Powered Money: Designing, Developing & Delivering Community Currencies*, New Economics Foundation, p.34

62 Ibid, p.34

63 Castlemilk Timebank, 2013, 'About us', Castlemilk Timebank ↗

64 Chakrabortty, A, 2018, 'In 2011 Preston hit rock bottom. Then it took back control', *The Guardian*

65 Chakrabortty, A, 2019, 'In an era of brutal cuts, one ordinary place has the imagination to fight back', *The Guardian*

66 Burns, H, 2016, 'The Digital Society & Health Inequalities' at Net Benefits: Is the Digital Society Good for Us?, University of Strathclyde

67 Dunlop, S et al, 2012, *Oxfam Humankind Index: The new measure of Scotland's Prosperity*, Oxfam Scotland, p.4

68 Barcelona en Comú et al, 2019, *Fearless Cities: A Guide to the Global Municipalist Movement*, New Internationalist

69 Peters, A, 2019, 'What happened when Oslo decided to make its downtown basically Car-free?', Fast Company ↗

70 McCarthy, AM, 2018 'Paris is going completely car-free one day every month', Lonely Planet ↗

71 Bol, D, 2019, 'Edinburgh's monthly car-free days at a glance', *Edinburgh Evening News*

72 Cross, L, 2019, 'Car-free Sundays are the norm in Colombia's capital city, Bogotá', Inhabitat ↗

73 Begg, D et al, 2018, *Connecting Glasgow: Creating an Inclusive, Thriving, Liveable City*,

CHAPTER 11 CONTINUED

Glasgow City Council, p.4
74 This could perhaps finally be a good use for
the controversial M74 extension, which cost
£692 million and opened in June 2011 despite
the sustained Jam 74 campaign run by local
people and against the recommendations of the
public inquiry.
75 MVRDV, Austin-Smith:Lord, 2018, *(Y)our
Broomielaw: Glasgow City Centre District
Regeneration Frameworks Broomielaw District*,
Glasgow City Council, p.78–9
76 MVRDV, 2016, (Y)our City Centre, MVRDV ↗
77 Alsarras, N, 2010, 'Festival transforms
autobahn into world's longest street party',
DW ↗
78 Inhabitat, 2014, 'How the Cheonggyecheon
River Urban Design Restored the Green Heart
of Seoul', Inhabitat ↗
79 Mairs, J, 2017, 'MVRDV transforms 1970s
highway into "plant village" in Seoul',
Dezeen ↗

Chapter 12

1 Poverty Alliance, 2019, 'Making Scotland's
transport system work for everyone' at
Get Heard Scotland, Glasgow Caledonian
University
2 This is a quote from Pascale Robinson,
public transport campaigner at Better Buses for
Greater Manchester, who has been fighting for
the re-regulation of their city's whole public
transport network, which has suffered the
same problems as Glasgow since deregulation
in 1986. The 'Our Network' plan for a fully-
integrated public transport network (described
in Chapter 10), released by Transport for
Greater Manchester in June 2019 is a massive
step forward for them. Greater Glasgow
must now follow their lead. See: ABC, 2019,
'The Fight for Greater Manchester's Buses',
Association of British Commuters, YouTube ↗
3 Schumacher, EF, 1973, *Small is Beautiful:
A Study of Economics as if People Mattered*,
Abacus, p.17
4 Beadie, B, 2017, 'The Glasgow Effect', Kiltr ↗
5 Schumacher, EF, 1973, *Small is Beautiful*,
Abacus, p.30
6 Ibid, p.140
7 Begg, D et al, 2018, *Connecting Glasgow:
Creating an Inclusive, Thriving, Liveable City*,
Glasgow City Council, p.7
8 Sloman, L et al, 2018, *We need fare-free
buses! It's time to raise our sights*, Transport for
Quality of Life
9 Ward, B, Lewis, J, 2002, *Plugging the
Leaks: Making the most of every pound that
enters your local economy*, New Economics

Foundation, p.19
10 Harkins, C, 2019, 'The public health
implications of low income and debt' at
Money, Debt & Health, Glasgow's Healthier
Future Forum 23
11 Trebeck, K, Williams, J, 2019, 'What
about decoupling' in *The Economics of
Arrival: Ideas for a Grown Up Economy*,
Policy Press, p.91–3
12 Fellow Travellers, 2019, 'Introducing the
Frequent Flyer Levy', A Free Ride ↗
13 Stephen, W, 2004, *A Vigorous Institution:
The Living Legacy of Patrick Geddes*, Luath
Press
14 Craig, C, 2010, *The Tears that Made
the Clyde: Well-being in Glasgow*, Argyll
Publishing, p.335
15 Eurostat, 2018, 'Air transport statistics',
Statistics Explained ↗
16 Schumacher, EF, 1973, *Small is Beautiful*,
Abacus, p.56
17 Ibid, p.57–8
18 Ibid, p.58
19 Wallace-Wells, D, 2019, *The
Uninhabitable Earth: A Story of the Future*,
Allen Lane, p.131
20 Ibid, p.133
21 King, E, 2014, 'UK has made largest
contribution to global warming says study',
Climate Home News ↗
22 Rosen, E, 2017, 'As Billions More Fly,
Here's How Aviation Could Evolve', National
Geographic ↗
23 Colau, A, 2019, 'Transforming fear
into hope' in *Fearless Cities: A Guide to
the Global Municipalist Movement*, New
Internationalist, p.145
24 Ibid, p.146
25 Ibid, p.147
26 Hansen, B et al, 2019, 'Sanctuary
cities' in *Fearless Cities: A Guide to the
Global Municipalist Movement*, New
Internationalist, p.137–44
27 Fearless Cities, 2018, 'About', Fearless
Cities ↗
28 Colau, A, 2019, 'Transforming fear
into hope' in *Fearless Cities: A Guide to
the Global Municipalist Movement*, New
Internationalist, p.147
29 The Economist, 2017, 'A new study
details the wealth hidden in tax havens', The
Economist ↗
30 GCPH, 2018, 'Population Overview', The
Glasgow Indicators Project ↗
31 Craig, C, 2010, *The Tears That Made the
Clyde*, Argyll Publishing, p.27
32 Even Darren McGarvey himself admits he
is now 'becoming middle class', with his own
'social mobility' being the core theme of his
2019 show 'Scotland Today' at Edinburgh

Festival Fringe.

33 Mould, O, 2018, *Against Creativity: Everything you have been told about creativity is wrong*, Verso, p.156

34 Ibid, p.98

35 Ibid, p.160

36 McGarvey, D, 2016, 'Dear Creative Scotland', *Loki The Scottish Rapper*, YouTube ↗

37 Mould, O, 2018, *Against Creativity*, Verso, p.183

38 This is of course debatable given that it's clear we must move away from GDP-driven economies fixated with 'growth' and avoid the 'internal contradiction' of urbanisation which David Harvey highlights (in Chapter 4).

39 Forrest, A, 2018, 'Estonia Is About To Roll Out Free Public Transport Across The Whole Country', Huffington Post ↗

40 Friends of the Earth, 2019, 'Why you shouldn't need a ticket to ride: free bus travel makes for a climate-friendly transport', Medium ↗

41 SPT, 2007, 'Survey shows SPT "in tune" with the public', Strathclyde Partnership for Transport ↗

42 Boyle, J, McDonald, T, 2016, 'Sick man of Europe gets sicker: Scots life expectancy lower than former communist states', *The Sunday Post*

43 Elledge, J, 2019, 'Greater Manchester mayor Andy Burnham has unveiled the region's new "tube map"', CityMetric ↗

44 Reid, J, 1972, 'Alienation Speech' at University of Glasgow, YouTube ↗

45 Coote, A et al, 2010, *21 Hours: Why a shorter working week can help us all to flourish in the 21st century*, New Economics Foundation

46 Black, I et al, 2015, *From 'I' to 'We': Changing the narrative in Scotland's relationship with consumption*, Common Weal, p.11

47 Schumacher, EF, 1973, *Small is Beautiful*, Abacus, p.83

48 Ibid, p.30

49 Ibid, p.83

50 Walsh, D et al, 2016, *History, Politics & Vulnerability*, Glasgow Centre for Population Health, p.56

51 Reid, J, 1972, 'Alienation Speech' at University of Glasgow, YouTube ↗

52 Shaull, R, 1972, 'Foreword' in *Pedagogy of the Oppressed*, Penguin, p.13–4

53 Ibid, p.14

54 Veiel, A, 2017, *Beuys*, Zero One Film

55 Mould, O, 2018, *Against Creativity*, Verso, p.53

56 p.46

57 Schumacher, EF, 1973, *Small is Beautiful*, Abacus, p.30

58 Barcelona en Comú, 2016, *How to Win Back the City En Comú: Guide to Building a Citizen Municipal Platform*, Barcelona en Comú, p.1

59 Barcelona en Comú et al, 2019, *Fearless Cities: A Guide to the Global Municipalist Movement*, New Internationalist, p.14

60 Gray, A, 2012, 'Settlers and Colonists' in *Unstated: Writers on Scottish Independence*, Word Power Books, p.106

61 Craig, C, 2010, *The Tears That Made the Clyde*, Argyll Publishing, p.100

62 MVRDV, 2016, (Y)our City Centre, MVRDV ↗

63 Barcelona en Comú et al, 2019, *Fearless Cities: A Guide to the Global Municipalist Movement*, New Internationalist, p.14

64 Beuys, J, 1973, 'I Am Searching for Field Character' in *Joseph Beuys in America: Energy Plan for the Western Man*, Da Capo Press, p.22

65 Craig, C, 2010, *The Tears That Made the Clyde*, Argyll Publishing, p.204–5

66 Ewan, R et al, 2012, *The Glasgow Schools*, The Common Guild, p.8

67 Ibid, p.10

68 Ibid, p.8

69 Ibid, p.10

70 Ibid, p.23

71 Appiah, KA, 2016, 'Mrs May, we're all citizens of the world', BBC News ↗

72 Barcelona en Comú, 2016, *How to Win Back the City En Comú: Guide to Building a Citizen Municipal Platform*, Barcelona en Comú

73 This was quoted by one of the audience members at *The Glasgow Effect: A Discussion* event at the Glad Café on 3 February 2016. Also quoted in Hames, S et al, 2012, *Unstated: Writers on Scottish Independence*, Word Power Books, p.127

74 'Making ourselves at home' is the metaphor which Trebeck and Williams use for an economy which is already GDP-rich and so now must now stop 'growing'. See: Trebeck, K, Williams, J, 2019, *The Economics of Arrival: Ideas for a Grown Up Economy*, Policy Press, p.7

75 McCormack, C, 1995, 'War Without Bullets', *Cathy McCormack*, YouTube ↗

76 Said, EW, 1984, 'Reflections on Exile' in *Reflections on Exile: And Other Literary & Cultural Essays*, Granta, p.180

Luath Press Limited

committed to publishing well written books worth reading

LUATH PRESS takes its name from Robert Burns, whose little collie Luath (*Gael.*, swift or nimble) tripped up Jean Armour at a wedding and gave him the chance to speak to the woman who was to be his wife and the abiding love of his life. Burns called one of the 'Twa Dogs' Luath after Cuchullin's hunting dog in Ossian's *Fingal*. Luath Press was established in 1981 in the heart of Burns country, and is now based a few steps up the road from Burns' first lodgings on Edinburgh's Royal Mile. Luath offers you distinctive writing with a hint of unexpected pleasures.

Most bookshops in the UK, the US, Canada, Australia, New Zealand and parts of Europe, either carry our books in stock or can order them for you. To order direct from us, please send a £sterling cheque, postal order, international money order or your credit card details (number, address of cardholder and expiry date) to us at the address below. Please add post and packing as follows: UK – £1.00 per delivery address; overseas surface mail – £2.50 per delivery address; overseas airmail – £3.50 for the first book to each delivery address, plus £1.00 for each additional book by airmail to the same address. If your order is a gift, we will happily enclose your card or message at no extra charge.

Luath Press Limited

543/2 Castlehill
The Royal Mile
Edinburgh EH1 2ND
Scotland
Telephone: +44 (0)131 225 4326 (24 hours)
email: sales@luath. co.uk
Website: www. luath.co.uk